SOLID STATE PHYSICS

VOLUME 42

Founding Editors

FREDERICK SEITZ

DAVID TURNBULL

SOLID STATE PHYSICS

Advances in
Research and Applications

Editors

HENRY EHRENREICH

DAVID TURNBULL

Division of Applied Sciences
Harvard University, Cambridge, Massachusetts

VOLUME 42

ACADEMIC PRESS, INC.

Harcourt Brace Jovanovich, Publishers

Boston   San Diego   New York
Berkeley   London   Sydney
Tokyo   Toronto

ACADEMIC PRESS, INC.
1250 Sixth Avenue
San Diego, California 92101

United Kingdom Edition published by
ACADEMIC PRESS INC. (LONDON) LTD.
24–28 Oval Road, London NW1 7DX

Cover art for the paperback edition of *Solid State Physics, Vol. 42* is reprinted from *Phys. Rev. B* **37**(13):7961, May 1988, and *LANSCE* **4,** Winter 1988, LA LP 88-6. Reprinted by permission of The American Physical Society and Los Alamos National Laboratory.

LIBRARY OF CONGRESS CATALOG CARD NUMBER: 55-12200

ISBN   0-12-607742-8 (hardcover)
ISBN   0-12-607790-8 (paperback)

PRINTED IN THE UNITED STATES OF AMERICA
89 90 91 92   9 8 7 6 5 4 3 2 1

Contents

Polytetrahedral Order in Condensed Matter

D. R. NELSON AND FRANS SPAEPEN

Physical Properties of the New Superconductors

M. TINKHAM AND C. J. LOBB

The Structure of $Y_1Ba_2Cu_3O_{7-\delta}$ and Its Derivatives

R. BEYERS AND T. M. SHAW

Electronics Structure of Copper-Oxide Superconductors

K. C. HASS

Electron Correlations in Two Dimensions

A. ISIHARA

Contributors to Volume 42

Numbers in parentheses indicate the pages on which the authors' contributions begin.

R. BEYERS (135), *IBM Research Division, Almaden Research Center, San Jose, California 95120-6099*

K. C. HASS (213), *Research Staff, Ford Motor Company, Dearborn, Michigan 48121-2053*

A. ISIHARA (271), *Department of Physics, State University of New York at Buffalo, Buffalo, New York 14214*

C. J. LOBB (91), *Physics Department and Division of Applied Sciences, Harvard University, Cambridge, Massachusetts 02138*

D. R. NELSON (1), *Lyman Laboratory of Physics, Harvard University, Cambridge, Massachusetts 02138*

T. M. SHAW (135), *IBM Research Division, Thomas J. Watson Research Center, Yorktown Heights, New York 10598-0218*

FRANS SPAEPEN (1), *Division of Applied Sciences, Harvard University, Cambridge, Massachusetts 02138*

M. TINKHAM (91), *Division of Applied Sciences, Harvard University, Cambridge, Massachusetts 02138*

Preface

This volume deals with three topics at the forefront of interest in solid state physics: (1) condensed matter having polytetrahedral order; important examples include liquids, crystals, glasses, Frank-Kasper phases and quasi-crystals; (2) aspects of high-temperature superconductivity; and (3) the electronic behavior of two dimensional systems, most notably the integral and fractional quantum Hall effects. These subjects represent qualitatively new departures for the condensed matter sciences, and authoritative reviews in a single volume are greatly welcomed. Moreover, these reviews are copiously referenced and designed to be as comprehensive and accessible as possible. Much of this material will ultimately find its way into standard solid state textbooks, and, indeed, may be regarded as a complement to those books that are presently available.

The article by Nelson and Spaepen deals with polytetrahedral order in condensed matter. Why this type of order is so frequently encountered is worth a few words. The lowest energy configuration of four atoms interacting by pairwise central potentials is a regular tetrahedral one. Twenty such tetrahedra packed together with slight distortions form an icosahedral group in which twelve atoms are uniformly distributed around a central one. Frank showed that clusters of 13 atoms should be more stable in this configuration than in either of the crystalline close-packed configurations, which include the more energetic half octahedral configurations. However, space cannot be filled by packing regular tetrahedra, and infinite systems composed of one kind of atom, interacting by non-directed forces, generally exhibit crystalline close-packed structures. By contrast, in non-periodic structures such as liquids or glasses, the preferred atomic short range order may be of the polytetrahedral type. Further, when two kinds of atoms, with appropriate size ratios, are packed together, the preferred crystalline structures can be of the Frank–Kasper type in which space is filled by systematically distorted tetrahedral configurations. Presumably, polytetrahedral configurations are prominent in the recently discovered alloy structures (quasicrystals) which exhibit icosahedral point group symmetry. Spaepen and Nelson review the evidence for polytetrahedral order for a variety of systems and show in their comprehensive and incisive article that it is a powerful unifying concept in providing a theoretical description.

Various aspects concerning the new high-temperature superconductors are considered in three of the articles. Tinkham and Lobb review the physical properties of these superconductors. They emphasize the properties of YBCO as a typical example because it was the first of the 90 K materials to be discovered and because it has been studied most thoroughly. The discussion in this and the remaining superconductivity articles emphasizes those features about which a reasonable degree of concensus has emerged. The Tinkham–Lobb article develops a new point of view for understanding the disappointing resistive properties of these materials in strong magnetic fields, which is based on activated phase-slip or fluxon motion. Such slippage varies rapidly with increasing operating temperature. For example, assuming an activation energy of 1000 K, one finds an astronomical lifetime ($\sim 10^{100}$ years) at 4 K, but a lifetime of only one *hour* at 100 K. This sobering conclusion undelines that room temperature operation at high magnetic fields would be more problematic even if materials with requisitely high transition temperatures can be found.

The diversity of topics addressed in this important article is made accessible to the non-specialist through its frequent specific references to Tinkham's classic *Introduction to Superconductivity* (see this volume, p. 92, ref. 5.), which explains the basic ideas underlying this subject in a clear, pedagogical manner.

The second of the three articles, by Beyers and Shaw, surveys the structure of YBCO and its derivatives and provides an overview of the structural studies that have been performed. The basic atomic arrangements and how these change with processing are discussed at the outset. This discussion is followed by an enumeration of commonly found defects and the role they play in controlling superconducting properties. The studies of elemental substitutions are described. These studies have both practical and theoretical interest because they provide clues to the roles played by various structural features in the microscopic mechanisms responsible for high-temperature superconductivity. The article also contains suggestions for future structural studies and a delineation of the generally agreed upon features at the time of writing. The textual material is augmented by many illustrations, which, given the complexity of these structures, will be helpful to the reader. An elaborate set of references to the literature is also provided.

The electronic structure of YBCO and the La-based systems is reviewed by K. C. Hass in the third superconductivity paper. The article summarizes the state-of-the-art band calculations and results based on simpler, more idealized model Hamiltonian treatments that include electron correlation effects, each designed to capture the essential features that give rise to superconductivity. The calculated electronic structures are also compared to the results of a wide range of spectroscopic data including photoemission, inverse photoemission, Auger spectroscopy, X-ray absorption, electron energy loss, and optical experiments. According to the author's view, two conclusions that now seem to be well established are that for both La and YBCO materials the behavior is dominated by the same microscopic physics and that the crucial Cu 3d and O 2p electrons are sufficiently strongly interacting that conventional band theoretic ideas are inapplicable.

It should be emphasized that this article is *not* designed to present a fundamental theory describing the superconductivity in these systems. This subject is still sufficiently controversial that a comprehensive and succinct review would be premature. However, some general boundaries on the ingredients of a proper theory are now beginning to emerge. We hope to return to this subject in future volumes.

The third topic that has recently stimulated great interest in the physics community concerns two-dimensional systems, which are commonly encountered in semiconductor heterostructures and superlattice interfaces. To be sure, there has been important applications-oriented physics which needed to be developed. However, the real impetus for fundamental investigations came in 1980 with the discovery of the quantum Hall effect. A. Ishihara's review of the subject is not encyclopedic, but rather is limited to an explication of the low-temperature properties of two-dimensional electron systems with emphasis on the role of electron correlation effects. He presents a systematic exposition of the subject beginning with fundamentals concerning the ground state, phase transitions involving the Wigner lattice, magnetoconduction and localization. He ends with an account of both the integral and fractional quantum Hall effect, which is sufficiently detailed that it should be useful to a broad range of scientists, from graduate students to experts. Such complete accounts in the review literature have been very rare, if at all present.

Henry Ehrenreich
David Turnbull

SOLID STATE PHYSICS, VOLUME 42

Polytetrahedral Order in Condensed Matter

DAVID R. NELSON

Lyman Laboratory of Physics
Harvard University
Cambridge, Massachusetts

FRANS SPAEPEN

Division of Applied Sciences
Harvard University
Cambridge, Massachusetts

I. Introduction

The physics, crystallography, and materials science of condensed matter would in many ways be much simpler in a two-dimensional world.

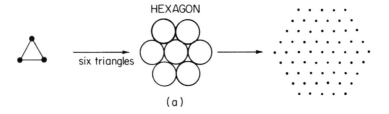

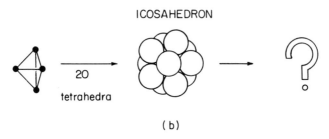

FIG. 1. (a) Particle packing in two dimensions: equilateral triangles are preferred locally and pack naturally to form a close-packed triangular lattice. (b) Particle packing in three dimensions: although tetrahedra are preferred locally and combine with slight distortions to form a regular icosahedron, the fivefold symmetry axes of the icosahedron preclude a simple space-filling lattice.

For identical particles interacting with simple pair potentials, liquids would have the same short-range order as crystals, crystals would always form a triangular lattice, and undercooling liquids fast enough to form a glass would be virtually impossible. The reason for this state of affairs lies in the geometry of 2-D particle packings: As shown in Fig. 1a, triplets of particles will tend to form equilateral triangles to minimize the energy or maximize the density. Six such triangles pack naturally to form a hexagon, which should be the dominant motif characterizing short-range order in a dense liquid. Such a hexagon can be extended very easily to form a triangular (i.e., hexagonal close-packed) lattice, which is the expected ground state for classical particles with a wide variety of pair potentials. A liquid with hexagonal short-range order automatically contains many nuclei of the stable crystal, which prevents the undercooling necessary to form a glass.

The situation is quite different in three dimensions, again for elementary geometrical reasons: Four hard spheres form a dense tetrahedral packing, in which each sphere is in contact with the three others.

Similarly, the most stable configuration of a cluster of 4 atoms interacting by pairwise central potentials is a tetrahedral one, in which the 6 interatomic distances can all be optimized. As shown in Fig. 1b, 20 such tetrahedra pack naturally, with slight distortions, to form an icosahedron. Unlike the hexagon in two dimensions, however, the icosahedron has (a total of 6) fivefold symmetry axes and cannot be periodically extended to fill space. The inability of perfect tetrahedra to fill space already appears in the geometry of the icosahedron, where the distance between atoms on the surface is about 5% longer than the distance of surface atoms to the center. The incompatibility of the preferred short-range tetrahedral order with a simple global packing of tetrahedra is known in condensed matter physics as "frustration."

The tetrahedron is nevertheless the natural unit for the study of short-range order, and its relation to long-range order, in condensed phases of "spherelike" atoms. Nature solves the frustration problem in a variety of ways. Because there is no space-filling lattice of perfect tetrahedra, we must typically consider crystals composed of very distorted tetrahedra (bcc lattices) or peculiar mixtures of perfect tetrahedra and octahedra (fcc and hcp lattices) when searching for ground states of simple materials. When the frustration is relaxed by combining particles of two different sizes (see below), Frank–Kasper phases[1] composed entirely of slightly distorted tetrahedra become stable. Undercooled liquids and metallic glasses can be viewed as spacing-filling arrays of distorted tetrahedra via the Voronoi construction.[2] Discussions of polytetrahedral order have recently received new impetus from the experimental discovery of quasicrystals,[3] which have a macroscopic icosahedral symmetry and which, in many cases, are closely related to Frank–Kasper compounds.

Attempts to understand frustrated polytetrahedral particle packings can be traced back to the pioneering work of Frank,[1,4] Bernal,[5] and Coxeter[6–8] in the 1950s. Although the work of Frank and Bernal was widely appreciated at the time, the contributions of the mathematician

[1]F. C. Frank and J. Kasper, *Acta Crystallogr.* **11**, 184 (1958); *Acta Crystallogr.* **12**, 483 (1959).

[2]See, for example, R. Collins in "Phase Transitions and Critical Phenomena" (C. Domb and M. S. Green, eds.). (New York, Academic Press, 1972.)

[3]D. Shechtman, I. Blech, D. Gratias, and J. W. Cahn, *Phys. Rev. Lett.* **53**, (1984).

[4]F. C. Frank, *Proc. Royal Soc. London* **215A**, 43 (1952).

[5]J. D. Bernal, *Proc. Royal Soc. London Ser. A***280**, 299 (1964).

[6]H. S. M. Coxeter, *Illinois J. of Math.* **2**, 746 (1958).

[7]H. S. M. Coxeter, "Regular Polytopes." Dover, New York, 1975.

[8]H. S. M. Coxeter, "Introduction to Geometry." Wiley, New York, 1969.

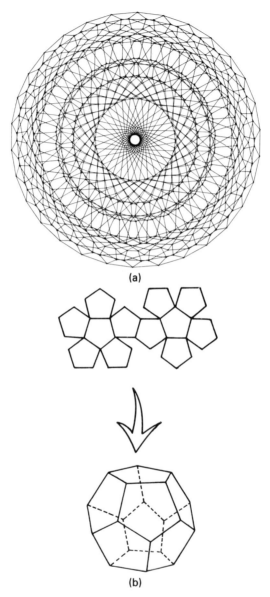

(a)

(b)

FIG. 2. (a) Projection of the Wigner–Seitz cells of polytope $\{3, 3, 5\}$. Frustration is absent because these regular dodecahedra can perfectly tile the curved three-dimensional surface of a sphere embedded in four dimensions. This lattice of dodecahedra is called "polytope" $\{5, 3, 3\}$. (b) Two-dimensional analogue of Fig. 2a: Twelve pentagonal Wigner–Seitz cells can tile without frustration when wrapped around the two-dimensional surface of an ordinaty sphere to form a regular dodecahedron.

Coxeter are less well known. Coxeter introduced a mean field theory of dense-random packing called the "statistical honeycomb model" and emphasized the importance of "polytope $\{3, 3, 5\}$," a regular four-dimensional Platonic solid that provides a regular tetrahedral tiling of S^3, the curved, three-dimensional surface of a sphere embedded in four dimensions. The 120 atoms of polytope $\{3, 3, 5\}$ form a regular lattice where each atom sits in an identical icosahedral coordination shell and serves as a paradigm for the kind of global order prevented by frustration in flat space. The 120 unit cells of polytope $\{3, 3, 5\}$, each one of which is a perfect dodecahedron, are shown in projection in Fig. 2a. Such perfect order is only possible in a curved space, just as the curvature of an ordinary sphere allows 12 pentagons to be packed without frustration to form a perfect dodecahedron—see Fig. 2b.

Theoretical interest in polytetrahedral packings revived in the late 1970s with work by Kleman and Sadoc,[9] who discussed defects in the context of polytope $\{3, 3, 5\}$, and by Sadoc,[10] who suggested polytope $\{3, 3, 5\}$ as a model of metallic glasses. Recently, the ideas of Frank and Bernal about polytetrahedral order have been combined with the curved space icosahedral paradigm provided by polytope $\{3, 3, 5\}$ to produce a theory of the statistical mechanics of geometrical frustration.[11–14] In this paper we review the experimental and simulation evidence for polytetrahedral order in small clusters, liquids, glasses, and periodic or quasiperiodic crystals and show that it is a powerful unifying concept in their theoretical description.

II. Liquids, Glasses, and Polytetrahedral Crystals

1. CLUSTERS

Polytetrahedral structures of small clusters of atoms interacting with pairwise central potentials are energetically favored. Figures 3 and 4 show how two particularly favorable configurations are formed from

[9] M. Kleman and J. F. Sadoc, *J. Phys. (Paris) Lett.* **40**, L569 (1979).

[10] J. F. Sadoc, *J. Phys. (Paris) Colloq.* **41**, C8-326 (1980).

[11] D. R. Nelson, *Phys. Rev. Lett.* **50**, 982 (1983); *Phys. Rev.* **B28**, 5515 (1983).

[12] J. Sethna, *Phys. Rev. Lett.* **51**, 2198 (1983); *Phys. Rev.* **B31**, 6278 (1985).

[13] D. R. Nelson and M. Widom, *Nucl. Phys.* **B240**, [FS12], 113 (1984).

[14] S. Sachdev and D. R. Nelson, *Phys. Rev. Lett.* **53**, 1947 (1984); *Phys. Rev.* **B32**, 1480 (1985).

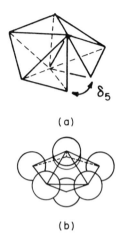

(a)

(b)

FIG. 3. (a) Packing of five regular tetrahedra around a common edge. The gap angle is $\delta_5 = 7.35°$. (b) Seven-atom cluster, forming a regular pentagonal bipyramid.

individual identical regular tetrahedra. Since the dihedral angle between 2 tetrahedral faces is $\alpha = \cos^{-1}(1/3) = 70.53° = 2\pi/5.104$, 5 tetrahedra fit *almost* perfectly around a common edge (Fig. 3a). If atoms are placed at the vertices of the corresponding regular pentagonal bipyramid, a low-energy 7-atom cluster is formed. The 7.35° gap in Fig. 3a, although small, is the basis of the *frustration* that becomes important in larger structures. It is also the reason why the interatomic distance between the "capping" atoms in the cluster of Fig. 3b is slightly larger than the one between nearest neighbors in the ring of 5.

As discussed in the Introduction, it is possible to put together 20 regular tetrahedra that share a common central vertex. Since the solid angle subtended by 3 tetrahedral planes is $(3\alpha - \pi) = 0.551 = (4\pi/22.8)$ steradians, it is clear that a substantial "gap" remains in the 20-tetrahedron configuration (see Fig. 4a). If, however, the tetrahedra are slightly distorted, so that the length of the edges containing the central vertex is $(\tau\sqrt{5}/4)^{1/2} = 0.951$ times the length of the others (τ is the golden ratio, $(\sqrt{5} + 1)/2 = 1.6180$), the triangles on the surface close up to form a regular icosahedron (Fig. 4b). If identical atoms are placed at the vertices and the center of this icosahedron, a very favorable 13-atom cluster is formed (Fig. 4c). Note how it is made up of 12 interpenetrating pentagonal bipyramids of Fig. 3b. The "gap" in Fig. 4c is now spread out more uniformly than in Figure 4a. Both configurations of gaps are another illustration of the "frustration" associated with the three-dimensional packing of regular tetrahedra.

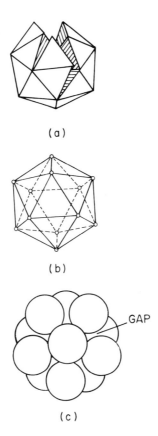

FIG. 4. (a) Twenty regular tetrahedra sharing a common vertex. Note the 0.551 steradian gap. (b) Regular icosahedron. (c) Thirteen hard spheres in an icosahedral cluster. Note the gaps left by the frustration.

It was first pointed out by F. C. Frank in 1952 that energy of an icosahedral 13-atom cluster relaxed in a Lennard–Jones potential was 8.4% lower than that of 13-atom clusters corresponding to the nearest-neighbor environments in close-packed (face-centered cubic or hexagonal) crystals.[4] Figure 5 shows that the latter are made up of tetrahedra and half octahedra (associated with the square faces). Since then, a number of systematic studies have been made of the relative energies of small clusters of identical particles interacting through pair potentials. Using static relaxation, Hoare and Pal[15] identified several growth sequences for the formation of different polytetrahedral isomers. They

[15]M. R. Hoare and P. Pal, *Adv. Phys.* **20**, 161 (1971).

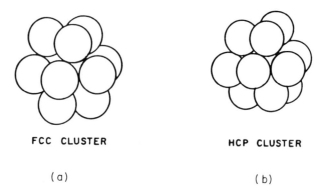

FCC CLUSTER   HCP CLUSTER

(a)   (b)

FIG. 5. Two 13-atom clusters corresponding to the nearest-neighbor environment in the close-packed hard sphere structures. Note the squares on the surface of each cluster, which, with the central sphere, form half octahedra. (a) Nearest-neighbor environment for the face-centered dubic structure (cuboctahedral cluster). (b) Nearest-neighbor environment for the hexabonal close-packed structure.

found that up to the maximum size they considered (64 atoms) the energy of the (noncrystalline) polytetrahedral packings was still lower than that of the corresponding close-packed crystalline clusters. Molecular dynamic studies[16,17] on slowly cooled Lennard–Jones clusters, in which the atoms can sample many configurations, also showed that structures with pentagonal, and especially icosahedral symmetry predominate in clusters of about 40 atoms or less. Honeycutt and Andersen[16] observed that above 40 atoms structures leading to the Mackay icosahedron (55 atoms) are formed, but that a transition to face-centered cubic structures did not occur until the number of atoms exceeded several thousand. Electron diffraction experiments by Farges and collaborators[18] on noble gas cluster beams produced by free jet expansions show that clusters containing less than 150 atoms indeed have a polytetrahedral (icosahedral) structure.

Although pair potentials are satisfactory for describing the interactions in noble gas clusters, they are not necessarily adequate for metallic clusters. More sophisticated approaches, such as self-consistent pseudo-potential local-spin-density (LSD) calculations,[19] combinations of pseudopotential and density-functional methods,[20] or a generalized valence

[16]J. D. Honeycutt and H. C. Andersen, *J. Phys. Chem.* **91,** 4950 (1987).
[17]T. L. Beck and R. S. Berry, *J. Chem. Phys.* **88,** 3910 (1988).
[18]J. Farges, M. F. De Feraudy, B. Raoult, and G. Torchet, *Surf. Sci.* **106,** 95 (1981); J. Farges, B. Raoult, and G. Torchet, *J. Chem. Phys.* **59,** 3454 (1973).
[19]J. L. Martins, J. Buttet, and R. Car, *Phys. Rev. B***31,** 1804 (1985).
[20]J. Flad, H. Stoll, and H. Preuss, *J. Chem. Phys.* **71,** 3042 (1979).

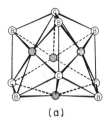 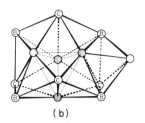 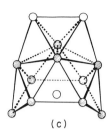

(a)　　　　　　　(b)　　　　　　　(c)

FIG. 6. Three polytetrahedral clusters found in generalized valence bond calculations to have lower energies than a regular tetrahedral cluster. Cluster (a) consists of a central tetrahedron (shaded atoms, dotted lines), four capping atoms (C) and five bridging atoms (B). Cluster (b) differs from (a) in the position of only one atom. The shaded atoms of cluster (c) from a portion of an icosahedral surface. All three clusters have multiple local fivefold symmetry axes. [From M. H. McAddon and W. A. Goddard, *J. Non-Cryst. Sol.* **75**, 149 (1985).]

bond (GVB) approach,[21] give some surprising results for the smallest clusters. For example, in all approaches it is found that the most stable configuration of 4 neutral monovalent metal atoms (Na, Li) is a *planar* one, with the atoms at the vertices of a rhombus (with short diagonal 6 to 14% shorter than the edges, depending on the method).

The most stable neutral 7-atom cluster, however, is found to be again the pentagonal bipyramid structure of Fig. 3b, with the distance between the caps 1.6% larger than the other nearest-neighbor distances in the cluster, according to the LSD method. On the 13-atom cluster the methods disagree. The LSD work indicated that a slightly distorted cuboctahedron (f.c.c. environment; Fig. 5a) has a lower energy than the icosahedron. A configuration interaction calculation,[22] however, found the icosahedron to be the most stable configuration. The GVB work found the icosahedron to be more stable than the crystalline configurations of Fig. 5, but also led to the discovery of 3 configurations that are even more stable (Fig. 6) than the icosahedron. These are polytetrahedral structures, similar to the nearest-neighbor environments found in the γ-brass structure. Note that they contain several pentagonal bipyramid configurations.

Although more theoretical and experimental work is obviously needed to clarify the structure of small metallic clusters, the present understanding seems to support the idea that polytetrahedral short-range order is also strongly favored in metallic systems. In fact, the structural rules

[21]M. H. McAddon and W. A. Goddard, *J. Non-Cryst. Sol.* **75**, 149 (1985); *J. Phys. Chem.* **91**, 2607 (1987); *Phys. Rev. Lett.* **55**, 2563 (1985).
[22]G. Pacchione and J. Koutecký, *J. Chem. Phys.* **81**, 3588 (1984).

derived from the GVB approach explicitly state that optimum structures are those that maximize the number of tetrahedra because optimum bonding involves valence electrons that prefer to localize in tetrahedral interstices. Energy differences between such polytetrahedral structures are then the result of different arrangements of these tetrahedra.

2. POLYTETRAHEDRAL CRYSTALS

As discussed above, and illustrated in Figs. 3 and 4, it is impossible to fill space by the close packing of regular tetrahedra. It is possible to do so, however, if some distortion of the tetrahedra is allowed. In fact, there exists an unambiguous procedure to divide up a structure into (irregular) tetrahedra.[2] The Voronoi (or Dirichlet) cell of an atom in a structure is defined as the part of space that is closer to that atom than to any other. It is the innermost polyhedron formed by the bisecting planes of the lines between the atom and all others. Connecting the nearest neighbors, i.e.,

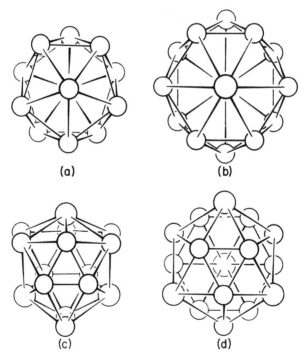

FIG. 7. Coordination polyhedra in the Frank–Kasper phases. (a) $Z = 12$. (b) $Z = 14$. (c) $Z = 15$. (d) $Z = 16$. The central atoms are omitted.

atoms whose Voronoi cells share a face, divides space into tetrahedra. The center of each tetrahedron is a common vertex of the Voronoi cells of the 4 atoms. These tetrahedra pack closely around each atom and define a triangulated coordination polyhedron of Z nearest neighbors.

Although this procedure is entirely general, it is most useful if the tetrahedra are only slightly distorted, as in complex alloy structures of atoms that are similar in size. In that case, the triangles in the coordination polyhedra are approximately equilateral, and a $Z12$ coordination polyhedron closely resembles the icosahedron of Fig. 4b. Note that a 10% size difference between the atoms corresponds to perfect icosahedral coordination, as discussed above. Frank and Kasper[1] have analyzed the structure of such alloys by identifying 3 other coordination polyhedra ($Z14$, $Z15$ and $Z16$) with approximately equilateral triangles as faces. The four Frank–Kasper coordination polyhedra are shown in Fig. 7. Consistent with its faces being nearly equilateral, each coordina-

TABLE I

NAME	EXAMPLES	ATOMS PER UNIT CELL	SPACE GROUP	$Z = 12$	$Z = 14$	$Z = 15$	$Z = 16$	$\bar{Z}$	$\bar{q}$
$A15$	$\beta - W$, Cr_3Si	8	$Pm\bar{3}n$	2	6	0	0	13.500	5.1111
σ	$Cr_{46}Fe_{54}$, $\beta - U$	30	$P4_2/mnm$	10	16	4	0	13.467	5.1089
H	Complex	30	$Cmmm$	10	16	4	0	13.467	5.1089
K'	Complex	82	$Pmmm$	28	42	12	0	13.463	5.1087
F	Complex	52	$P6/mmm$	18	26	8	0	13.462	5.1086
J	Complex	22	$Pmmm$	8	10	4	0	13.455	5.1081
ν	$Mn_{81.5}Si_{18.5}$	186	$Immm$	74	80	20	12	13.441	5.1072
Z	Zr_4Al_3	7	$P6/mmm$	3	2	2	0	13.428	5.1064
P	$Mo_{42}Cr_{18}Ni_{40}$	56	$Pbnm$	24	20	8	4	13.428	5.1064
δ	$MoNi$	56	$P2_12_12_1$	24	20	8	4	13.428	5.1064
K	$Mn_{77}Fe_4Si_{19}$	220	$C2$	100	76	16	28	13.418	5.1057
R	$Mo_{31}Co_{18}Cr_{51}$	159	$R\bar{3}$	81	36	18	24	13.396	5.1042
μ	Mo_6Co_7	39	$R\bar{3}m$	21	6	6	6	13.385	5.1034
—	K_7Cs_6	26	$P6_3/mmc$	14	4	4	4	13.385	5.1034
$p\sigma$	$W_6(Fe, Si)_7$	26	$Pnam$	14	4	4	4	13.385	5.1034
M	$Nb_{48}Ni_{39}Al_{13}$	52	$Pnam$	28	8	8	8	13.385	5.1034
I	$V_{41}Ni_{36}Si_{23}$	228	Cc	132	24	24	48	13.369	5.1024
C	$V_2(Co, Si)_3$	50	$C2/m$	30	4	4	12	13.360	5.1018
T	$Mg_{32}(Zn, Al)_{49}$	162	$Im\bar{3}$	98	12	12	40	13.358	5.1017
X	$Mn_{45}Co_{40}Si_{15}$	74	$Pnnm$	46	4	4	20	13.351	5.1012
—	Mg_4Zn_7	110	$C2/m$	70	4	4	32	13.345	5.1008
$C14$	$MgZn_2$	12	$P6_3/mmc$	8	0	0	4	13.333	5.1000
$C15$	$MgCu_2$	24	$Fd\bar{3}m$	16	0	0	8	13.333	5.1000

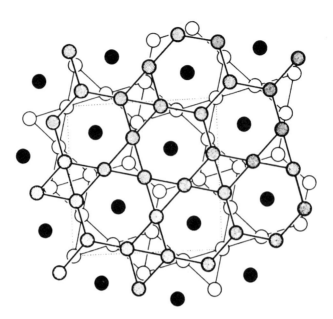

FIG. 8. Projection along the c-axis of the atoms in the σ-phase (tetragonal). The black atoms have coordination number $Z = 14$ and are connected by a set of Frank–Kasper skeleton lines parallel to the c-axis. Note the rings of six neighbors common to pairs of black atoms, as required by the coordination polyhedron of Figure 7(b). The gray or white atoms have coordination numbers $Z = 12$, 14, or 15, and are connected by skeleton lines that lie in the planes perpendicular to the c-axis.

tion polyhedron has 12 vertices with surface coordination of 5, and $(Z - 12)$ vertices with surface coordination 6; none of the 6-coordinated vertices are adjacent.

Frank and Kasper then define a "skeleton" of lines that connect the atoms that have 6 neighbors in common. This skeleton therefore contains all the $Z14$, $Z15$ and $Z16$ atoms, and since each of their coordination shells contains at least 2 (i.e., $Z - 12$) atoms with sixfold surface coordination, none of the lines can terminate in the structure. The $Z15$ and $Z16$ atoms are, respectively, three- and fourfold nodes in the skeleton (see Fig. 22, below). Table I lists the major types of polytetrahedral, or "Frank–Kasper" phases.[23] The skeleton provides a simple and characteristic description of the structure of each type of alloy. For example, in the $\beta - W$ (A15) structure, the skeleton consists of sets of

[23]D. P. Shoemaker and C. B. Shoemaker, *in* "Aperiodicity and Order, Vol. 1," (M. V. Jaric, ed.). Academic Press, Boston, 1987.

straight lines parallel to the cubic axes. In the tetragonal σ-phases (Fig. 8), it consists of straight lines parallel to the c-direction, and planar networks perpendicular to them. In the $MgCu_2$ Laves phase it forms a diamond-cubic like structure. Note that this phase only has $Z = 12$ and $Z = 16$ polyhedra. The latter are the fourfold nodes of the diamond-cubic skeleton.

3. Liquids

Frank's original argument showing that an icosahedral coordination shell (Fig. 4c) is energetically favored over the close-packed crystalline coordination shells (Fig. 5) was prompted by Turnbull's experiments that demonstrated that many simple metallic liquids could be undercooled far below their thermodynamic melting point T_M. His most complete study was made on mercury,[24] which could be undercooled to as much as $(2/3)$ T_M.

Up to that time, it had been thought that the short-range order in these liquids was similar to that in the respective crystal. After all, the densities of the two phases are very similar, and x-ray diffraction had shown the average coordination distances and coordination numbers in the liquid to be very similar to those in the close-packed crystals. This naturally led to the formulation of microcrystalline-type models of the liquid, and the presence of this embryonic crystallinity seemed to explain the ease of crystal nucleation and the lack of undercooling.

Turnbull's work definitively discarded this picture. By subdividing the liquid into very small micrometer-size droplets he was able to isolate the ubiquitous heterogeneous nuclei that normally prevented substantial undercooling of the liquid and to demonstrate the occurrence of homogeneous crystal nucleation at large undercooling. Figure 9 shows this with a dilatometric measurement on a Hg emulsion: the contraction upon crystallization occurs 60 K below the equilibrium melting point, marked by the expansion. In other Hg emulsions he observed undercoolings as large as 80 K. Frank's suggestion showed that atoms with the same coordination number and coordination distance (and, hence, similar atomic volume) can still have fundamentally different short-range order. The atoms at the center of the configurations of Figs. 4c and 5 each have 12 nearest neighbors at the same distance. What makes the polytetrahedral short-range order of the icosahedron fundamentally different from that of the tetrahedral–octahedral crystalline configuration, however, is the *orientational* part of the coordination. Due to the orientational

[24]D. Turnbull, *J. Chem. Phys.* **20,** 411 (1952).

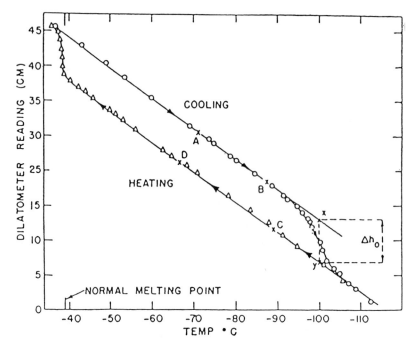

Fig. 9. Dilatometric measurements on a mercury emulsion, demonstrating the large undercooling that is required for homogeneous crystal nucleation from the melt. [From D. Turnbull, *J. Chem. Phys.* **20,** 411 (1952).]

averaging, this kind of difference in the coordination could, of course, not be observed in diffraction experiments on bulk liquids. Although direct observations of orientational order in liquids are still lacking, modeling and simulation have been providing increasing evidence that polytetrahedral order is the basis of the fundamental difference between simple liquids and their corresponding crystals. This will be elaborated on in Section 7.

Turnbull could analyze his measurements of the homogeneous nucleation frequency, I, with classical nucleation theory,[25] which states:

$$I = I_0 \exp(-16\pi\sigma^3/3(\Delta G_v)^2 k_B T), \qquad (3.1)$$

where I_0 is a prefactor that depends on the atomic jump frequency and the density, σ is the crystal–liquid interfacial tension, and ΔG_v is the

[25]D. Turnbull, *in* "Solid State Physics" (F. Seitz and D. Turnbull, *eds.*), vol. 3, p. 225. Academic Press, N.Y., 1965.

difference in free energy, per unit volume, between the crystal and liquid. Since ΔG_v is proportional to the undercooling $T_M - T$, the observation of the first measurable nucleation at temperatures as low as $(2/3)$ T_M can only be accounted for by a large solid–liquid interfacial tension. For mercury,[24] the interfacial tension obtained by fitting the nucleation data is as large as 0.62 times the heat of fusion per atom in the crystal plane.

The large experimental value of the interfacial tension initially came as a surprise. Since the densities of the two phases are similar, no contribution from "bond breaking" (as in the liquid–vapor and crystal–vapor interfaces) was expected. In an *intercrystalline* boundary, the interfacial tension consists mainly of the excess energy associated with the density deficit arising from the strong crystallographic constraints on the structure of the boundary; in a crystal–liquid boundary, however, these constraints are relaxed, and the density deficit and the attendant energetic contribution to the interfacial tension were, correctly, expected to be small.

Turnbull later pointed out that the interfacial tension therefore had to be largely *negentropic* in origin.[26] The relaxation of the liquid necessary to eliminate the density deficit leads to ordering of the liquid in a few layers near the crystal surface. These layers have no translational symmetry in the directions parallel to the interface but are localized to some degree in the perpendicular direction. As a result, the configurational entropy of the atoms in these layers is lowered, which constitutes a *positive* contribution to the interfacial excess free energy, $F = U - TS$. This effect will be discussed further using a simple model of the crystal–melt interface in Part III.

4. GLASSES

A glass is obtained by cooling a liquid, without crystallization, below the temperature where its viscosity exceeds, conventionally, 10^{14} Pa.s. It is then solidlike, and atomic transport is slow enough to prevent transformation to the stable crystalline state. Although glasses of covalent network formers (e.g., silicates) or hydrogen-bonded molecules (e.g., glucose, glycerine, etc.) have been known for a long time, the existence of glasses of simple, spherelike atoms such as metals is a fairly recent discovery. On the basis of their theory for the temperature dependence of atomic transport in simple liquids, Turnbull and Cohen

[26]D. Turnbull, *in* "Physics of Non-Crystalline Solids" (J. A. Prins, ed.), p. 41. North-Holland, Amsterdam, 1964.

predicted in 1959 that glass formation should be a universal phenomenon: in the absence of crystallization, any liquid, including the metals, should exhibit a glass transition.[27] That such a temperature could, in principle, be reached with a liquid metal had already become clear from Turnbull's undercooling experiments. They also pointed out that alloy liquids with a composition close to a deep eutectic were the best candidates for metallic glass formation.[28] These predictions were confirmed by Duwez' 1960 discovery of glassy Au_4Si (close to the Au-Si eutectic) obtained by rapid quenching of the melt.[29]

In the next few years a very large number of metallic glass alloys were found, and often the deep eutectic criterion was the guide to their discovery. The largest single category among them are alloys consisting of about 80% noble or late transition metals (Au, Ni, Pd, . . .) and about 20% metalloids (Si, B, P, . . .). The next large category are alloys of early transition metals with late transition or noble metals (Ni-Nb, Cu-Zr, Cu-Ti, . . .), which can usually be prepared over a fairly broad range of compositions around the equiatomic one. A particularly interesting category are the "nearly-free electron" glasses, such as Mg_2Zn; they typically form around a eutectic composition that is located symmetrically from a Laves (Frank–Kasper) phase, such as $MgZn_2$.[30]

The availability of metallic glasses provided an extraordinary opportunity for studying the structure of simple liquids, since some of the thermal effects that obscure structural details are diminished. The most important discovery was made by Cargill,[31,32] who showed that the radial distribution function he had obtained from diffraction experiments on amorphous Ni-P alloys could be fit almost perfectly by the one determined by Finney[33] for the dense-random packing of hard spheres (see Fig. 10).

The dense-random packings of hard spheres (DRP) were originally mechanical models obtained by placing a large number of identical hard

[27]M. H. Cohen and D. Turnbull, *J. Chem. Phys.* **31,** 1164 (1959).

[28]D. Turnbull and M. H. Cohen, *J. Chem. Phys.* **34,** 120 (1961); M. H. Cohen and D. Turnbull, *Nature* **189,** 131 (1961).

[29]W. Klement, R. H. Willens, and P. Duwez, *Nature* **187,** 869 (1960).

[30]Reviews of the structure and composition ranges of metallic glasses can be found in D. Turnbull, *J. de Physique* **35,** C-4, 1 (1974); H. A. Davies, *in* "Amorphous Metallic Alloys," (F. E. Luborsky, ed.), p. 8. Butterworths, London, 1983; C. N. J. Wagner (ibid. p. 58); K. Suzuki (ibid. p. 74); J. Hafner, *in* "Glassy Metals I, Topics in Applied Physics" **46,** ed.), p. 93. Springer, Berlin, 1981.

[31]G. S. Cargill III, *J. Appl. Phys.* **41,** 12 and 2248 (1970).

[32]G. S. Cargill III, *in* "Solid State Physics." (F. Seitz and D. Turnbull, eds.), vol. 30, p. 227. Academic Press, N.Y., 1975.

[33]J. L. Finney, *Proc. Roy. Soc. London* **319A,** 497 (1970).

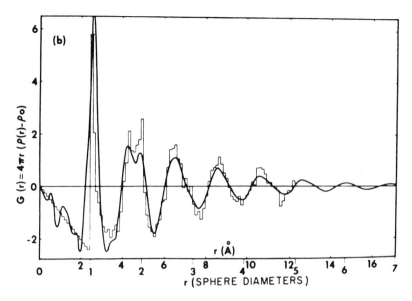

Fig. 10. Comparison of the reduced radial distribution functions for amorphous $Ni_{76}P_{24}$ (smooth line) and Finney's dense-random packing of hard spheres (histogram). [From G. S. Cargill III, *J. Appl. Phys.* **41**, 2248 (1970).]

spheres (such as ball bearings) in a container, the walls of which were roughened to prevent "nucleation" of close-packed crystals, and mechanically densifying this assembly as much as possible. The structure was then frozen by allowing paint or wax to fill the interstices. After being frozen it could be taken apart sphere by sphere for analysis. The first systematic studies of this kind were made by Scott[34] and by Bernal.[5,35] Bernal determined the nearest-neighbor bonds between the spheres, on the basis of an approximate maximum separation criterion and analyzed the assembly of polyhedra outlined by those bonds. He found that 86% of them were tetrahedra, 6% were octahedra, and the remainder were larger polyhedra, many of which were "deltahedra." These are polyhedra that in their ideal shape have identical equilateral triangles as faces.[36-38] Some of them have already been encountered above: the tetrahedron, the triangular bipyramid (double tetrahedron), the octahedron, the pentagonal bipyramid (Fig. 3b; counted as 5 tetrahedra by Bernal), and

[34]G. D. Scott, *Nature* **188**, 908 (1960); *Nature* **194**, 956 (1962).
[35]J. D. Bernal, *Nature* **188**, 910 (1960).
[36]H. M. Cundy, *Math. Gaz.* **36**, 263 (1952).
[37]H. Freudenthal and B. L. van der Waerden, *Simon Stevin* **25**, 115 (1947).
[38]M. F. Ashby, F. Spaepen, and S. Williams, *Acta Met.* **26**, 1647 (1978).

the icosahedron (Fig. 4b; a deltahedron large enough to contain a central sphere and thus be counted as 20 tetrahedra), which are the deltahedra with 4, 5, 6, 7, and 12 vertices, respectively. The deltahedra that form "Bernal holes" in the DRP are the dodeca-deltahedron, the capped triagonal prism and the capped Archimedian antiprism, which have 8, 9, or 10 vertices, respectively. (They are illustrated in Fig. 23 in Section 6.)[39] They can be considered as interstitial spaces between the polytetrahedral configurations that make up most of the structure. Frost[40] has reanalyzed the hole structure in the DRP by using the coordinates of Finney's 9000 sphere model and a rigorous maximum nearest-neighbor distance of $\sqrt{12/7} = 1.3093$ sphere diameters. This value was chosen to prevent interpenetration of the tetrahedra formed by the nearest-neighbor bonds. His analysis confirmed the preponderance of tetrahedral configurations (86% by number) but found that the interstitial space contained, besides the Bernal holes, some very large, nonconvex deltahedra (2% by number; 42% by volume). It seems possible, however, that they can be divided up into smaller deltahedra around the larger interstitial sphere positions by using a more flexible nearest-neighbor definition that more closely approximates Bernal's original criterion.

The basic polytetrahedral character of the DRP is also apparent from the coordination polyhedra. Finney[33] determined the Voronoi cells of all the spheres in his model and found that five-sided faces, corresponding to rings of five tetrahedra as in Fig. 3b, predominate. His actual distribution of Voronoi faces is shown in Fig. 11. This feature was already known from simple experiments in which very soft spheres (e.g., clay,[41,34] or peas[42]) are initially randomly packed and then compressed, which makes them take the shape of their Voronoi cells. The resulting polyhedra also had many five-sided faces. Occasionally perfect pentagonal dodecahedra are found, corresponding to a perfect icosahedral coordination.

A striking natural example is provided by the calcite oolite shown in Fig. 12.[41] These oolites are polycrystalline aggregates that form originally

[39]Bernal's original list of the polyhedra in the dense-random packing consisted of the tetrahedron, the octahedron, the half octahedron, the trigonal prism, the Archimedian antiprism, and the dodecadeltahedron. Often, however, the square faces of the trigonal prism and the Archimedian antiprism are capped with the half octahedra, thus forming the deltahedra of primary interest here. Note that, if one used the Voronoi construction to assign near neighbors, all the more complex Bernal holes (except for a set of measure zero) would be partitioned into (distorted) tetrahedra.

[40]H. J. Frost, *Acta Med.* **30**, 889 (1982).

[41]E. V. Shannon, *J. Wash. Acad. Sci.* **17**, 409 (1927). We thank D. J. Milton for bringing this paper to our attention.

[42]Stephen Hales, "Vegetable Staticks: An Account of Some Statistical Experiments on the Sap in Vegetables." Innys and Woodward, London, 1727.

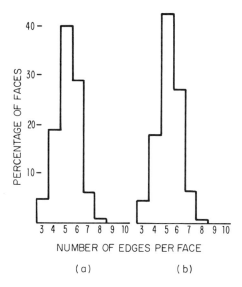

FIG. 11. Distribution of the number of edges per face of the Voronoi polyhedra of two unrelaxed dense-random packings of hard spheres. (a) Finney model; (b) Scott model. [From J. L. Finney, *Proc. Roy. Soc.* **319A,** 497 (1970).]

FIG. 12. Calcite oolite with pentagonal dodecahedral morphology corresponding to its Voronoi polyhedron, formed by continued growth from interstitial solution in a dense-random packing of initially spherical oolites.

as spheres from solution. When the spheres reach a certain size they settle out in a randomly packed assembly similar to the DRP. Since calcite continues to be precipitated from the interstitial solution, flat faces are formed between initially touching spheres, thus leading to formation of a natural "Voronoi polyhedron" for each sphere.

The packing fraction (the fraction of the volume taken up by the spheres) of the DRP is very well established at 0.636,[33,43] albeit only empirically. It should be noted that this value is smaller than that of the crystalline close-packed (face-centered cubic, or hexagonal) (0.740) or body-centered cubic structure (0.680). (See Table III in Section 7).

For reasons of ease of formation and thermal stability, all metallic glasses are alloys. Some nearly pure amorphous metals have been formed by vapor deposition on substrates at liquid helium temperature,[44,45] but some impurity is still required to prevent collision-limited crystallization at low temperatures.[46] Their radial distribution function is fit well by the DRP. The relation of the structure of the heavily alloyed glasses to the DRP is somewhat less straightforward. In the metal–metalloid glasses, x-ray diffraction is dominated by scattering from the majority component of heavy metal atoms. Cargill's success in fitting the radial distribution function (RDF) of Ni-P glasses with a DRP[31] prompted Polk[47] to suggest that the metal atoms form a DRP skeleton, with the metalloids occupying the larger holes in this structure. Later detailed investigations of the size of the interstitial holes in the DRP[40] showed that they are somewhat too small to accommodate the metalloids: in order to accommodate 20% interstitial metalloids (a typical composition for these glasses), the metalloid/metal size ratio should be less than 0.74. Since for all glass formers this ratio is substantially larger[48] (0.93–0.83), some distortion of the DRP structure to accommodate the metalloids is required.

The qualitative implications of the Polk "hole-filling" model nevertheless remain important in the understanding of the structure of alloy glasses. They are: (i) the metalloid atom has only metal atoms as nearest neighbors, and (ii) the coordination of the metalloid atoms is lower than that of the metal atoms. Subsequent diffraction work[30,49] has confirmed

[43]E. J. LeFevre, *Nature Physical Science* **235**, 20 (1972).

[44]W. Buckel, *Z. Phys.* **138**, 136 (1954).

[45]T. Ichikawa, *Phys. Sta. Sol.* **19**, 707 (1973).

[46]For a review see F. Spaepen and D. Turnbull, *in* "Laser Annealing of Semiconductors" (J. M. Poate and J. W. Mayer eds.) p. 15. Academic, N.Y., 1982.

[47]D. E. Polk, *Scripta Met.* **4**, 117 (1970); *Acta Met.* **20**, 485 (1972).

[48]F. Spaepen, *Proc. 3rd Int. Conf. on Rapidly Quenched Metals*, vol. 2, (B. Cantor, ed.), p. 235. The Metals Society, London, 1979.

[49]J. F. Sadoc and J. Dixminer, *Mat. Sci. Eng.* **23**, 187 (1976); Y. Waseda, *Prog. Mat. Sci.* **26**, 1 (1981). T. M. Hayes, J. W. Allen, J. Tauc, B. C. Giessen, and J. J. Hauser, *Phys. Rev. Lett.* **40**, 1282 (1978).

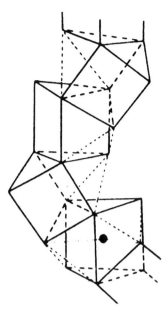

FIG. 13. Illustration of how, in a stereochemically defined model, trigonal prisms are combined to form the nine-vertex coordination polyhedra (the capped trigonal prism of Fig. 23(b)) of the metalloid atoms in metal-metalloid glasses. [From P. H. Gaskell, in "Glassy Metals II, Topics in Applied Physics" **53**, edited by H. Beck and H.-J. Güntherrodt, Springer, Berlin (1983), p. 23.]

this. In fact, the chemical ordering in analogous crystalline intermetallic compounds has exactly the same qualitative features.[50] For example, in crystalline Ni_3P each phosphorus atom is surrounded by 9 nickel atoms in a capped trigonal prism configuration,[51] which is one of the Bernal holes discussed above. These sorts of observations have led to the development of the "crystal chemically defined models" for metallic glasses,[52,53] in which trigonal prisms with a metalloid atom at the center are connected to form an amorphous structure (see Fig. 13).

Attempts have also been made to incorporate the high degree of chemical ordering in the computer algorithms for generating binary dense random packings. The most detailed work of this type has been done by Boudreaux,[54] who has built the largest models of this type and has paid

[50] S. Rundqvist, *Ark. Kemi* **20**, 67 (1962).
[51] S. Rundqvist, *Acta Chem. Scand.* **16**, 995 (1962).
[52] P. H. Gaskell, *in* "Glassy Metals II, Topics in Applied Physics" **53** (Springer, Berlin, 1983) p. 5.
[53] J. M. Dubois and G. LeCaer, *Acta Met.* **32**, 2101 (1984).
[54] D. S. Boudreaux and J. M. Gregor, *J. Appl. Phys.* **48**, 152 (1977); *J. Appl. Phys.* **48**, 5057 (1977).

special attention to problems of isotropy and compositional homogeneity. By relaxing the clusters in appropriately defined Lennard–Jones-type potentials, he obtained detailed agreement with the experimental radial distribution functions and structure factors.

III. Statistical Geometry

5. FRUSTRATION IN TWO DIMENSIONS

As discussed in the Introduction, the obvious analogue of the icosahedron in two dimensions occurs when 6 triangles combine to form a hexagon. Unlike the icosahedron, however, the resulting cluster of 7 identical disks can be packed neatly to form a space-filling triangular lattice. Although there is no frustration for identical particles on a flat surface, illuminating analogies with 3-D polytetrahedral order can be obtained by introducing frustration artificially into 2-d "polytriangular" particle arrays.

In two dimensions, liquids can be regarded as regions of "hexatic"[55–57] order interrupted by disclinations, which are local points of five- and sevenfold symmetry in an otherwise sixfold medium. A microscopic definition of point disclinations results from applying the Dirichlet construction to a 2-D particle configuration[2] and looking for coordination numbers that deviate from 6. This construction uses a two-dimensional version of the more familar 3-D Voronoi or Wigner–Seitz cells to break 2-D space into disjoint regions surrounding each particle. Its effect here is to "triangulate" the medium by assigning a set of nonoverlapping bonds connecting "nearest neighbors," defined as particles whose cells share a common edge. Disclinations, defined geometrically as the deviation of the local coordination number from 6, provide a vivid way of thinking about the statistical mechanics of undercooling. As shown in Fig. 14, a high-temperature liquid can be viewed as a dense "plasma" of 5's and 7's, to which we assign "charges" of +1 (diamonds) and −1 (asterisks), respectively. As the liquid cools, the triangles connecting nearest neighbors become less distorted as the five- and sevenfold disclinations begin to pair off and annihilate.

[55]B. I. Halperin and D. R. Nelson, *Phys. Rev. Lett.* **41,** 121 and 519(E) (1978); *Phys. Rev.* **B19,** 2457 (1979).

[56]D. R. Nelson, *Phys. Rev.* **B25,** 269 (1982). D. R. Nelson and B. I. Halperin, *Phys. Rev.* **B19,** 2457 (1979).

[57]D. R. Nelson, M. Rubinstein, and F. Spaepen, *Phil. Mag.* **A46,** 105 (1982).

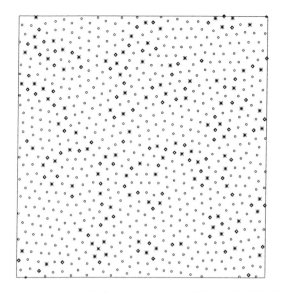

FIG. 14. Dirichlet construction applied to computer simulation by R. Morf of particles in a dense liquid. Particles with five neighbors are indicated by diamonds and those with seven neighbors are denoted by asterisks. All other particles have six neighbors.

A famous theorem of Euler asserts that, in a triangulation such as that provided by the Dirichlet construction,

$$F - E + V = 0, \qquad (5.1)$$

where F is the number of cells, E is the number of edges, and V is the number of vertices.[8] The right-hand side of this equality is zero (rather than 2, the value for a triangulation of sphere) because we imagine particles with periodic boundary conditions, i.e., on a surface with the topology of a torus. To use this theorem to derive a constraint on the average coordination number, note first that particles occupy vertices in a Dirichlet triangulation, so $V = N$, where N is the total number of particles. Since each cell contributes 3 edges, and each edge is associated with 2 triangular cells, we have $E = 3F/2$, which can be used to eliminate F from Eq. (5.1). The total number of edges, however, is $E = (1/2) \sum_{i=1}^{N} Z_i$, where Z_i is the coordination number of the ith vertex and factor 1/2 eliminates double counting. These facts can be combined to show that the average coordination number is exactly 6,

$$\bar{Z} = \frac{1}{N} \sum_{i=1}^{N} Z_i = 6. \qquad (5.2)$$

Supposing for simplicity that the fluid is dense enough so that only 5-, 6- and 7-coordinated particles are present; we see from Eq. (5.2) that there must be exactly as many fivefold disclinations as sevenfold ones. The "plasma" of Fig. 14 is therefore charge-neutral, and the pairing for 5's and 7's envisaged above can be continued with decreasing temperatures until only sixfold particles are present in a defect-free triangular lattice. This topological point of view suggests that freezing can be a fairly gradual process in two dimensions. Indeed, the disclination–dislocation theory of 2-d freezing predicts two continuous phase transitions,[55,56] with the latent heat of freezing spread out in an intermediate, hexatic phase. The pairing of disclinations can only be carried to completion because of the absence of frustration.

To make this topological picture more typical of undercooling in three dimensions we must add frustration to prevent easy crystallization into a triangular lattice. One could, for example, consider cooling miscible binary liquids of two different sizes: To minimize the energy or maximize the density, we must now pack a mixture of equilateral, isosceles and other kinds of triangles. Experimental[57,58] and theoretical[59] studies of such systems show that, although it is easy to destroy translational order with this sort of frustration, orientational order is much more robust. One often finds hexatic "glasses" with extended orientational correlations at low temperatures, instead of the isotropic behavior expected for a three-dimensional undercooled liquid or glass. We expect, moreover, that the geometric frustration of tetrahedral packings in three dimensions can, in principle produce glasses even for *identical* particles; we would like to find a way to mimic this phenomenon in two dimensions.

To produce frustration leading naturally to glasses for identical particles in two dimensions, it is instructive to consider first dense random packing on the curved surface of a sphere. If we use spherical polar coordinates (r, ϕ), where r is the distance to the north pole and ϕ is the azimuthal angle, the metric on a sphere of radius $R_0 = \kappa^{-1}$ is[8]

$$d^2s = d^2r + \frac{\sin^2 \kappa r}{\kappa^2} d^2\phi. \qquad (5.3)$$

Now, slightly fewer than six particles will pack around a central one. It is easy to program a computer to pack hard disks on a sphere by using this metric. An equilateral triangle of disks is centered on the north pole, and successive disks are brought into contact with the growing cluster so that

[58]M. Rubinstein and D. R. Nelson, *Phys. Rev.* B**26**, 6254 (1982).
[59]D. R. Nelson, *Phys. Rev.* B**27**, 2902 (1983).

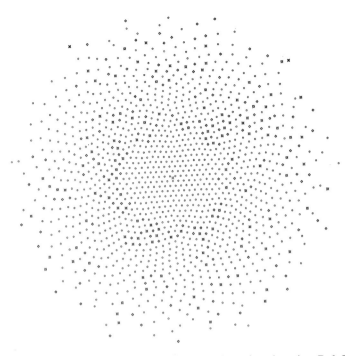

FIG. 15. Polar projection of particles packed on the surface of a sphere (see Ref. 58). Disks are deposited in a compact array starting at the north pole. The curvature-induced frustration eventually forces disclinations (asterisks and diamonds) into the initially sixfold particle packing.

they are as close as possible to the center. A very similar construction, which starts with a tetrahedron of hard spheres, has been used by Bennett[60] to obtain dense-random packing models of metallic glasses in three dimensions.

In 2-D flat space; i.e., as the radius of the sphere tends to infinity, this algorithm produces a triangular closed-packed lattice. As shown in Fig. 15, the result for finite R_0 is initially a triangular lattice. The frustration builds up with increasing r, however, until disclinations must be introduced at a radius of order κ^{-1}.[58] Note that frustration introduces a characteristic length scale (in the case, the radius of the sphere) beyond which the locally preferred hexagonal order can no longer maintain itself. Euler's theorem on a sphere reads[8]

$$F - E + V = 2, \qquad (5.4)$$

[60]C. H. Bennett, *J. Appl. Phys.* **43**, 2727 (1972).

which can be used to show for a spherical Dirichlet triangulation that

$$\sum_{i=1}^{N} (6 - Z_i) = 12, \tag{5.5}$$

i.e., the triangulations contain 12 more fivefold disclinations than sevenfold ones. One plausible candidate for the ground state is a "superlattice" of these 12 disclinations arrayed at the vertices of a regular icosahedron. A familar example is the 32–panel surface of a soccar ball. If we think of each panel as the Voronoi cell of particle, we find a "defect lattice" of 12 anomalous 5-coordinated particles immersed in a sea of 20 hexagonal sites. As we shall see the Frank–Kasper phases play a similar role in three-dimensional flat space.

An attractive and theoretically interesting alternative to spherical packings is to pack the particles on an infinite surface of constant *negative* Gaussian curvature.[61] The properties are defined by a metric which reads, in hyperbolic polar coordinates,[8]

$$d^2s = d^2r + \frac{\sinh^2 \kappa r}{\kappa^2} d^2\phi. \tag{5.6}$$

The quantity κ^{-1} is a tunable frustration length scale that tends to infinity as a space becomes flat ($\kappa \to 0$). Physically, κ^{-1} is the length scale on which frustration effects become important on this "surface of saddle points." When κ is nonzero, cracks open up between 6 particles symmetrically arranged around a central atom to form a hexagon, in analogy with a similar property of the icosahedron (see Fig. 16). One then has the same difficulty in filling space with equilateral triangles as for tetrahedra in 3 D flat space. The ratio l/d is a measure of the frustration in both cases.

It is straightforward to integrate up the metric in Eq. (5.6) and find the distance l_{ab} between two particles with polar coordinates (r_a, ϕ_0) and (r_b, ϕ_b):

$$\cosh(\kappa l_{ab}) = \cosh(\kappa r_a) \cosh(\kappa r_b) - \sinh(\kappa r_a) \sinh(\kappa r_b) \cosh(\phi_a - \phi_b). \tag{5.7}$$

This formula is closely related to the distance between 2 points on a sphere with imaginary radius $R_0 = i/\kappa$. Let us apply Eq. (5.7) to the

[61]D. R. Nelson, in "Topolocial Disorder in Condensed Matter" (T. Ninomiya and F. Yonezawa, eds.). Springer, Berlin, 1983.

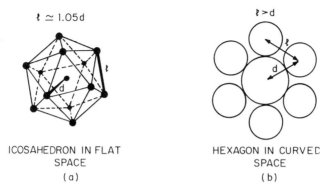

ICOSAHEDRON IN FLAT   HEXAGON IN CURVED
SPACE   SPACE
(a)   (b)

FIG. 16. Analogy between frustration of (a) an icosahedron in three-dimensional flat space and (b) a hexagon embedded in a two-dimensional hypobolically curved surface. (The apparent size difference of the disks is an artifact of projecting the hexagon onto a flat surface.) In both cases (a) and (b), frustration opens up cracks between the particles on the surface.

arrangement of 7 disks shown in Fig. 16b, taking the origin of our polar coordinate system to be the center of the middle disk. The apparent size difference in the disks is due to projecting a small portion of curved space onto a flat piece of paper; the disks all have an identical diameter d. Assuming the disks on the perimeter are symmetrically arranged so as to just touch the disk at the center, we find that the ratio of the geodesic distance l joining their centers to disk diameter d is

$$\frac{l}{d} = \frac{1}{\kappa d} \cosh^{-1}\left[\frac{1}{2}\cosh^2(\kappa d) + \frac{1}{2}\right]. \qquad (5.8)$$

As shown in Fig. 17, l exceeds d for $\kappa > 0$, so there is indeed extra space between the perimeter disks, just as in the case of the spheres in Fig. 4c.

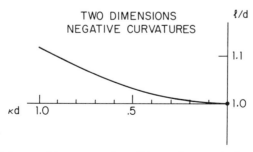

FIG. 17. Ratio of distance l between surface disks to distance d of a surface disk to the center of the curved space hexagon shown in Fig. 16(b). This measure of the frustration is shown as a function of κd, where $-\kappa^2$ is the Gaussian curvature of the surface.

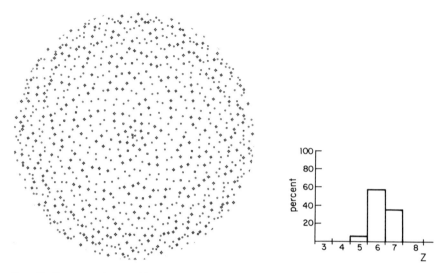

Fig. 18. Dense-random packing on a surface of constant negative curvature. As in Fig. 14, diamonds and asterisks indicate 5- and 7-coordinated particles. In contrast to Fig. 14, frustration induces a net excess of sevenfold disclinations as indicated in the coordination number Z histogram shown in the inset.

Figure 18 shows a hard-disk particle configuration obtained via a deterministic packing algorithm for hard disks with $\kappa d = 0.5$, where d is the disk diameter.[58] Figure 18 is a 2-D analogue of dense random packing. The coordination number histogram in the inset shows that most particles have coordination number 6, with an asymmetry between the remaining 5- and 7-coordinated particles (diamonds and asterisks, respectively). If the 5's and 7's are regarded as microscopic point disclination changes, we see from the histogram in the inset that introducing frustration now produces a bias toward sevenfold disclinations.

It is easy to see more quantitatively that a finite net disclination charge density must characterize a surface of constant negative curvature. We imagine triangulating an array of particles on such a manifold with a set of geodesic lines joining neighboring atoms. The integral curvature of one of the geodesic triangles obtained in this way is[8]

$$\int_{\Delta}\int_{ABC} -\kappa^2 \, dS = A + B + C - \pi, \qquad (5.9)$$

where A, B, and C are the three angles characterizing the triangle. Summing over all triangles i in a large region, and ignoring edge effects, we obtain

$$-\kappa^2 S = \sum_i (A_i + B_i + C_i - \pi) = 2\pi V - \pi F, \qquad (5.10)$$

where V is the number of vertices (particles) in the net, F is the number of triangular faces, and S is the total surface area of the region. Since the number of edges (bonds) in the net is $E = 3F/2$, and we also have $E = \bar{Z}V/2$, Eq. (5.10) can be reexpressed in the form

$$\bar{Z} = 6 + 3s\kappa^2\pi, \qquad (5.11)$$

where s is the surface area per particle. This formula quantifies the small excess of sevenfold disclination shown in Fig. 18.

Although the ground state for particles on a hyperbolic manifold is probably a Frank–Kasper-like superlattice of 7's embedded in an otherwise 6-coordinated medium, one might expect a glassy configuration like Fig. 18 when particles are cooled rapidly on a hyperbolic manifold. It is in any event clear from Eq. (5.11) that the pairing of 5's and 7's that leads to a triangular lattice can never be carried to completion in the presence of geometrical frustration.

6. Three Dimensions

a. *Polytope* {3, 3, 5}

As shown in Fig. 16, the cracks on the surface of an icosahedral cluster of atoms also appear on the surface of a hexagon embedded in a surface of constant negative curvature. This frustration vanishes, however, as the curvature tends to zero. It turns out to be useful as well to vary the frustration in *three*-dimensional particles packings, and, if necessary, tune it to zero. One appealing way to change the frustration is to imagine particles embedded on the curved three-dimensional surface S^3 of a four-dimensional sphere.

Points $\mathbf{x}$ on the surface of a four-dimensional sphere of radius κ^{-1} are conveniently parameterized by using polar coordinates,

$$\mathbf{x} = \kappa^{-1}[\cos\phi \sin\theta \sin(\kappa r), \; \sin\phi \sin\theta \sin(\kappa r), \; \cos\theta \sin(\kappa r), \; \cos(\kappa r)], \qquad (6.1)$$

where (r, θ, ϕ) play a role similar to ordinary polar coordinates in flat space. It is easy to show by using this four-dimensional embedding of S^3

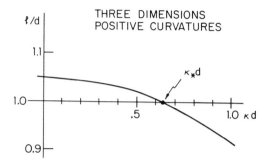

FIG. 19. Frustration as a function of curvature in three dimensions. The ratio l/d defined in Fig. 16(a) for an icosahedron embedded in the surface of four-dimensional sphere of radius $R = \kappa^{-1}$ is shown as function of κd.

that the geodesic distance l_{ab} between two points (r_a, θ_a, ϕ_a) and (r_b, θ_b, ϕ_b) is given by

$$\cos(\kappa l_{ab}) = \cos(\kappa r_a) \cos(\kappa r_b) + \sin(\kappa r_a) \sin(\kappa r_b)$$
$$\times [\cos \theta_a \cos \theta_b + \sin \theta_a \sin \theta_b \cos(\phi_a - \phi_b)]. \qquad (6.2)$$

The 20 tetrahedra in a symmetric, icosahedral cluster of 13 particles divide the 4π of solid angle surrounding the central particle into 20 equal pieces. This is true in both flat and curved space. It follows that angles between the 12 radial bonds are the same in S^3 as in flat space. Thus, we can take for angular coordinates of 2 neighboring radial bonds $(\theta_a, \phi_a) = (0, 0)$ and $(\theta_b, \phi_b) = (\cos^{-1}[(1 + 3\sqrt{5}/5)/2], 0)$, just as in flat space.[62] If the length of the bonds is d, we find from Eq. (6.2) that the geodesic distance between surface atoms l is related to d by

$$\frac{l}{d} = \frac{1}{\kappa d} \cos^{-1}[1 - (1 - \sqrt{5}/5) \sin^2 \kappa d]. \qquad (6.3)$$

The relationship between frustration and curvature is summarized in Fig. 19. When κ tends to zero, we recover the flat space ratio l/d defined in Fig. 16(a),

$$l/d = \frac{4}{\sqrt{10 + 2\sqrt{5}}} \approx 1.051462. \qquad (6.4)$$

[62]A useful source of geometrical information on the icosahedron is "CRC Standard Mathematical Tables" (S. M. Selby, ed.), pp. 15 and 16. The Chemical Rubber Co., Cleveland, 1970.

As illustrated in Fig. 19, however, one can adjust l/d to unity by taking $\kappa = \kappa^*$, where

$$\kappa^* d = \cos^{-1}((1 + \sqrt{5})/4) = 0.638318. \qquad (6.5)$$

Equation (6.16) defines a special "commensurate" sphere radius $(\kappa^*)^{-1}$ such that the frustration vanishes. It is precisely at this radius that Coxeter[6–8] shows how 600 perfect tetrahedra can be imbedded without frustration on the surface of a four-dimensional sphere to form a regular lattice of 120 particles. Every one of the vertices in this four-dimensional Platonic solid (or "polytope") sits in an identical, 12-particle isosahedral coordination shell. A two-dimensional projection[7] is shown in Fig. 20. This icosahedral crystal is called "polytope $\{3, 3, 5\}$" because it is composed of tetrahedra (denoted by the symbol $\{3, 3\}$ because 3 equilateral *tri*angles meet at every vertex) with 5 tetrahedra wrapped around every bond. The 7.5° crack associated with the pentagonal bond spindle of Fig. 3a has closed up completely. The Wigner–Seitz cells associated with the vertices of polytope $\{3, 3, 5\}$ are shown in Fig. 2a. The role of $\{3, 3, 5\}$ in describing dense-random packing of hard spheres

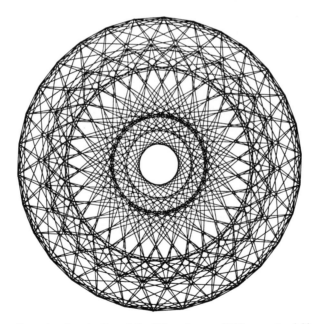

FIG. 20. Two-dimensional projection of the 120 vertices and 720 nearest-neighbor bonds of polytope $\{3, 3, 5\}$. The Wigner–Seitz cells of this perfect icosahedral lattice are shown in projection in Fig. 2a.

in flat space was recognized by Coexter in 1958.[6] It was suggested as a model for metallic glasses by Sadoc in 1980.[10] Kleman and Sadoc[9] considered a variety of tetrahedral particle arrays in both spherical and hyperbolic spaces and suggested that defects like cut surfaces and disclinations would be necessary to map such configurations into flat space.

Here we take a related but different point of view: We shall define disclinations microscopically directly in flat space by a 3-D analogue[11] of the 5–7 construction discussed in Section 5. As we shall see, disclinations are inevitable except in the special commensurately curved space S^3 that admits polytope $\{3, 3, 5\}$ as a ground state. Polytope $\{3, 3, 5\}$ will be used explicitly, however, only in Part IV, where we use it to construct an order parameter theory of the statistical mechanics of glass.[13,14] Instead of projecting polytope $\{3, 3, 5\}$ onto flat space,[9,10] we shall project a local flat space particle configuration onto polytope $\{3, 3, 5\}$ to define an order parameter.[12-14]

b. Defect Lines

A microscopic construction for disclination lines in any three-dimensional particle configuration is summarized in Fig. 21. Following early work by Frank and Kasper,[1] we first partition space into tetrahedra by assigning near-neighbor bonds via the Voronoi construction.[2] Links of plus or minus disclination line are assigned to bonds, depending on their local environment. A "disclination-free" bond is surrounded by 5 tetrahedra (Fig. 21) and plays a role analogous to the 6-coordinated

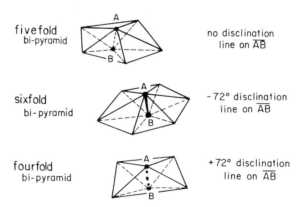

FIG. 21. Microscopic definition of defect lines associated with polytetrahedral order. Sixfold and fourfold bond spindles play the roles, respectively, of the asterisks and diamonds in Fig. 18. Polytope $\{3, 3, 5\}$ consists of fivefold bond spindles only.

particles in Section 5. The excitations analogous to 7's and 5's in two dimensions are the bonds surrounded by 6 and 4 tetrahedra. All bonds in polytope $\{3, 3, 5\}$ are spindles for fivefold bipyramids, composed of perfect tetrahedra. Although bipyramids of perfect tetrahedra are impossible in flat space, less distortion is required for fivefold bipyramids than for four- and sixfold bipyramids. As discussed in Frank and Kasper,[1] sixfold bipyramids require less distortion than fourfold ones. Other kinds of bipyramids represent more exotic excitations.

Frank and Kasper give a proof that a link of sixfold line cannot simply terminate at an atom whose remaining bonds are fivefold. Such an atom

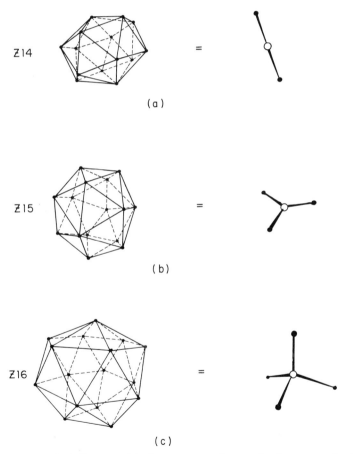

(a)

(b)

(c)

FIG. 22. The canonical Kasper polyhedra, represented as nodes in a network of $-72°$ disclinations (see also Fig. 7). The particles at the centers of these polyhedra are not shown for clarity.

would have coordination number $Z = 13$.[1] (There are other ways to make coordintion shells with $Z = 13$; see below.) It is easy to repeat the Frank–Kasper argument and show that a fourfold link terminating at a particle whose remaining bonds are fivefold (corresponding to $Z = 11$) is also impossible.

A number of standard coordination polyhedra have simple representations in terms of this construction. The canonical Kasper polyhedra[1] are shown in Fig. 22 (see also Fig. 7), together with their representation as nodes for $-72°$ disclination lines. The antidefects of the canonical Kasper polyhedra, which are nodes of $+72°$ disclination line, are shown in Fig. 23. The resulting $Z = 10$, 9, and 8 coordination shells are the canonical hole polyhedra discussed by Bernal (see Section 4).[2,5]

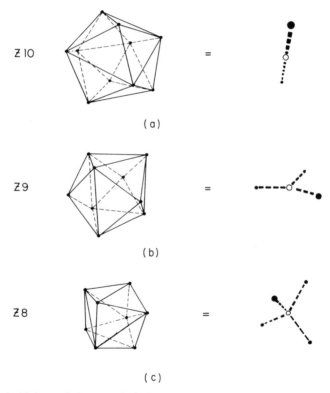

$Z\ 10$

$=$

(a)

$Z\ 9$

$=$

(b)

$Z\ 8$

$=$

(c)

FIG. 23. Antidefects of the canonical Kasper polyhedra. If particles (*e.g.*, the smaller metalloid atoms in a binary metallic glass) are placed in the centers of these polyhedra, they become nodes in a network of $+72°$ disclinations. In the absence of central particles, the polyhedra are the canonical "holes" discovered in dense random packing by Bernal (see Refs. 2 and 5).

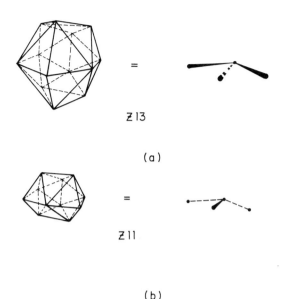

(a)

(b)

FIG. 24. (a) Z13 and (b) Z11 coordination polyhedra in a polytetrahedral particle packing. With the insertion of central particles, these can be used to terminate the "bubbles" shown in Fig. 25.

Upon filling the free volume at the surface of an icosahedron with an extra atom, one obtains a particle with $Z = 13$, characterized by 10 fivefold bonds, and 1 fourfold bond interposed between 2 sixfold ones, as shown in Fig. 24a. It is natural to suppose that $Z = 13$ atoms will often occur in pairs, terminating the bubbles of two solid lines and one dashed line shown in Fig. 25. A small bubble corresponds to an interstitial (actually a split interstitial) in polytope $\{3, 3, 5\}$. The antidefect of the bubble consisting of 2 $Z = 13$ atoms formed by removing one particle from an icosahedral coordination shell to form a $Z = 11$ (Fig. 24b). The particle at the center then has 8 fivefold bonds, together with a sixfold bond interposed between 2 fourfold ones. A bubble terminated by two $Z = 11$ particles is shown in Fig. 25. A small bubble of this kind can be viewed as a vacancy in polytope $\{3, 3, 5\}$.

Lines of six- and fourfold links are microscopically defined wedge disclination lines in an otherwise icosahedral medium. By following the rotations of pentagonal bond spindles surrounding the 2 colinear sixfold bonds of a $Z = 14$ Kasper polyhedron in Fig. 22, one sees that these sixfold bonds are indeed segments of $-72°$ disclination line. A similar construction shows that the 2 colinear fourfold bonds in the $Z = 10$ coordination polyhedron in Fig. 23 are segments of $+72°$ disclination line.

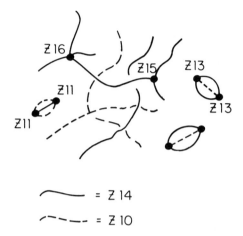

FIG. 25. Undercooled liquid represented as a tangled mass of disclination lines obtained from the topological building blocks shown in Figs. 22–24. Icosahedra, or fragments of icosahedra in the form of fivefold bond spindles, occupy the regions free of disclination lines. In this sense, this figure is the dual of Fig. 33, which shows the icosahedra present in a computer simulation of an undercooled liquid.

Figure 25 is intended to represent a polytetrahedral uncooled liquid, with regions of short-range order broken up by a network of disclination lines.

c. Dense Random Packing and the Frank–Kasper Phases

The microscopic defect construction described above can be applied to dense random packing models. As discussed in Part II, dense-random packing models are believed to capture the intrinsic structure of liquids just above the glass transition. Figure 26 shows the distribution of edges per face for Voronoi polygons in a dense-random packing model[63] constructed by Ichikawa[64] for amorphous iron films using Bennett's packing algorithm.[60] Also shown are results for this model after relaxation in a soft potential (see also Fig. 11).[63] It is easy to show that the number of edges on a particular face of a Voronoi polygon is the same as the number of tetrahedra surrounding the bond bisected by that face. Thus, Fig. 26 is a direct measure of the distribution of defect lines in the packing. As originally emphasized by Bernal,[5] fivefold bipyramids dominate in dense-random packing. Note, however, that there are more

[63]R. Yamamoto, H. Shibuta, T. Mihar, K. Haga, and M. Doyama, *in Proceedings of the International Conference On Rapidly Quenched Metals,* (T. Masumoto and K. Suzuki eds.), Vol. 1. Japan Institute of Metals, Sendai, 1982.
[64]T. Ichikawa, *Phys. Status Solidi* **19,** 707 (1973).

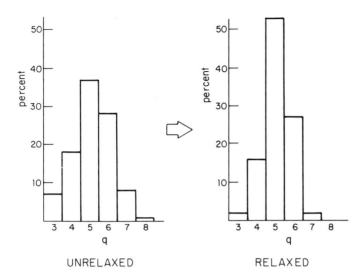

FIG. 26. Edge-per-face Voronoi cell statistics for dense-random packing. The number of edges of a face is also the number of tetrahedra wrapped around the bond piercing of that face. Fivefold bond spindles predominate in the unrelaxed state, and their number increases upon relaxation. The asymmetry between four- and sixfold disclination lines is evidence for an unpairable excess of $-72°$ disclinations induced by frustration. Note the similarity to the inset of Fig. 18.

links of sixfold than of fourfold disclination line. This is a direct result of the frustration in 3-D of flat space and is quite similar to the asymmetry between 7's and 5's displayed in Fig. 18. When the original model is relaxed, the number of fivefold bonds increases, at the expense of the number of nonicosahedral bonds. The relaxation can be viewed as an annihilation four- and sixfold links of disclination line. The more exotic three- and sevenfold links also participate in this pairing process.

Table II compares the results displayed in Fig. 26 with the same

TABLE II.   PERCENTAGES OF q-FOLD BONDS IN RELAXED AND UNRELAXED ICHIKAWA–BENNETT AND FINNEY DENSE-RANDOM- PACKING MODELS (ALL VALUES IN %)

	n_3	n_4	n_5	n_6	n_7	n_8
Unrelaxed Ichikawa–Bennett	7	18	37	28	8	1
Relaxed Ichikawa–Bennett	2	16	53	27	2	0
Unrelaxed Finney	5	19	40	29	6	1
Relaxed Finney	2	20	43	32	3	0
Liquid argon near the triple point	7	20	36	27	8	2

information for Finney's unrelaxed (Fig. 7)[33] and relaxed[65] ball-bearing dense-random packing model. Although the changes are not as large as in the Bennett–Ichikawa model, the trend toward increasing the number of fivefold bonds is the same. It seems plausible that Bennett's model has more fivefold bonds (and hence more icosahedral order) because the method of construction is biased toward perfect tetrahedra.

A bcc crystal can also be analyzed in this way. Although there are no fivefold bonds, there is again an asymmetry in the eight sixfold and the six fourfold bipyramids associated with every Wigner–Seitz cell. Every atom has coordination number $Z = 14$. Close-packed fcc crystals are a special degenerate case. The existence of perfect octahedra of six particles makes the assignment of bonds via the Voronoi construction ambiguous. There are similar ambiguities with the Dirichlet construction for square lattices in two dimensions. In both cases, the ambiguity is removed by applying an infinitesimal shear strain. The coordination number of an fcc crystal then becomes 14, with the same fraction of six- and fourfold bonds as in the bcc case.

Because of the asymmetry between $-72°$ and $+72°$ disclination lines, the pairing process suggested by Fig. 26 cannot be carried to completion. There must be residual excess of unpairable sixfold bipyramids. This is precisely the situation found by Frank and Kasper in their pioneering study of complex crystalline alloys.[1] A number of crystalline phases with large unit cells, particularly in transition metal alloys, consist of atoms with $Z = 12$ icosahedral coordination shells interrupted by atoms with $Z = 14$, 15, and 16. The atoms with $Z > 12$ combine to form a contiguous disclination network threading through an otherwise icosahedral medium. Table I summarizes the properties of some typical Frank–Kasper phases. As already mentioned in Section 2, in A15 compounds, the disclination network consists of three orthogonal grids of $Z = 14$ atoms. The σ phase of, for example, CoCr consists of parallel lines of $Z = 14$ atoms threading perpendicular planar networks of $Z = 14$ and $Z = 16$ particles, while the 162-atom unit cell of Mg_{32} $(Al, Zn)_{49}$ is cominated by a disclination network consisting of interconnected dodecahedra.

The information in Table I is taken from the Frank–Kasper papers[1] and from the review by Shoemaker and Shoemaker.[23] The average number of edges per face of the Voronoi polygons $\bar{q}$ in these phases was obtained from the average coordination number $\bar{Z}$ using the formula

$$\bar{q} = 6 - 12/\bar{Z}. \qquad (6.6)$$

Note from Table I that $\bar{q}$ is always about 2% larger than 5 in the

[65]J. A. Barker, J. L. Finney, and M. R. Hoare, *Nature* **257**, 120 (1975).

Frank–Kasper phases. This excess is a direct measure of the number of anomalous sixfold bonds.

To prove Eq. (6.6), we shall use a generalization of the Euler relation, Eq. (5.1), applicable to any polytetrahedral network inscribed on the surface S^3 of a sphere in four dimensions. This generalization reads[8]

$$T - F + E - V = 0, \qquad (6.7)$$

where T, F, E, and V are, respectively, the number of tetrahedral cells, triangular faces, edges, and vertices in the network. Since this relation is independent of sphere radius R_0, it will apply in particular to the limit $R_0 \to \infty$, i.e., to flat space.

We start with the identifications $V = N$ and $E = N_b$, where N and N_b are the total number of particles and bonds, and the definitions

$$\bar{Z} = \frac{1}{N} \sum_{i=1}^{N} Z_i, \qquad (6.8a)$$

$$\bar{q} = \frac{1}{N_b} \sum_{j=1}^{N_b} q_j, \qquad (6.8b)$$

where Z_i is the coordination number of the ith vertex and q_j is the number of tetrahedra wrapped around the jth bond. It is clear that N and N_b are related, $N_b = \frac{1}{2} N \bar{Z}$. Since each tetrahedron has four faces, shared between two tetrahedra, we also have $F = \frac{1}{2}(4T) = 2T$. Using these relations to eliminate F and $E = N_b$ from Eq. (6.7) we have

$$N(\tfrac{1}{2}\bar{Z} - 1) = T. \qquad (6.9)$$

The total number of tetrahedra T can be written using (6.8b) as $T = \frac{1}{6} N_b \bar{q}$, where the factor $1/6$ appears because every tetrahedron has 6 bonds. Reexpressing T in terms of N gives $T = \frac{1}{12} \bar{Z} \bar{q}$, which, when substituted into Eq. (3.21), leads to the desired result,

$$\bar{Z} = \frac{12}{6 - \bar{q}}. \qquad (6.10)$$

Note that when $\bar{q} = 5$, as it is for polytope $\{3, 3, 5\}$, $\bar{Z} = 12$, the coordination number of the icosahedron. Since Eq. (6.10) is independent of the Gaussian curvature of the four-dimensional sphere, we expect that it is also valid for negatively curved three-dimensional hyperbolic manifolds: One could also imagine relaxing the frustration of flat space by going to surface with a *negative* Gaussian curvature such that exactly 6, instead of 5, perfect tetrahedra could be wrapped around every bond.[9] Equation (6.10) shows, however, that the average coordination number would then be infinite, so that method of eliminating the frustration leads to extremely unphysical particle packings, at least for metallic systems.

d. *Relaxation to an Ideal Glass*

An interesting "mean field" or "effective medium" approach for predicting the statistics of flat space tetrahedral particle packings has been discussed by Coxeter.[6,8] Coxeter argued that some of the properties of dense-packed hard spheres could be modeled by a fictitious space-filling "statistical honeycomb" polytope $\{3, 3, q\}$, where in the simplest of several models discussed by him,

$$\bar{q}_{\text{ideal}} = 2\pi/\cos^{-1}(1/3) = 5.104299. \tag{6.11}$$

By definition, this polytope is composed of perfect tetrahedra, with a fractional number, $\bar{q}_{\text{ideal}}$, wrapped around every bond. Since the dihedral angle of a tetrahedron is $\cos^{-1}(1/3)$, one clearly needs $2\pi/\cos^{-1}(1/3)$ of them to fill the 2π radians of angle round a bond. The fractional part, $\bar{q}_{\text{ideal}} - 5 \approx 0.104299$, is a measure of the size of the crack in Fig. 3a.

The statistical honeycomb model also occupies a special place in the theory of packing fractions. The packing fraction f for an arrangement of identical hard spheres is defined to be the volume of a sphere divided by the volume per particle. The packing fractions for some common crystalline lattices are shown in Table III. Also shown is the packing fraction of the statistical honeycomb model (see Appendix) obtained by analytically continuing the result for structures with integral numbers of tetrahedra per bond off the integers—the Wigner–Seitz cell for the statistical honeycomb model has $Z = 13.4$ identical faces! As is evident from Table III, the statistical honeycomb model is denser than all the

TABLE III

LATTICE	PACKING FRACTION
diamond	$\sqrt{3}\pi/16 \doteq 0.34$
simple cubic	$\pi/6 \doteq 0.52$
body-centered cubic	$\sqrt{3}\pi/8 \doteq 0.68$
face-centered cubic	$\sqrt{2}\pi/6 \doteq 0.74$
statistical honeycomb	$\dfrac{1}{\sqrt{2}}\left[6\cos^{-1}\left(\dfrac{1}{3}\right) - 2\pi\right] \doteq 0.78$

common regular crystalline lattices. Although the statistical honeycomb model is an abstraction, and not a real lattice, its packing fraction f_{ideal} is in fact a rigorous upper bound for all possible hard sphere particle configurations, disordered or crystalline.[66,67] It plays the same role as the packing fraction $f = \pi/2\sqrt{3} \approx 0.907$ of the triangular lattice does in the packing of hard disks.[8]

It is, of course, impossible to have a fractional number of perfect tetrahedra associated with the bonds of real physical system. It is more realistic to allow for some dispersion in the near-neighbor bond lengths. One can show, however, that the result (6.11) is, in fact, exact for a special polytetrahedral configuration we shall call an "ideal glass."[11]

Consider first an arbitrary array of identical particles interacting via a simple pair potential in three-dimensional flat space. Although the bonds joining near neighbors divide space into tetrahedra, the curvature incommensurability insures that not all tetrahedra will have equal edges. The strains imbedded in disordered particle configurations can be relaxed by allowing motion into an extra dimension. The disclinations shown in Fig. 14, for example, could clearly lower their energy by buckling out of the two-dimensional plane. Imagine a similar relaxation into a fourth dimension for three-dimensional particle configurations, under the constraint of constant coordination number topology. No near-neighbor bonds, as defined by the Voronoi construction, can be broken. The result wll be a crinkled, three-dimensional surface, with local regions of positive and negative curvature. The bonds in this relaxed configuration will be more nearly equal than in the initial, flat arrangement. We define an "ideal glass" to be that configuration in flat space that is able to equalize *all* near-neighbor distances via the above relaxation process. Although the initial particle configuration is frustrated, the frustration is removed after relaxation, since all particles then sit at the minima of their neighbor's pair potentials.

The value of $\bar{q}$ for this special configuration follows from a formula developed to make discrete approximations to the formulas of general relativity, namely[68]

$$\sum_j l_j \delta_j = \frac{1}{2} \int {}^{(3)}R\sqrt{g}\, d^3x, \tag{6.12}$$

where l_j is the length of the jth bond, and the integral is over the scalar

[66]C. A. Rogers, *Proc. London Math. Soc.* **8**, 609 (1958).
[67]N. J. A. Sloane, *Scientific American,* January 1984, p. 116.
[68]C. W. Misner, K. S. Thorne, and J. A. Wheeler, "Gravitation." W. H. Freeman, San Francisco, 1971.

curvature $^{(3)}R$ of the three-dimensional surface. The quantity δ_j is the deficit angle associated with the jth bond, and g is the determinant of the metric tensor. Equation (6.12) is a tetrahedral discretization procedure for integrals on curved manifolds that becomes more and more accurate with increasing numbers of mesh points. The space enclosed by a tetrahedron is regarded as flat, and the curvature is concentrated on the bonds.[68] For an ideal glass in flat space, both the scalar curvature and all the deficit angles in Eq. (6.12) are zero. After relaxation, all the l_j are equal, and the δ_j vary in sign and magnitude from bond to bond. We expect that the *integral* curvature, at least in the limit of very large system size, *remains* zero. The assumption is that the initial flat space integral curvature is encoded into to bond topology, and that as many regions of positive as negative curvature are generated by the relaxation process.

The angular deficit δ_5 associated with 5 perfect tetrahedra packed around a bond is shown in Fig. 3a. This is the mismatch that results when the 5 tetrahedra are taken apart and reassembled into flat space. One clearly has

$$\delta_5 = 2\pi - 5y, \tag{6.13}$$

where y is the dihedral angle of a perfect tetrahedron,

$$y = \cos^{-1}(1/3). \tag{6.14}$$

The result for a j-fold bond is

$$\delta_j = 2\pi - jy. \tag{6.15}$$

Inserting this result into Eq. (3.24), we have

$$2\pi d \sum_j F_j - yd \sum_j jF_j = 0, \tag{6.16}$$

where F_i is the number of i-fold bonds, and d is the common value of all bond lengths l_j determined by, say, the minimum in the pair potential. Because

$$\bar{q} = \sum_j jF_j / \sum_j F_j, \tag{6.17}$$

Equation (6.11) follows immediately from Eq. (6.16). Since the relaxation process preserves the bond topology, this is also the value of $\bar{q}$ for an ideal glass in flat space.

Figure 27 shows the values of $\bar{q}$ for the extensive list of Frank–Kasper phases shown[23] in Table I, as a function of the size of the unit cell. All values of $\bar{q}$ are remarkably close to the ideal glass result, with an accuracy of a few parts in 10^4. As the number of atoms in a unit cell gets larger, these crystals seem better able to approximate the irrational value (6.11).

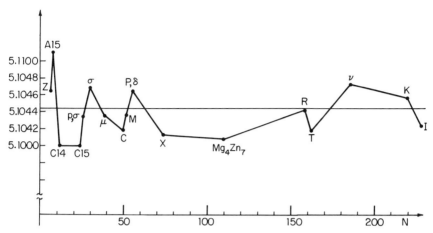

FIG. 27. Average number $\bar{q}$ of edges-per-face for the Voronoi cells of the Frank–Kasper phases as function of the unit cell size N. These values closely approximate Coxeter's "ideal glass" result $\bar{q} = 2\pi/\cos^{-1}(1/3)$. The average is pulled above 5 by the excess sixfold disclinations.

In this sense, we can regard an ideal glass as a Frank–Kasper phase with an infinitely large unit cell.

Figure 28 summarizes results for $\bar{q}$ and $\bar{Z}$ in a variety of different flat space systems. Starting with a computer simulation of liquid argon near the triple point,[69] these systems become progressively more ordered as $\bar{q}$ decreases. Note that the relaxed Ichikawa–Bennett model is further down on the curve that a perfect bcc crystal ($\bar{q} = 5\frac{1}{7}$, $\bar{Z} = 14$)! The Frank–Kasper phases are clustered about the ideal glass value, with

$$\bar{Z}_{ideal} = \frac{6}{12 - \bar{q}_{ideal}} \approx 13.4. \tag{6.18}$$

This progression can be viewed as a gradual pairing of plus and minus disclination lines, until only $-72°$ disclinations are left in the "Frank–Kasper limit". A similar progression (without intervening crystalline phases) should occur as one undercools a liquid toward the glass transition. We expect that increasing the number of 4–6 pairs will increase $\bar{Z}$ (and, hence, $\bar{q}$), because there are many more coordination shells possible with multiple four- and sixfold bonds for $Z \geq 14$ than for $Z < 14$. If one excludes cooperative arrangements of the lines, there need not be a sharp phase transition associated with this process. The ideal,

[69]A. Rahman, *J. Chem. Phys.* **45**, 2585 (1966).

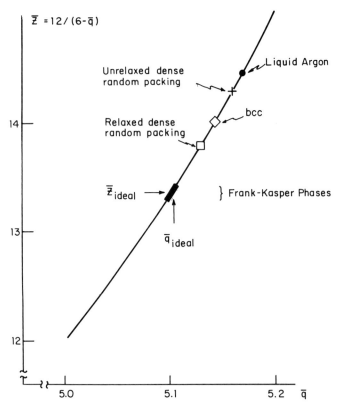

FIG. 28. Average coordination number $\bar{x}$ vs. $\bar{q}$ for progressively more polytetrahedrally ordered systems. The Frank–Kasper phases are clustered within the solid bar close to the "ideal glass" values.

"fully-paired" structure is presumably quite similar locally to a Frank–Kasper phase. It need not be crystalline, however.

IV. Computer Simulations, the Glass Transition, and Frustration in Binary Mixtures

7. MEASUREMENTS OF SHORT-RANGE ICOSAHEDRAL ORDER
   IN LIQUIDS

We have given a variety of physical arguments for frustrated polytetrahedral order in undercooled liquids, leading to short-range order with an icosahedral symmetry. As shown in Fig. 26, short-range icosahedral order does indeed appear in relaxed dense random packing models, as evidenced by the large number (over 50%) of pentagonal bond spindles. Although there are relatively few *complete* icosahedra in dense-random packing, there are many icosahedral fragments.[70] If this order is a feature of metallic glasses (which dense-random packing is supposed to model), it should also appear in undercooled liquids in metastable equilibrium just above the glass transition. Although it is hard to detect short-range icosahedral order directly in the laboratory, microscopic measurements of local order are possible by using computer simulations.

Quantitative measures of short-range icosahedral order in liquids were first developed in 1981 by Steinhardt *et al.*[71] in molecular dynamics simulations of 864 particles interacting via a Lennard–Jones pair potential

$$V(r) = 4\varepsilon\left[\left(\frac{\sigma}{r}\right)^{12} - \left(\frac{\sigma}{r}\right)^6\right],$$

with periodic boundary conditions. The runs were carried out for $n^* \equiv n\sigma^3 = 0.97$, where n is the particle density. This trajectory is shown on the phase diagram for Lennard–Jonesium in Fig. 29. The maximum run times during cooling were of order $30{,}000\tau_0$, where τ_0 is a microscopic Lennard–Jones collision time, of order 10^{-14} sec for argon. The cooling rate was roughly 10^{12} degrees per second in argon units, relatively slow compared to many computer simulations, but much faster that presently attainable in the laboratory. Such cooling rates are necessary to avoid nucleating an fcc crystal when the liquid is cooled below the liquid–solid phase boundary.

These simulations focused on bond-orientational order, defined by considering clusters of nearest neighbor "bonds" surrounding every atom like that shown in Fig. 30. To quantify the orientational order embodied in a particular bond cluster, the density of bonds $\rho(\mathbf{r}, \Omega)$ on a unit sphere surrounding one of N atoms at position $\mathbf{r}$ can be expanded in spherical harmonics.

$$\rho(\mathbf{r}, \Omega) = \sum_{l=0}^{\infty} \sum_{m=-l}^{l} Q_{lm}(\mathbf{r})Y_{lm}(\Omega). \tag{7.1}$$

---

[70] If f is the fraction of fivefold bond spindles, the fraction of complete icosahedra should be roughly f^{12}, which is small even if f is 50% or more. Thus, short-range icosahedral order is not incompatible with the relatively few complete icosahedra observed in the dense random packing studies of Finney (Refs. 33 and 65) and in the simulations of F. H. Stillinger and R. A. La Violette, *Phys. Rev.* **A25**, 5136 (1986).

[71] P. J. Steinhardt, D. R. Nelson, and M. Ronchetti, *Phys. Rev. Lett.* **47**, 1297 (1981); *Phys. Rev.* **B28**, 784 (1983).

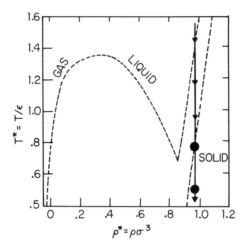

FIG. 29. Constant density undercooling trajectory for a Lennard–Jones liquid. The bond orientational order was measured at the points indicated before the undercooled liquid had crystallized into the equilibrium solid phase.

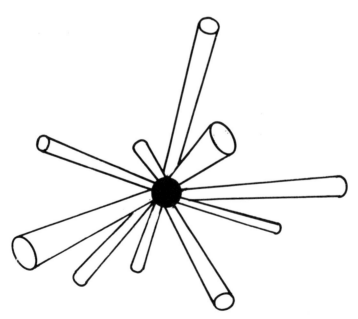

FIG. 30. Cluster of near-neighbor bonds used to define the bond orientational order parameters discussed in the text.

In an *isotropic* liquid, all spatially averaged expansion coefficients

$$Q_{lm} = \frac{1}{N} \sum_{i=1}^{N} Q_{lm}(\mathbf{r}_i), \tag{7.2}$$

except the trivial term Q_{00} will vanish identically. For a bond cluster like that associated with Fig. 5a, which could be the nucleus of an fcc crystal, there are additional nonzero $\langle Q_{lm} \rangle$'s characteristic of cubic symmetry with $l = 4, 6, 8, \ldots$. Q_{lm}'s with odd l vanish due to inversion symmetry. If the symmetry of the cluster is icosahedral, on the other hand, the only nontrivial allowed spherical harmonics occur for $l = 6, 10, 12, \ldots$.[72]

Steinhardt *et al.*[71] measured the set of rotationally invariant bond-orientational order parameters

$$Q_l = \sqrt{\frac{4\pi}{2l+1} \sum_{m=-l}^{l} |Q_{lm}|^2}. \tag{7.3}$$

Figure 31 shows the Q_l's for liquids in metastable equilibrium at temperature $T^* \equiv k_B T / \varepsilon = 0.72$ and $T^* = 0.51$, i.e., at the points indicated by dots on Fig. 27.[73] There is an increase in the value of Q_6 upon undercooling.

Since this increase could be attributed either to an increase in the number of fcc clusters or to an increase in the number of icosahedral clusters, a more discriminating test for icosahedral order is necessary.[71] Such a test is provided by the cubic invariant

$$\hat{W}_6 = \frac{\sum_{m_1, m_2, m_3} \left(\begin{smallmatrix} 6 & 6 & 6 \\ m_1 & m_2 & m_3 \end{smallmatrix} \right) Q_{6m_1} Q_{6m_2} Q_{6m_3}}{[\sum_{m=-l}^{l} |Q_{lm}|^2]^{3/2}}, \tag{7.4}$$

where the $\left(\begin{smallmatrix} 6 & 6 & 6 \\ m_1 & m_2 & m_3 \end{smallmatrix} \right)$ are Wigner $3j$ symbols.[74] For a cluster with cubic

---

[72] See Appendix C of Ref. 14

[73] As shown by H. Jonsson and H. C. Andersen (see Ref. 80), the large Q_6 values at low temperatures shown in similar histograms in Ref. 71 may be due to an anomalous run in which the liquid was gradually crystallizing. The data shown in Fig. 31, however, came from a run uncontaminated by crystallization. The third order invariant $\hat{W}_6$ discussed in Ref. 71, which is more discriminating in distinguishing icoshedral from incipient fcc order, was obtained from this uncontaminated run. We are grateful to H. C. Andersen for conversations on this point and to P. J. Steinhardt for supplying the data used in Fig. 31.

[74] L. D. Landau and E. M. Lifshitz, "Quantum Mechanics." Pergamon, Oxford, 1965 p. 106.

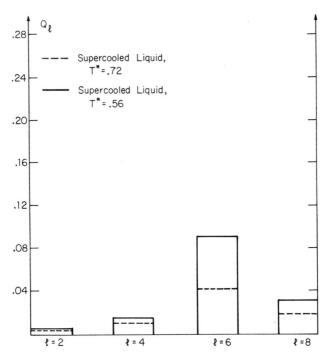

Fig. 31. Rotationally invariant bond orientational order parameters at the two reduced temperatures indicated in Fig. 29.

symmetry, it can be shown that

$$|\hat{W}_6^{\text{cubic}}| = \frac{4}{\sqrt{92378}} \simeq 0.01311,$$ (7.5)

while icosahedral symmetry leads to

$$|\hat{W}_6^{\text{icos}}| = \frac{11}{\sqrt{4199}} = 0.169754.$$ (7.6)

The *sign* of $\hat{W}_6$ can distinguish whether a small cluster of particles has an overall icosahedral or dodecahedral shape.[71] The meaning of the sign is less clear in large systems, where it may even change with time. At large time-averaged absolute value of $\hat{W}_6$, however, is indicative of icosahedral order in the sample, provided the Q_{6m}'s are large enough for the ratio (7.4) to be meaningful. It was conjectured by Steinhardt *et al.*[71] and proven by Jaric[75] that the value (7.6) for a perfect icosahedron is in fact a *rigorous upper bound* on all possible configurations of particles, disordered or crystalline.

At $T^* = 0.72$, the small Q_{6m}'s lead to large fluctuations in the ratio (7.4). At $T^* = 0.56$, however, the large Q_{6m}'s lead to a well-defined third-order invariant $\hat{W}_6 = 0.060939$, more than four and a half times the value indicative of fcc symmetry, and 36% of the maximum possible icosahedral value. During the cooling that produced the growth of these predominantly icosahedral Q_6's, the correlation length for orientational order grew from less than an interparticle spacing to 3 or 4 interparticle distances.[71] Nonzero Q_6's (presumably with an icosahedral symmetry), intermediate between those of a normal liquid and a cubic crystal, have also been reported by Mountain and Brown[76] in undercooled liquids that form a glass. Much smaller Q_6's, however, were found in simulations of 50–50 binary soft sphere mixtures by Thirumalai and Mountain.[77]

The simulations of Steinhardt *et al.*[71] are indicative of icosahedral order which grows with decreasing temperature. The discussion of statistical geometry in Part III, however, suggests that frustration will prevent *infinite* range icosahedral correlations, at least in simulations of identical particles (icosahedral quasicrystalline alloys are an important exception to this rule; see Part VI). Long-range icosahedral bond-orientational order is, of course, a feature of polytope $\{3, 3, 5\}$; "bonds" between neighboring atoms are replaced by geodesics on S^3 connecting nearest neighbors, and spherical harmonics associated with the tangents to these geodesics at a particular atom can be defined with respect to a special, singularity-free coordinate system. If points on S^3 are denoted by (x_0, x_1, x_2, x_3), $\sum_{i=0}^{3} x_i^2 = \kappa^{-2}$, this local coordinate system is defined by the orthonormal triad

$$\hat{e}_1 = \kappa^{-1}(-x_1, x_0, -x_3, x_2), \tag{7.7a}$$

$$\hat{e}_2 = \kappa^{-1}(-x_2, x_3, x_0, -x_1), \tag{7.7b}$$

$$\hat{e}_3 = \kappa^{-1}(-x_3, -x_2, x_1, x_0), \tag{7.7c}$$

which is everywhere orthogonal to the unit radius vector

$$\hat{e}_0 = \kappa^{-1}(x_0, x_1, x_2, x_3). \tag{7.8}$$

[75]M. V. Jaric, *Nucl. Phys.* B265[FS], 647 (1986).
[76]R. D. Mountain and A. C. Brown, *J. Chem. Phys.* **80**, 2730 (1984).
[77]D. Thirumalai and R. D. Mountain, *J. Phys.* C20, L399 (1987); R. D. Mountain and D. Thirumalai, *Phys. Rev.* A36, 3300 (1987).

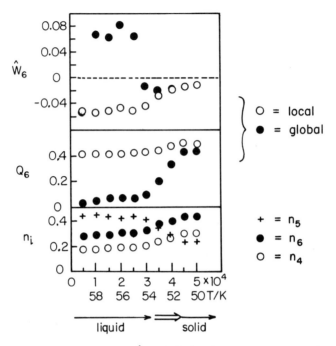

FIG. 32. Local and global invariants $\hat{W}_6$ and Q_6 for the constant pressure simulations of Nosé and Yonezawa (*J. Chem. Phys.* **84**, 1804 (1986)), for a Lennard–Jones liquid undercooled in one-degree steps with 10^4 Monte Carlo iterations at each temperature. Also shown are the relative proportions n_4, n_5, and n_6 of four-, five-, and sixfold bond spindles as a function of temperature.

Straley[78] has measured the bond-orientational order parameters in a simulation of 120 particles on S^3, and finds that the $\{Q_l\}$ associated with perfect icosahedral order become nonzero below the temperature at which this system crystallizes into polytope $\{3, 3, 5\}$.

Because frustration tends to twist up icosahedral order in flat space, it is probably better to define Q_6 and $\hat{W}_6$ *locally* for a particular bond cluster rather than to average over the entire system. This point has been made in an interesting paper by Nose and Yonezawa,[79] who measured Q_6, $\hat{W}_6$, and the numbers of four-, five-, and sixfold bond spindles in a constant pressure simulation study of crystallization in an undercooled Lennard–Jones liquid. Local Q_6's and $\hat{W}_6$'s were obtained from Eqs. (7.3) and (7.4), using Q_{lm}'s obtained from a weighted average

[78]J. P. Straley, *Phys. Rev.* **B34**, 405 (1986) and private communication.
[79]S. Nosé and F. Yonezawa, *J. Chem. Phys.* **84**, 1803 (1986).

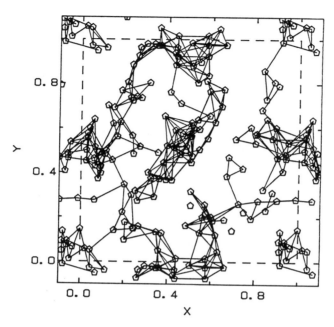

Fig. 33. Icosahedral order in the binary glass (20% large atoms) studied by Jonsson and Andersen (*Phys. Rev. Lett.* **60**, 2295 (1988)). Completed icosahedra are represented by pentagons that are connected by neighboring icosahedra by the solid lines.

over the Voronoi bonds associated with a particular particle. Equations (7.3) and (7.4) were then averaged over the entire sample. Figure 32 shows these quantities as a function of temperature, in argon units. Also shown along the temperature axis is the number of Monte Carlo time steps. In the range 59 to 54° K, the predominance of fivefold bond spindles is indicative of short-range icosahedral order. This is accompanied by large values of both the local Q_6 and $\hat{W}_6$. When the system crystallizes, which happens over the temperature interval 54 to 52° K, a drop in the number of fivefold bonds is accompanied by a reduction in $\hat{W}_6$ from a large icosahedral value to a much smaller value appropriate to local fcc order. The value of Q_6, however, changes only slightly, showing that Q_6 by itself cannot distinguish between fcc and icosahedral order. Note that the globally averaged symmetry indicator $\hat{W}_6$ is also indicative of icosahedral order in the liquid phase.

A close connection between the development of local icosahedral order and the glass transition has been established in some striking simulations of undercooled Lennard–Jonesium by Jonsson and Andersen.[80] These

[80]H. Jonsson and J. C. Andersen, *Phys. Rev. Lett.* **60**, 2295 (1988).

authors use the structural relaxation method of Stillinger and Weber[81] to prove the intrinsic structure (undistorted by thermal fluctuations) of the liquid as a function of temperature. At high temperatures, this structure is temperature independent, but, slightly above the glass transition temperature T_g, a structural relaxation involving an increase in the number of fivefold bond spindles is clearly identified. At T_g, the relaxation is slow and leads to hysteresis upon heating. Icosahedral order is particularly pronounced in a simulation leading to a two-component glass, consisting of 20% larger atoms. As shown in Fig. 33, there are many regions of fivefold symmetry, including 150 complete icosahedra in the 1500-atom sample. In a sense, Fig. 33 is the dual of Fig. 25, which shows defects threading through an otherwise icosahedral medium. Fewer icosahedra appear in smaller samples, suggesting that periodic boundary conditions tend to suppress icosahedral order.

A physical picture of glass transiton based on the idea of frustrated short-range icosahedral order will be developed in the next subsection.

8. A Scenario for the Glass Transition

Ideas about the statistical geometry of polytetrahedral order can be combined with the computer evidence for short-range icosahedral order in undercooled liquids to produce a qualitative argument for the existence of a glass transition. Of prime practical importance are metallic glasses, which are typically obtained by rapidly cooling a molten binary mixture. We share with Turnbull and Cohen,[25-27] however, the conviction that even simple one-component liquids would form glasses if it were possible to cool them rapidly enough in the laboratory. This belief is borne out in numerous computer simulations of glass transition,[82] and was recently confirmed experimentally by Pusey and van Megen[83] for an essentially monodisperse colloidal suspension of 3400 Angstrom poly-(methylmethacrylate) spheres. Crystals do not have time to nucleate in colloids because particle relaxation times are many orders of magnitude longer than they would be in a conventional atomic or molecular liquid.

We can think of an undercooled liquid as shown schematically on the right side of Fig. 34: a tangled mass of ±72° wedge disclination lines thread through a dense liquid with a high degree of short-range icosahedral order. When this liquid cools, some of the plus and minus disclination lines (four- and sixfold bond spindles) annihilate, increasing

---

[81]F. H. Stillinger and T. A. Weber, *Phys. Rev.* **A25**, 978 (1982).
[82]C. A. Angell, J. H. R. Clarke, L. V. Woodcock, *Adv. Chem. Phys.*, **48**, 397 (1981).
[83]P. N. Pusey and W. van Megen, *Phys. Rev. Lett.* **59**, 2083 (1987).

FRANK - KASPER   MOLTEN LIQUID/GLASS
PHASE

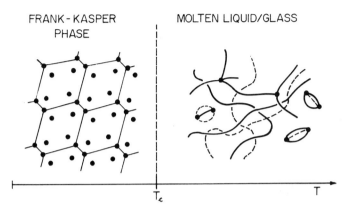

FIG. 34. Schematic of ordered and disordered polytetrahedral phases. An undercooled liquid can be viewed as a tangled mass of disclination lines permeating an otherwise icosahedral medium (compare with Fig. 33). If the frustration is not too large, there may be a temperature T_c below which the lines order into a Frank–Kasper phase. Forming the Frank–Kasper ground state requires disentangling the defect lines present at higher temperatures.

the amount of short-range icosahedral order. The geometrical frustration of three-dimensional flat space, however, insures that there will be an unpairable excess of sixfold disclination lines. If the cooling process were infinitely slow, the coordination number and disclination statistics would approach the "ideal glass" values discussed in Section 6.d and illustrated in Fig. 28. As we have seen, the Frank–Kasper phases provide an excellent approximation to an "ideal glass." If crystallization into an fcc crystal is prevented, there may ultimately be a transition into a crystalline Frank–Kasper phase, shown on the left side of Fig. 34.

If the cooling is more rapid, the entangled disclination lines will be unable to find the ordered Frank–Kasper crystal and will drop out of equilibrium at a glass transition. As discussed in Part V, there are topological reasons which make it difficult for disclination lines to cross. These entanglement constraints become more pronounced with increasing short-range icosahedral order. The entanglement that leads to the glass transition is intimately connected with the excess of negative wedge disclinations forced in by the topological frustration.

As shown by Straley,[84] it is very easy to avoid the glass transition entirely when particles are embedded in the frustration-free surface of a four-dimensional sphere. Straley used Monte Carlo simulations to cool 120 particles on S^3, as well as 108 particles with periodic boundary

[84]J. P. Straley, *Phys. Rev.* B**30**, 6592 (1984).

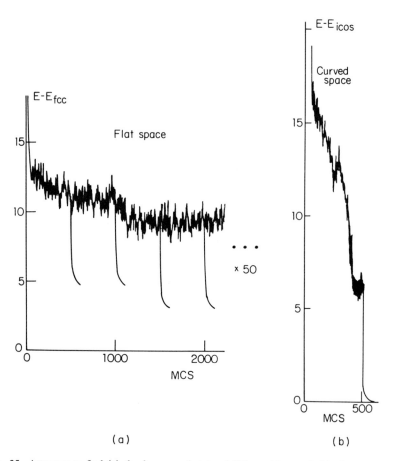

FIG. 35. Attempts to find (a) the fcc ground state of 108 particles cooled in flat space with periodic boundary conditions and (b) the icosahedral polytope {3, 3, 5} ground state of 120 particles cooled on the curved surface of a four-dimensional sphere. This figure is taken from Monte Carlo simulations by J. P. Straley (*Phys. Rev.* **B30,** 6592 (1984)). The ground state is much easier to find in curved space, where frustration is absent.

conditions in flat space, with a repulsive $1/r^{12}$ potential. Figure 35a shows the energy in flat space at low temperature minus the energy of the stable fcc crystal as a function of the number of Monte Carlo time steps. Every 500 steps, the thermal energy is drained out of the system, and the particles attempt to find the equilibrium fcc crystal. If the resulting energy is higher than the fcc value, the system is returned to its configuration before the quench, and the simulation is continued. Approximately 200 quenches spaced over 100,000 Monte Carlo time steps failed to find the ground state. Crystallization is much easier in the non-Euclidean simulation: as illustrated in Fig. 35b, perfect "crystallization" of 120 particles, as measured by the energy relative to a polytope $\{3, 3, 5\}$ ground state, occurred after only 500 Monte Carlo steps in 3 out of 11 trials. The energies of defects (including vacancies and interstitials) that might characterize metastable states in the 8 trials that failed to find the ground state were studied by Straley.[85] Easy crystallization (now to a triangular lattice) also occurred in two dimensions, again due to the absence of frustration.

9. FRUSTRATION AND GLASS FORMATION IN BINARY ALLOYS

Some understanding of glass-forming tendencies in binary alloys can be obtained by considering how two different particle sizes influence the frustration. A particularly simple example is provided by mixtures of magnesium and calcium, which we approximate by hard spheres with diameter ratio 1.33. A dilute concentration of large particles (in this case, magnesium) *relaxes* the frustration by providing preferred sites for the topologically required excess of sixfold bond spindles: large atoms typically sit inside the $Z14$, $Z15$, and $Z16$ coordination shells shown in Fig. 22. A dilute concentration of large atoms increases the number of complete icosahedra, as observed in the simulations of Jonsson and Andersen.[80] Although the large atom itself will not be icosahedral, its extra volume can "fill in the cracks" in 12 of the coordination shells of its nearest neighbors to produce $Z12$ icosahedral coordination shells, as happens in the Frank–Kasper phases. The optimal concentration of large atoms can be estimated by assuming that all coordination shells are either $Z12$ or $Z16$, with all large atoms at the $Z16$ sites. A concentration x of $Z16$'s such that the average coordination number is the "ideal glass" result $Z = 13.4$ requires that

$$16x + 12(1 - x) = 13.4. \tag{9.1}$$

The frustration is minimized when $x = 35\%$, very close to the maximum in the melting curve shown in Fig. 36, which occurs at the stochiometry $x = 33\%$ of a (Frank–Kasper) Laves phase.

Conversely, introducing a dilute concentration of small atoms into a matrix of larger ones will *increase* the frustration. One would expect, for example, *fewer* complete icosahedra if the binary mixture simulations of Jonsson and Andersen[80] were repeated with 20% *smaller* particles. One

[85]J. P. Straley, *Materials Science Forum* **4**, 93 (1985). For earlier characterizations of defects in polytope $\{3, 3, 5\}$, see R. Mosseri, Saclay Ph.D. Thesis (unpublished).

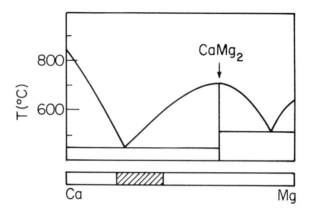

FIG. 36. Schematic phase diagram for calcium and magnesium showing the Laves (*i.e.*, Frank–Kasper) phase CaMg$_2$ at the maximum in the melting curve and the deep eutectic, which forms when the proportions of calcium and magnesium are reversed.

might hope to maximize the frustration by interchanging the proportions of large and small atoms predicted by Eq. (7.9). One does indeed find a deep eutectic in Fig. 36 near the predicted concentration $x = 65\%$. Glass formation is possible in the laboratory only near this eutectic, as delineated by the bar in the figure. If the cooling rate could be increased further, however, we would expect the region of glass formation to expand away from the eutectic, until it encompassed the entire phase diagram.

Chemical short-range order, which goes beyond simple hard sphere models, also plays an important role in metallic glass formation. In metal–metalloid glasses, the minority metalloid constituents are believed to be surrounded by compact 8- and 9-particle coordination shells (see Section 4). In amorphous Ni-P, for example, every phosphorous atom is surrounded by a 9-atom cobalt coordination shell. These coordination shells are, in fact, just the 3-line nodes for +72° disclinations shown in Fig. 23. Thus, seeding cobalt with phosphorus atoms amounts to seeding the material with nodes for disclination lines with the "wrong" sign— lines, which would annihilate with their antidefect in an icosahedral model without chemical short-range order. Sachdev[86] has analyzed the large unit cell crystalline ground state of Ni$_3$P and shown that the ratios of four-, five- and sixfold bond spindles are very similar to dense random packing. This suggests that the primary difference between amorphous and crystalline Ni$_3$P is in the entanglement of the defect lines.

[86]S. Sachdev, Ph.D. Thesis, Harvard University.

V. Statistical Mechanics of Polytetrahedral Order

The geometrical arguments of Parts III and IV must be supplemented with a statistical mechanical approach to obtain a complete description of undercooled liquids near the glass transition. In Section 10, we show how the mismatch between the order in an fcc crystal and a polytetrahedral liquid gives rise to a large, temperature-dependent interfacial tension, that accounts for the large experimentally observed undercoolings. In Section 11, we construct an order parameter theory of polytetrahedral liquids that explains the experimentally observed structure function near the glass transition.

10. STATISTICAL ESTIMATE OF THE CRYSTAL-MELT INTERFACIAL
    TENSION

As discussed in Section 3, the large value of the crystal-melt interfacial tension implicit in the large undercooling obtained for simple metallic liquids came initially as a surprise, since the similarity in density between the two phases and the absence of strong configurational constraints in the liquid make the energetic contributions to the interfacial tension quite small. Turnbull's suggestion[26] of a *negentropic* origin of the interfacial tension, resulting from the entropy loss in the liquid as it becomes more localized in one dimension near the crystal surface, can be made explicit with a simple static hard sphere model[87,88] based on the fundamental structural difference between a polytetrahedral liquid and close-packed crystal.

A static model of a phase boundary can be created by adding atoms on a terminating plane of one phase (here the crystal), in such a way that the structure evolves toward that of the second phase. Since in a static model the two phases are obviously not in equilibrium, it is necessary to identify characteristic structural features for each, so that by changing the construction rules a transition from one phase to the other can be forced. In the case of the hard sphere crystal–liquid interface the difference in orientational short-range order discussed in Section 3 and illustrated in Figs. 4 and 5 can be used for this purpose.

By explicitly excluding octahedral configurations (characteristic of the crystal, as shown in Fig. 5), maximizing the number of tetrahedra (characteristic of the liquid) and the density, three two-dimensional nearest-neighbor configurations are formed on the close-packed plane

[87]F. Spaepen, *Acta Met.* **23,** 729 (1975).
[88]F. Spaepen and R. B. Meyer, *Scripta Met.* **10,** 257 (1976).

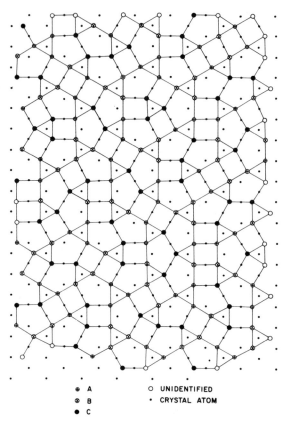

⊕ A   ○ UNIDENTIFIED
⊗ B   · CRYSTAL ATOM
● C

FIG. 37. Structure of the first interfacial layer in a static hard sphere model of the interface between a melt and close-packed crystal plane. The layer is fully localized in the direction perpendicular to the interface and has no translational symmetry in the lateral directions. [From F. Spaepen, *Acta Met.* **23,** 729 (1975).]

that can be connected, as shown in Fig. 37, to form the first interfacial layer. As can be seen, this layer has no translational symmetry parallel to the interface and is fully localized in the perpendicular direction (i.e., all atoms are in one plane). The atoms of the next interfacial layer are located at the centers of the polygons of the first layer to maximize the density; as a result, they are found at different distances from the interface, resulting in a lesser degree of localization (see Fig. 38). It was assumed that all the interfacial localization had taken place in these two layers. Recent molecular dynamics simulations[89] and density functional

[89]J. H. Sikkenk, J. O. Indekeu, J. M. J. van Leeuwen, and E. O. Vossnack, *Phys. Rev. Lett.* **59,** 98 (1987).

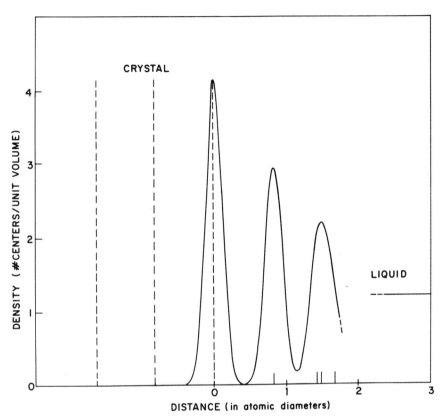

FIG. 38. Density of atom centers as a function of distance perpendicular to the interface for the model of Fig. 37. The atom positions are marked, and broadening typical for a metal at its melting point is applied.

calculations [90,91,92] confirm that the transition layer is indeed quite narrow.

It is interesting to note here that the presence of a planar crystalline surface introduces correlations that extend throughout the bulk of the liquid. Figure 39 shows the projection of the centers of the hard spheres in a model system,[93] created by pouring identical spheres into a box with roughened sides to prevent crystallization. On the bottom of this box was a close-packed crystal plane that had the first interfacial layer of Fig. 37 glued in place. Glueing was necessary to prevent crystallization. This

[90]C. Ebner and W. F. Saam, *Phys. Rev. Lett.* **38,** 1486 (1977).

[91]W. A. Curtin, *Phys. Rev. Lett.* **59,** 1228 (1987).

[92]W. E. McMullen and D. W. Oxtoby, *J. Chem. Phys.* **88,** 1967 (1988).

[93]F. Spaepen, Ph. D. Thesis, Harvard University (1975).

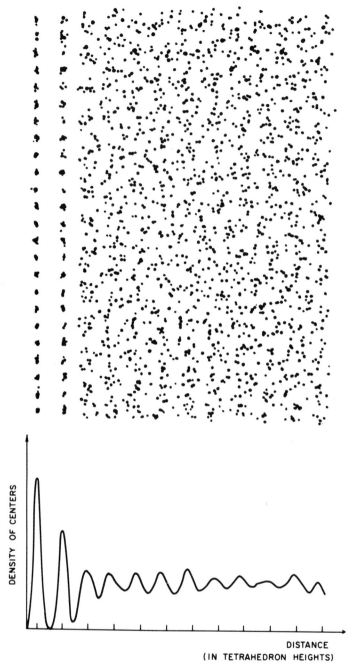

FIG. 39. Projection of the centers of a model system in which spheres are poured randomly onto a crystal plane and the first interfacial layer of Fig. 37. The projection is parallel to the crystal plane. The smoothed density of centers is shown as a function of distance perpendicular to the interface.

assembly was then densified by shaking. The resulting structure was analyzed by filling the interstitial spaces with plastic and slicing the model after hardening. It is striking that positional correlations with a period equal to the tetrahedral height persist throughout the sample in the direction perpendicular to the interface. Similar correlations have been observed in dense-random packings of hard spheres created with other boundary conditions.[94] They are probably accentuated by the hard-sphere nature and the essentially low-temperature state of these static models, since in the simulations with softer potentials they appear to die out more quickly.

The change in the thermodynamic quantities as a function of position normal to the interface is illustrated in Fig. 40. The number of interface atoms, i.e., the number of atoms whose configurational entropy is determined by the configurational entropy of the structure of Fig. 37, is 1.10 times the number of atoms in the crystal plane. This includes all the atoms in the first interfacial layer and 46% (corresponding to the degree of localization) in the second layer. The configurational entropy of the structure of Fig. 37 can be calculated by observing[88] that the disorder in this structure is equivalent to that of placing dimers on a honeycomb structure (Figs. 41a–d). Elser[95] recently pointed out that this problem is also equivalent to that of plane partitions, as illustrated by Figs. 41e and 41f, for which he presented exact solutions. The configurational entropy is 0.338314 per dimer (or per triangle in Fig. 41b), which corresponds to a configurational entropy of 0.077241 per interface atom. The resulting value for the surface tension (at the melting point) is then 0.85 times the heat of fusion per atom in the crystal plane. Thompson[96] has made a similar estimate for the interfacial tension between the (110) plane in the body-centered cubic structure and the melt and found a value of 0.71 times the heat of fusion per atom in the crystal plane.

If the interfacial tension is negentropic in origin it should increase with temperature. For a flat interface (i.e., under equilibrium conditions along the p-T coexistence line) it is straightforward to show that, if the interfacial negentropy is constant, the interfacial tension is proportional to temperature. This proportionality is in agreement with very recent density functional calculations by Curtin[91] and by McMullen and Oxtoby.[92] The value of the surface tension calculated by the latter is 0.87 times the heat of fusion per atom in the crystal plane. The temperature dependence of the surface tension at constant pressure (i.e., using a curved interface to establish equilibrium below the melting point), which

[94]R. Alben, G. S. Cargill, and J. Wenzel, *Phys. Rev. B***13,** 835 (1976).
[95]V. Elser, *J. Phys. A***17,** 1509 (1984).
[96]C. V. Thompson, Ph.D. Thesis, Harvard University (1981).

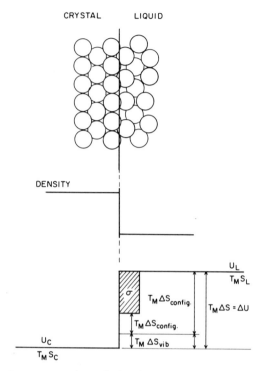

FIG. 40. Schematic representation of the density, energy, vibrational entropy, and configurational entropy as a function of distance perpendicular to the interface of Fig. 37. The negentropic origin of the interfacial tension, σ, is indicated.

is needed to account for the values measured in the nucleation experiments, still needs to be calculated. It is worth noting, however, that the lower value (0.62 times the heat of fusion per atom in the crystal plane) measured for mercury at $2/3\ T_M$ is consistent with a positive temperature coefficient. Turnbull also shows in his analysis[97] of the nucleation kinetics of some of his mercury emulsions that a positive temperature coefficient of the surface tension may be needed to keep the preexponential in Eq. (3.1) at a physically reasonable value.

11. ORDER PARAMETER THEORY OF POLYTETRAHEDRAL STATISTICAL MECHANICS

In this section, we show how to describe a liquid with short-range icosahedral order using a set of uniformly frustrated order

[97]D. Turnbull, *Prog. Mat. Sci.*, Chalmers Anniversary Volume 269 (1981).

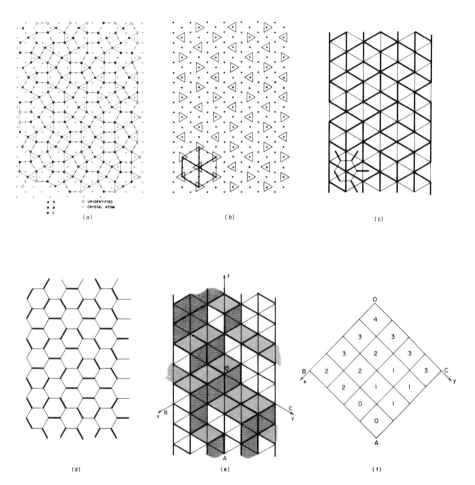

FIG. 41. Equivalent structures used in the calculation of the interfacial entropy. (a) Structure of the layer of Fig. 37. (b) The triangles of the layer of part (a) are shown to fall on a superlattice. The translational symmetry is broken by not allowing these neighboring triangles to have the same orientation. (c) Network obtained by connecting the centers of the triangles in part (b). Triangles with the same orientation are connected by thin lines; triangles with different orientation are connected by heavy lines. The sketch in the lower left corner shows the relation to part (d). (d) Network obtained by connecting the centers of the triangles of part (c), illustrating the pairing. Heavy lines connect the centers of paired triangles; thin lines connect the centers of unpaired triangles. All vertices have exactly one heavy line. This corresponds to dimerization on the honeycomb lattice. (e) Illustration of how the structure of part (c) can be regarded as a stacking of cubes. (f) Illustration of how the cube stacking of part (e) corresponds to a partitioning of the plane. The numbers in each square correspond to the number of cubes stacked on the OABC plane in part (e). [Adapted from F. Spaepen and R. B. Meyer, *Scripta Met.* **10,** 257 (1976) and V. Elser, *J. Phys.* **A17,** 1509 (1984).]

parameters.[13,14] Unlike the *orientational* order parameters discussed in Section 7, these order parameters will be sensitive to both the translational and orientational order exemplified by polytope $\{3, 3, 5\}$. Our approach is similar to Landau's theory[98] of a liquid just above a conventional freezing transition, where there is an order parameter for every reciprocal lattice vector in the incipient crystalline solid. As we shall see, non-Abelian matrices play the role of reciprocal lattice vectors in materials whose short-range order can be modeled by polytope $\{3, 3, 5\}$. It is their noncommutivity which leads to frustration, and makes the physics of glass a challenging problem.

We first review how to describe order in a conventional flat space crystalline solid. Transitional periodicity shows up when we Fourier expand the particle density,

$$\rho(\mathbf{r}) = \sum_{\mathbf{q}} \rho_{\mathbf{q}} e^{i\mathbf{q}\cdot\mathbf{r}}. \tag{11.1}$$

In a gas, or high-temperature liquid, one would expect all Fourier coefficients to occur with roughly equal weights. In a crystalline solid, however, the coefficients $\rho_{\mathbf{q}}$ such that $\mathbf{q}$ equals a special set of reciprocal lattice vectors $\{\mathbf{G}\}$ become very large. For a given crystal structure there is a well-known algorithm for determining the allowed $\mathbf{G}$'s, which assigns one reciprocal lattice vector to every set of Bragg planes.[99] In an x-ray scattering experiment, one measures the structure function $S(\mathbf{q})$, which is a thermal average,

$$S(\mathbf{q}) = \langle |\rho_{\mathbf{q}}|^2 \rangle. \tag{11.2}$$

In practice, one usually works with a sample composed of randomly oriented microcrystallites, and effectively averages over the different directions of $\mathbf{q}$. As shown schematically in Fig. 42, the angularly averaged function $\overline{S(q)}$ has sharp peaks at the reciprocal lattice vectors which characterize the crystalline solid.

Following Landau,[98] we can develop an order parameter theory of the freezing process by defining *local* Fourier components at the reciprocal lattice positions via

$$\rho_{\mathbf{G}}(\mathbf{r}) = \frac{1}{\Delta V} \int_{\Delta V} d^3 r' e^{-i\mathbf{G}\cdot\mathbf{r}} \rho(\mathbf{r}'), \tag{11.3}$$

[98] "The collected Papers of L. D. Landau" (D. ter Haar, ed.), p. 193. Gordon and Breach–Pergammon, New York, 1965; see also G. Baym, H. A. Bethe, and C. Pethick, *Nucl. Phys.* **A175**, 1165 (1971).

[99] N. W. Ashcroft and N. D. Mermin, "Solid State Physics." (Holt, Rinehart, and Winston, New York, 1976.

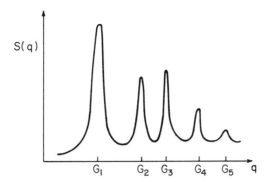

FIG. 42. Schematic powder diffraction pattern for a conventional flat space crystal, with peaks at the reciprocal lattice vectors G_i.

where ΔV is a hydrodynamic averaging volume centered at position r. In practice, one usually works with the Fourier modes corresponding to the smallest nonzero reciprocal lattice vectors. The free energy that describes the freezing transition is a rotationally and translationally invariant expansion in the ρ_G's,

$$F = \tfrac{1}{2}K \sum_{G} |(\nabla - i\mathbf{G} \times \mathbf{\theta})\rho_G|^2 + \tfrac{1}{2}r \sum |\rho_G|^2$$

$$+ w \sum_{G_1+G_2+G_3} \rho_{G_1}\rho_{G_2}\rho_{G_3} + O(\rho_G^4). \qquad (11.4)$$

The field $\mathbf{\theta}(\mathbf{r})$ measures the deviation of the local crystallographic axes from some reference orientation; its presence in the gradient term is required by rotational invariance.[100] The presence of a third-order term shows that, in the absence of strong fluctuations, the freezing transition is first order.

Before generalizing this approach to describe order in a flat space glass, it is helpful to first consider what happens when 120 particles are cooled on the surface S^3 of a four-dimensional sphere. This system is unfrustrated, and the ground state for simple pair potentials presumably consists of particles at the vertices of polytope $\{3, 3, 5\}$. Condensation into a $\{3, 3, 5\}$ "crystal" upon cooling reflects a broken SO(4) symmetry in the particle density $\rho(\hat{u})$, where $\hat{u}$ is a 4-D unit vector describing different positions on the surface of the sphere. Recall from Section 6a that the radius of the sphere necessary to accommodate a polytope with

[100]See, for example, D. R. Nelson and J. Toner, *Phys. Rev.* **B24**, 363, Sec III-D (1981).

geodesic distance d between the particles is

$$R = 1.591549d \equiv \kappa^{-1}. \tag{11.5}$$

When carrying out manipulations on the vertices of $\{3, 3, 5\}$, we can exploit the isomorphism between the group $SU(2)$ of 2×2 complex unitary matrices with unit determinant and the points on the surface of a four-dimensional sphere. This isomorphism is similar to the familiar identification of a circle with the set of complex numbers of unit modulus. Every point $\hat{u} = (u_o, u_x, u_y, u_z)$ is associated with a $SU(2)$ matrix as shown below,

$$\hat{u} \to \begin{pmatrix} u_0 + iu_z, & iu_x + u_y \\ iu_x + u_y, & u_0 - iu_z \end{pmatrix}. \tag{11.6}$$

Points on S^3 can now be multiplied together by using the multiplication rules of $SU(2)$ matrices. Many computations simplify because the 120 vertices of polytope $\{3, 3, 5\}$ can be oriented so that they form a group when combined in this way.[101] This group, which we shall call Y', is the lift of the 60-element icosahedral point group into $SU(2)$.

The density $\rho(\hat{u})$ for an arbitrary configuration of 120 particles on S^3 can be expanded in hyperspherical harmonics $Y_{n,m_1 m_2}(\hat{u})$,

$$\rho(\hat{u}) = \sum_{n=0}^{\infty} \sum_{m_1, m_2} Q_{n, m_1, m_2} Y_{n, m_1, m_2}(\hat{u}). \tag{11.7}$$

The coefficients $Q_{n, m_1 m_2}$ in this expansion are like the Fourier coefficients in Eq. (7.1). The nonnegative integer subscript n indexes $(n + 1)^2$-dimensional irreducible representations of the $4d$ rotation group $SO(4)$.[102] Unlike the usual three-dimensional spherical harmonics $Y_{lm}(\theta, \phi)$, there are *two* azimuthal quantum numbers, m_1 and m_2, which assume values with integer steps in the range

$$-n/2 \leq m_1, m_2 \leq n/2. \tag{11.8}$$

Summing over n is like summing over the *magnitudes* of the **q**-vectors in Eq. (11.1) (see below), and summing over m_1 and m_2 is like summing

[101]P. Du Val, "Homographies, Quaternions and Rotations." Oxford University Press, Oxford, 1964.

[102]L. C. Biedenharn, *J. Math. Phys.* **2**, 433 (1961): M. Bander and C. Itzykson, *Rev. Mod. Phys.* **38**, 330 (1966).

over their directions. Although unfamiliar to many physicists and materials scientists, these hyperspherical harmonics turn out to be proportional to the standard SU(2) Wigner matrices, so their properties can be readily determined.[102]

The flat space structure function displayed in Eq. (11.2) is both rotationally and translationally invariant when we average over the different directions of **q**. The analogous SO(4)-invariant correlation function for particles in thermal equilibrium in S^3 is[13]

$$S_n = \frac{1}{(n+1)^2} \sum_{m_1, m_2} \langle |Q_{n,m_1m_2}|^2 \rangle. \tag{11.9}$$

The quantity S_n becomes proportional to the powder average $\overline{S(q)}$ in the limit of a large number of particles imbedded in a sphere of infinite radius.[13] In analogy with the peaks in $\overline{S(q)}$ at the reciprocal lattice positions for crystalline solids (Fig. 42), we would expect that S_n can only be nonzero for certain special values of n for particles in the configuration $\{3, 3, 5\}$. The 7200-element group G of symmetry operations of polytope $\{3, 3, 5\}$ that can be continuously connected to the identity has the formal decomposition[101]

$$G = (Y' \times Y')/Z_2. \tag{11.10}$$

In Nelson and Widom,[13] these symmetries were exploited by using the trick embodied in Eq. (11.6) to show that the only allowed spherical harmonics for polytope $\{3, 3, 5\}$ occur for

$$n = 0,\ 12,\ 20,\ 24,\ 30,\ 32,\ 36,\ 40,\ 42,\ 48,\ 50,\ 52,\ 54,\ 56,\ 60, \tag{11.11}$$

and all even $n > 60$. This set of allowed n-values is the curved space analogue of the allowed magnitudes of reciprocal lattice vectors in a flat space crystal.

Formally, we can define wave vector magnitudes q in flat space in terms of the eigenvalues of the Laplacian operator acting on the complete set of functions in Eq. (11.1),

$$-\nabla^2 e^{i\mathbf{q}\cdot\mathbf{r}} = q^2 e^{i\mathbf{q}\cdot\mathbf{r}}. \tag{11.12}$$

The integers n play the role of wave vector magnitudes in curved space, in the sense that[13]

$$-\nabla_{\hat{u}}^2 Y_{n,m_1m_2}(\hat{u}) = n(n+2) Y_{n,m_1m_2}(\hat{u}), \tag{11.13}$$

when $\nabla_{\hat{u}}^2$ is the curved space Laplacian. Thus, we make the identification

$$q \leftrightarrow \sqrt{n(n+2)} \approx n, \tag{11.14}$$

and plot the structure function S_n of polytope $\{3, 3, 5\}$ just as we would $\overline{S(q)}$ for a flat space crystal. The "peaks" at special values of n are shown in Fig. 43, with intensities obeying the law[13] $S_n \sim 1/(n+1)$. These features should be compared with the peaks at special reciprocal lattice vectors in

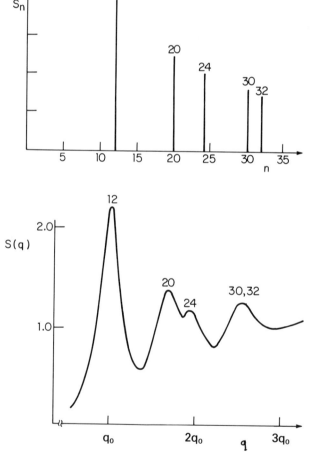

FIG. 43. Structure function S_n of polytope $\{3, 3, 5\}$ compared with the structure function of relaxed dense-random packing. The peaks for dense-random packing are very similar to those in S_n, except that they are broadened by frustration.

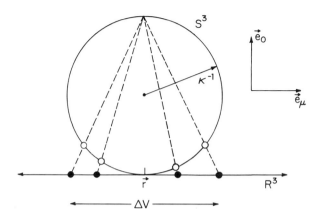

FIG. 44. Definition of the icosahedral order parameter. A local particle configuration in a volume ΔV is projected onto a tangent sphere of radius κ^{-1} given by Eq. (11.5), and the order parameter is then computed from Eq. (11.15b).

Fig. 42. The structure function of polytope $\{3, 3, 5\}$ is strikingly similar to the structure function for relaxed dense-random packing, also shown in Fig. 43. Peaks corresponding to $n = 12,\ 20,\ 40$ are clearly present, as well as a peak that appears to be a combination of $n = 30$ and 32. Even the peak intensities of polytope $\{3, 3, 5\}$ mimic those of dense-random packing!

Straley[103] has evaluated the function S_n for 120 particles interacting on a four-dimensional sphere via a repulsive $1/r^{12}$ potential and cooled rapidly via Monte Carlo computer simulation. Although the particle configuration that results is amorphous (all S_n are nonzero), there is a pronounced peak at $n = 12$, the smallest allowed nonzero value for $\{3, 3, 5\}$.

We now want to adapt these ideas to obtain an order parameter for liquids and glasses in flat space. The Fourier components for particles in S^3 are given by integrating the density over the surface of the sphere,

$$Q_{n, m_1 m_2} = \int d\Omega_{\hat{u}} \rho(\hat{u}) Y^*_{n, m_1 m_2}(\hat{u}). \tag{11.15a}$$

Fig. 44 illustrates how to define a local $n = 12$ order parameter via stereographic projection of a small flat space particle configuration onto a tangent sphere jof radius κ^{-1}. If $\Delta V'$ is the projection of the averaging volume ΔV onto S^3, and $\rho'(\hat{u})$ is the projected particle density, the

[103]J. P. Straley, private communication.

Fourier coefficients associated with the point $\mathbf{r}$ are

$$Q_{12,m_1m_2}(\mathbf{r}) = \frac{1}{\Delta V'} \int_{\Delta V'} d\Omega \hat{u} \rho'(\hat{u}) Y_{12,m_1m_2}(\hat{u}). \qquad (11.15b)$$

Equation (11.15b) replaces Eq. (11.3) in the usual Landau approach to flat space crystalline solids; it differs from Eq. (11.15a) only in the restricted range of the angular integration. We have focused on $n = 12$ because it corresponds to the smallest nonzero reciprocal lattice vector in a medium with local icosahedral order.

The symmetry of this order parameter is SO(4), modulo the symmetry group G of polytope $\{3, 3, 5\}$. An SO(4)-invariant Landau expansion for the free energy density reads

$$F_{12} = \tfrac{1}{2}K \, |D_\mu \mathbf{Q}_{12})|^2 + \tfrac{1}{2}r \, |\mathbf{Q}_{12}|^2$$

$$+ w \sum_{\substack{m_1m_2m_3 \\ m'_1m'_2m'_3}} \begin{pmatrix} 6 & 6 & 6 \\ m_1 & m_2 & m_3 \end{pmatrix} \begin{pmatrix} 6 & 6 & 6 \\ m'_1 & m'_2 & m'_3 \end{pmatrix}$$

$$\times Q_{12,m_1m'} Q_{12,m_2m'_2} Q_{12,m_3,m'_3} + O(Q_{12}^4). \qquad (11.16)$$

For notational convenience, we have replaced the 13×13 matrix Q_{12,m_1m_2} by a 169-component vector $\mathbf{Q}_{12}$ in the first two terms. Ordinary SO(3) Wigner 3j-symbols are used to make a rotationally invariant third-order term.[13] The frustration is embodied (see below) in the gradient operator D_μ, which is defined by

$$D_\mu \mathbf{Q}_{12} = (\partial_\mu - i\kappa L_{o\mu}^{(12)})\mathbf{Q}_{12}. \qquad (11.17)$$

Following a low-temperature continuum elastic treatment of glasses proposed by Sethna,[12] we require that neighboring particle configurations be related by rolling a $\{3, 3, 5\}$ template between them; the matrices $L_{o\mu}^{(12)}$ are the generators for the $n = 12$ representation of SO(4) for rolling in the (o, μ)-plane.[13] The quantity κ is the inverse radius of the tangent sphere in Fig. 44.

Nelson and Widom[13] argued that (for large, negative r) the polynomial part of Eq. (11.16) is minimized when the 169 numbers $\{Q_{12,m_1m_2}\}$ occur in the same proportions as for polytope $\{3, 3, 5\}$. Minimizing the polynomial part, however, only determines the order parameter up to a local SO(4) rotation. Minimizing the *gradient* term means that the order parameters at $\mathbf{r}$ and at $\mathbf{r} + \boldsymbol{\delta}$, where $\boldsymbol{\delta}$ is a small separation vector, are related,

$$\mathbf{Q}_{12}(\mathbf{r} + \boldsymbol{\delta}) = e^{i\kappa L_{o\mu}^{(12)}\delta^\mu} \mathbf{Q}_{12}(\mathbf{r}). \qquad (11.18)$$

Equation (11.18) is just a restatement of the requirement that neighboring textures are related by rolling. The quantity $\exp(i\kappa L_{o\mu}^{(12)})$ is an SO(4) Wigner matrix appropriate to the $n = 12$ representation. In a flat space crystal, the deviation of the density from its average value due to the reciprocal lattice vector $\mathbf{G}$ varies like

$$\Delta\rho(\mathbf{r} + \boldsymbol{\delta}) = e^{i\mathbf{G}\cdot\boldsymbol{\delta}}\,\Delta\rho(\mathbf{r}). \tag{11.19}$$

Comparison of Eqs. (11.18) and (11.19) makes it clear that the non-Abelian matrices

$$\{G_\mu\} = \{\kappa L_{o\mu}^{(12)},\ \kappa L_{o\mu}^{(20)},\ \kappa L_{o\mu}^{(24)}, \ldots\} \tag{11.20}$$

play the role of reciprocal lattice vectors in this theory.

Equation (11.18) means that the gradient term in the free energy (11.17) vanishes along the path connecting $\mathbf{r}$ and $r + \boldsymbol{\delta}$. The gradient term is frustrated, however, in the sense that it cannot be made to vanish everywhere. To see this, one need only multiply together the matrices corresponding to path shown in Fig. 45. One finds that the net change in the order parameter upon completing a circuit around this small plaquette is[13]

$$\mathbf{Q}_{12}(\mathbf{r}) \rightarrow (1 - \kappa^2 a^2 [L_{o\mu}^{(12)},\ L_{ov}^{(12)}])\mathbf{Q}_{12}(\mathbf{r}) \approx e^{-i\kappa a^2 L_{\mu v}^{(12)}}\mathbf{Q}_{12}(\mathbf{r}), \tag{11.21}$$

where $L_{\mu v}^{(12)}$ is the generator of rotations in the (μ, v)-plane. This change can only be accommodated by a density of negatively charged wedge disclination lines threading the plaquette, in agreement with the physical picture developed in Section 8.

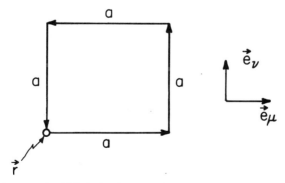

FIG. 45. Circuit around which it is, in general, impossible to lay down unfrustrated icosahedral order using polytope $\{3, 3, 5\}$ as a template.

It is interesting to compare the Landau theory described above with the continuum elastic approach suggested by Sethna.[12] Sethna worked with a "fixed length," $n = 1$ representation of SO(4). Upon applying a similar fixed length approximation to the Landau expansion in Eq. (11.16), one recovers Sethna's elastic free energy, except that his 4×4 SO(4) matrices are replaced by 169×169 ($n = 12$) representation matrices. (Both theories depend only on the 6 SO(4) Euler angles in this limit). Landau's approach is more general, however, because it allows for fluctuations in the *magnitude* of the order parameter and should be more useful at higher temperatures or in the presence of many defects. Since the magnitude of the order parameter vanishes at disclination cores, amplitude fluctuations will appear when one coarse grains over a volume containing several disclinations. If the physical picture developed in Section 8 is correct, this will be an appropriate description at all temperatures down to the glass transition temperature T_g.

Our expectation for the equilibrium statistical mechanics associated with the Landau theory is just that illustrated in Fig. 32. As usual, we assume that the quadratic parameter r changes sign with decreasing temperatures and that the other couplings are temperature independent. At high temperatures, we have a molten liquid, which can be regarded as a tangled mass of disclination lines. The low-temperature phase is a Frank–Kaser crystal conisting of a regular lattice of $-72°$ disclinations. The transition temperature T between these two phases is like the transition at $H_{c2}(T)$ in a type-II superconductor. The effect of (annealed) inhomogeneities in particle size (as would occur in binary mixtures) can be modeled by a renormalization of r.[14]

We shall now argue that many physical systems will drop out of equilibrium upon cooling, instead of condensing into a Frank–Kasper phase, due to entanglement of defect lines.[11] The order parameter discussed above has the symmetry SO(4)/G. The algebra of line defects in such a medium is given by homotopy group $\pi_1(SO(4)/G)$.[104] As shown in Nelson and Widom,[13] this group is just the direct product of Y' with itself,

$$\pi_1(SO(4)/G) = Y' \times Y'. \qquad (11.22)$$

This result means that every defect is characterized by a pair of SU(2) matrix charges, each chosen from the 120-element group Y'. The diagonal subgroup of elements of the form $(\alpha, \alpha) \equiv \alpha$ are the disclinations forced into the medium by the curvature incommensurability. A

[104]N. D. Mermin, *Rev. Mod. Phys.* **51,** 591 (1979).

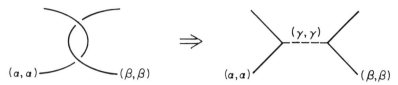

FIG. 46. Disclination entanglement in a predominatly icosahedral medium.

$-72°$ disclination about the $\hat{n}$-axis, for example is characterized by the "charge"

$$\alpha = e^{(1/2)i\theta\hat{n}\cdot\boldsymbol{\sigma}}, \tag{11.23}$$

where $\theta = -72°$ and $\boldsymbol{\sigma}$ is the vector of Pauli matrices. The combination laws of dislocations in conventional crystals are given by addition of Burgers vectors. Here, the defect combination laws are given by SU(2) matrix multiplication. The allowed nodes in networks of $-72°$ disclination lines predicted by this formalism are just those found in the Frank–Kasper phases.[11] As is shown in Fig. 46, it is very difficult for defects lines characterized by non-Abelian charges to cross. The lines cannot simply break and reform but must instead leave behind an "umbilical defect" with charge[104]

$$\gamma = \alpha\beta\alpha^{-1}\beta^{-1}. \tag{11.24}$$

This umbilical string ties the two defects together and impedes their crossing. As shown by Nelson,[11] two $\pm72°$ disclination lines corresponding to rotations about different symmetry axes always produce another $\pm72°$ umbilical defect when they try to cross. Thus, defect lines are prevented from crossing by an approximately linear potential proportional to the length of this umbilical defect.

One would expect the kinetic constraints associated with entanglement to scale with the amount of short-range icosahedral order, becoming more and more severe as a liquid is undercooled more and more below the equilibrium freezing transition T_m. [T_m marks the transition to a conventional (e.g. fcc) crystalline phase.] The transition of this metastable liquid to an ordered Frank–Kasper network of disclination lines should occur at a temperature T_c that is depressed well below T_m by frustration effects. Good glass-formers will drop out of equilibrium due to entanglement at a temperature $T_g < T_m$ before the transition to a Frank–Kasper phase actually occurs.

Just above T_g one can model the short-range icosahedral order in a dense liquid by a sum of free energies as in Eq. (11.16), one for each of the allowed values of $n = 0, 12, 20, 24 \ldots$. It probably suffices to truncate this expansion at quadratic order in the $\mathbf{Q}_n$, since the large

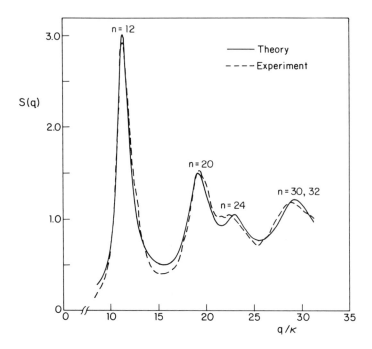

FIG. 47. Fit of the Landau theory of short-range icosahedral order to the structure function of vapor-deposited amorphous iron. Although the peak heights and widths are used to fit Landau parameters, the peak positions are entirely determined by the theory.

intrinsic density of defects forces these order parameters to be small. The particle density is given in terms of the Q_n's by[13,14]

$$\rho(\mathbf{r}) = \sum_{\substack{n=0,12,20 \\ 24,\dots}} \sum_{m_1 m_2} Q_{n_1 m_1 m_2}(\mathbf{r}) Y^*_{n,m_1 m_2}(-1), \qquad (11.25)$$

where the spherical harmonics are evaluated at the south pole of the tangent sphere in Fig. 44. It is straightforward to calculate the density-density correlations in the liquid by using Eq. (11.25) and this set of free energies. One obtains a remarkably good description[14] of the x-ray structure functions observed experimentally in a variety of metallic glasses, which exhibit peaks corresponding to $n = 12$, 20, and 24.

Figure 47 shows a fit[14] to the structure function of vapor deposited amorphous iron.[105] It is believed that a very similar essentially one-component glass would be obtained if iron could be cooled rapidly enough directly from the melt. Very similar structure functions are

[105]J. P. Lauriet, *J. Non-Cryst. Solids* **55**, 77 (1983).

TABLE IV

	THEORY	a-Co	a-Fe	MOLECULAR DYNAMICS
$\dfrac{q_{20}}{q_{12}}$	1.71	1.69	1.72	1.7
$\dfrac{q_{24}}{q_{12}}$	2.04	1.97	1.99	2.0

obtained from computer simulations, dense-random packing, and a wide variety of conventional binary metallic glasses. The good fit is not surprising, since the theory which produced it is based on polytope $\{3, 3, 5\}$, whose structure function shows great similarities to dense-random packing models of metallic glasses (see Fig. 43). The effect of frustration is to broaden the sharp peaks associated with the long-range icosahedral order of polytope $\{3, 3, 5\}$.

Although the peak widths and intensities are used to fit the parameters of the Landau theory, the relative peak positions are completely determined by the symmetry of the short-range icosahedral order. Table IV compares the peak positions with results from amorphous iron, cobalt and molecular dynamics simulations.[106]

The theory described here is still in an early stage of development. Once glassy x-ray structure functions are understood, one can try to understand other problems in solid state physics, like band structure and transport properties. The kinetics of entanglement presumably controls the viscosity and diffusion constants near T_g. The transition to the Frank–Kasper state may have something to do with an "intrinsic" but experimentally inaccessible glass transition at a temperature $T_0 < T_g$. A dynamical theory based on the icosahedral order parameter discussed here has been developed by Sachdev.[107]

VI. Quasicrystals

12. EXPERIMENTS AND THEORETICAL BACKGROUND

The fivefold rotational symmetry, characteristic of the short-range order in the polytetrahedral packings discussed in the previous sections,

[106]K. Kimura and F. Yonezawa, in "Topological Order in Condensed Matter" (F. Yonezawa and T. Ninomiya, eds.). Springer, Berlin, 1983.
[107]S. Sachdev, *Phys. Rev.* **B33,** 6395 (1986).

Fig. 48. Electron diffraction pattern taken along the fivefold symmetry axis of a rapidly quenched Ga-Mg-Zn quasicrystal. [W. Ohashi and F. Spaepen, *Nature* **330,** 555 (1987).]

is inconsistent with translational periodicity: it has long been known that periodic crystals can only have two-, three-, four-, or sixfold rotational symmetry.[108] The discovery in 1984 by Shechtman and coworkers[3] of an Al-Mn phase that exhibited a sharp electron diffraction pattern, till then commonly attributed to crystal periodicity, with full icosahedral (m $\bar{3}\bar{5}$) point symmetry, came therefore as a surprise to most workers in the field. Figure 48 shows an example of such a diffraction pattern, taken along the fivefold axis.

This apparent paradox is easily resolved when one realizes that the occurrence of sharp diffraction peaks does *not* necessarily imply simple periodicity.[109] In fact, as has been known by mathematicians since the

---

[108]L. D. Landau and E. M. Lifshitz, "Statistical Physics," Addison-Wesley, Reading, MA, 1970, p. 408.

[109]Reviews on quasicrytals can be found in D. R. Nelson and B. I. Halperin, *Science* **229,** 233 (1985); D. R. Nelson, *Sci. Am.* **255,** (2), 42 (1986); J. W. Cahn, *Mat. Res. Soc. Bull.,* March 1986, p. 9; C. L. Henley, *Comm. Condensed Mat. Phys.* **13,** 59 (1987); P. J. Steinhardt, *Science* **238,** 1242 (1987). For a readable account of the symmetries of polytope {3, 3, 5} and its connection with metallic glasses and quasicrystals, see M. Widom, *in* "Aperiodicity and Order: Introduction to Quasicrystals" (M. V. Jaric, ed.), Academic Press, Boston, 1987.

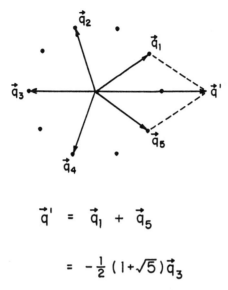

$$\vec{q}\,' = \vec{q}_1 + \vec{q}_5$$

$$= -\tfrac{1}{2}(1+\sqrt{5})\,\vec{q}_3$$

FIG. 49. The five basis vectors required to index the diffraction pattern of Fig. 48. Two of these vectors can be combined to give a vector that is the negative of the golden mean times of another element of the set.

end of the previous century, there exist many nonperiodic functions whose Fourier transform (i.e., diffraction pattern) is sharply peaked. A simple example is $f(x) = \cos x + \cos \sqrt{2}\, x$. The Fourier transform of this function has two sharp peaks at $q = 1$ and $q = \sqrt{2}$. If all the diffraction peak positions can be expressed as a sum of integer multiples of a finite number of vectors exceeding the dimensionality of the space, the function is called *quasiperiodic*. In the one-dimensional case above, 2 lengths are needed. To describe a two-dimensional quasiperiodic diffraction pattern, such as that of Fig. 48, 5 vectors $\mathbf{q}_i$ $(i = 1\text{–}5)$ are needed, as illustrated in Fig. 49. Three "generations" of linear combinations of these vectors are shown in Fig. 50; the similarity with the experimental pattern of Fig. 48 is clear. To span a three-dimensional quasiperiodic diffraction pattern, 6 vectors are needed; a number of schemes for his have been proposed.[110,111,112]

One type of quasiperiodic structure that is particularly useful in the study of these new materials is the Penrose pattern.[113] A two-dimensional

[110]A collection of reprints has been published in "The Physics of Quasicrystals," (P. J. Steinhardt and S. Ostlund eds.). World Scientific Publ. Co., 1987.
[111]V. Elser, *Phys. Rev.* **B32**, 4892 (1985).
[112]J. W. Cahn, D. Shechtman, and D. Gratias, *J. Mat. Res.* **1**, 13 (1986).
[113]R. Penrose, *Bull. Inst. Math. and its Appl.* **10**, 266 (1974).

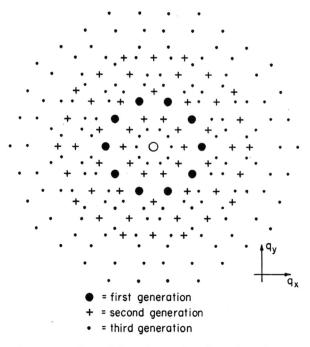

● = first generation
+ = second generation
• = third generation

FIG. 50. First three generations of intensity maxima in reciprocal space obtained by combining the vectors of Fig. 49. Second generation spots are the sum of two first generation basis vectors, while the third generation positions are obtained by adding together triplets of this first generation basis set. The similarity with the experimental pattern of Fig. 48 should be noted.

example, constructed by tiling the plane with 2 types of rhombuses according to certain matching rules,[114] is shown in Fig. 51. Since all the angles of the rhombuses are integral multiples of $2\pi/10$, all the edges lie along one of ten directions, giving the structure *perfect orientational order*. That the Penrose tiling is quasiperiodic was shown first by determining its optical diffraction pattern,[115] which turns out to be similar to that of Figs. 48 and 50. The translational periodicity corresponding to one of the main Bragg spots ("first generation" in Fig. 50) is indicated in Fig. 51 by a grid of evenly spaced ($d = 2\pi/q_i$) parallel lines through successive rows of tiles with parallel edges. De Bruijn[114] showed that these two-dimensional Penrose patterns can be obtained by projection of a slice of a five-dimensional simple cubic lattice along an irrational direction. This projection technique was generalized by Kramer and

[114]N. G. de Bruijn, *Proc. Ned. Akad. Wetensch. Ser.* **A43,** 39 and 53 (1981).
[115]A. L. Mackay, *Physics* **114A,** 609 (1982).

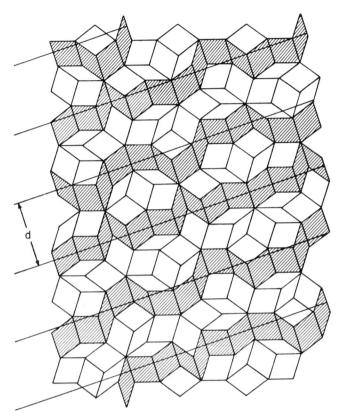

FIG. 51. Two-dimensional Penrose pattern, with a regularly spaced grid of lines superimposed on the shaded rows of the tiles with parallel sides. [From D. R. Nelson and B. I. Halperin, *Science* **229**, 233 (1985).]

Neri[116] to produce a three-dimensional Penrose pattern by projection of a six-dimensional lattice. The building blocks of this structure are then 2 rhombohedra with interior solid angles that are integral multiples of $4\pi/20$ steradians. Levine and Steinhardt[117] independently obtained a similar result, and were the first to propose using these blocks to model laboratory quasicrystals. Elser[111] and Duneau and Katz[118] used the projection technique to obtain excellent agreement between the experimental diffraction pattern and that of the three-dimensional Penrose tiling along the 2-fold, 3-fold, and 5-fold symmetry axes.

[116]P. Kramer and R. Neri, *Acta Cryst.* **A40**, 580 (1984).
[117]D. Levine and P. J. Steinhardt, *Phys. Rev. Lett.* **53**, 2477 (1984).
[118]M. Duneau and A. Katz, *Phys. Rev. Lett.* **54**, 2688 (1985).

Most of the quasicrystals that have been discovered so far are thermodynamically metastable. They are usually produced by nonequilibrium methods, such as rapid quenching from the melt,[3] transformation of the amorphous phase,[119] or ion mixing.[120] Upon heating they transform irreversibly to the periodic crystalline state, which can be observed as an exothermic reaction in the calorimeter.[121] A few stable quasicrystals have been discovered, however: Al_6CuLi_3,[122] $Ga_{21}Mg_{38}Zn_{41}$,[123] and $Cu_{20}Al_{65}Fe_{15}$,[124] They can be conventionally cast from the melt, and, upon heating, they transform endothermally in an equilibrium reaction (in all these cases involving the melt); below this transition temperature they can be held indefinitely. These stable quasicrystals have striking solidification morphologies and can be made up to several millimeters in size. Figure 52 shows the pentagonal dodecahedral morphology of the Ga-Mg-Zn quasicrystals. The Cu-Al-Fe quasicrystals are similar, whereas the Al-Li-Cu quasicrystals have a triacontahedral morphology. The presence of large facets on these stable quasicrystals tends to support the structural models discussed above, in which the icosahedral symmetry is homogeneously represented in all parts of the phase. It is difficult to see how microtwinning[125] would tend to expose planes with characteristic icosahedral symmetry, rather than the planes favored in the untwinned crystals. Similarly, if one tried to model the structure by a crystalline approximate with a large unit cell[126] (8 nm lattice parameter), the quasicrystal facets would be very high index planes, and it would again be difficult to explain why they would be exposed.

An intriguing question is to what extent the icosahedral symmetry observed in the quasicrystals is related to that discussed for polytetrahedral packing in the previous sections. There certainly seems to be a very strong link between the local icosahedral coordination and the formation of quasicrystalline phases. For example, the structural units of Elser and Henley's model for icosahedral Al-Mn-Si are the Mackay icosahedra that in a body-centered cubic arrangement make up most of

[119]S. J. Poon, A. J. Drehman, and K. R. Lawless, *Phys. Rev. Lett.* **55**, 2324 (1985).

[120]D. M. Follstaedt and J. S. Knapp, *J. Appl. Phys.* **59**, 1756 (1986).

[121]B. G. Bagley and H. S. Chen, *Mat. Res. Soc. Symp. Proc.* **57**, 451 (1986).

[122]P. Sainfort and B. Dubost, *J. de Phys.* **47**, C-3, 321 (1986); M. A. Marcus and V. Elser, *Phil. Mag.* **B54**, L101 (1986).

[123]W. Ohashi and F. Spaepen, *Nature* **330**, 555 (1987).

[124]A.-P. Tsai, A. Inoue, and T. Masumoto, *Jap. J. Appl. Phys.* **26**, L1505 (1987).

[125]L. Pauling, *Nature* **317**, 512 (1985); *Phys. Rev. Lett.* **58**, 365 (1987).

[126]K. M. Knowles and W. M. Stobbs, *in* "Phase Transformations '87," in press; K. M. Knowles, *in* "Quasicrystalline Materials," ILL/Codest Workshop, Grenoble (1988), in press.

FIG. 52. Stable Ga-Mg-Zn quasicrystals with pentagonal dodecahedral solidification morphology [W. Ohashi and F. Spaepen, unpublished results; similar to those in *Nature* **330**, 555 (1987).]

the stable crystalline phase.[127] More importantly, in light of the earlier discussion, many of the icosahedral phases are found at or near compositions that form polytetrahedral Frank–Kasper phases. For example, metastable icosahedral $(Al, Zn)_{49}Mg_{32}$ is found at the same composition as the stable Frank–Kasper phase (see Table I)[128] and the same tetrahedrally close-packed units (each decorated with $-72°$ disclination lines) can be used to construct both phases.[129] In the Ga-Mg-Zn system it was the known existence of the $GaMg_2Zn_3$ Frank–Kasper phase[130] that led to the discovery of the nearby stable icosahedral phase.[123] The stable Al-Li-Cu quasicrystals also occur near a Frank–Kasper phase isostructural with $(Al, Zn)_{49}Mg_{32}$.

[127]V. Elser and C. L. Henley, *Phys. Rev. Lett.* **55**, 2883 (1986); P. Guyot and M. Audin, *Phil. Mag.* **B52**, L15 (1985).
[128]G. V. S. Sastry and P. Ramachandrarao, *J. Mat. Res.* **1**, 247 (1986).
[129]C. L. Henley and V. Elser, *Phil. Mag, B***53**, L59 (1986).
[130]P. Villars and L. D. Calvert, "Pearson's Handbook of Crystallographic Data for Intermetallic Phases," Vol. 3, ASM. Metals Park, OH (1985), p. 2277.

In view of the discussion of sections 3 and 10, it can be expected that the similarity in short-range order between the liquid and the icosahedral phase would result in a lower interfacial tension, and hence to easier nucleation. This can only be definitively answered by performing undercooling experiments similar to those discussed in Section 3. There are, however, some indications that nucleation of the icosahedral phase from the melt is indeed quite easy. For example, rapidly quenched[131,132] Al-Mn alloys seem to have "micro-quasicrystalline" structure, with a grain size of no more than a few nm. This seems to indicate copious nucleation. It is also of interest to note that some of the Frank–Kasper phases themselves, such as the $MgZn_2$ Laves phase, seem to nucleate very easily in undercooling and casting experiments[133] in ternary systems. Finally, the structural similarity between the liquid and the icosahedral phases is also apparent when their structure factors are compared. The structure factors of most liquid and amorphous metals have many features in common: the relation between the peak positions and, if they can be resolved, the features of the second diffraction. It is therefore significant, as shown in the next section, that a typical quasicrystal structure factor can be calculated from a typical structure factor of a liquid near the glass transition.

13. THEORY OF POLYTETRAHEDRAL QUASICRYSTALS

A connection between the short-range order in metallic glasses and order in quasicrystals is suggested by Fig. 53, which shows schematically the diffraction pattern normal to the twofold symmetry axis for icosahedral crystals.[134] All these Bragg peaks can be written as integer linear combinations of 12 fundamental reciprocal lattice vectors of length q_0 pointing to the vertices of an icosahedron.[135] The four elements of this set that lie in the twofold symmetry plane are indicated by the large dots. Smaller dots mark the positions of peaks generated by linear combinations of 2, 3, and 4 members of the basis set. As indicated in Fig. 53, there are second-generation peaks at radii, which are 1.052, 1.701, and 2.0 times q_0. One might expect these peaks to be among the most intense Bragg spots. The positions of these peaks are correlated with the locations of the first 3 maxima in the typical metallic glass structure factors shown in Fig. 54. The pronounced first peak in glasses is broad

[131]L. A. Bendersky and S. D. Riddler, *J. Mat. Res.* **1**, 405 (1986).
[132]L. C. Chen and F. Spaepen, *Nature,* in press.
[133]S. Ebalard, W. Ohashi, and F. Spaepen, unpublished results.
[134]S. Sachdev and D. R. Nelson, *Phys. Rev. B***32**, 4592 (1985).
[135]D. R. Nelson and S. Sachdev, *Phys. Rev. B***32**, 689 (1985).

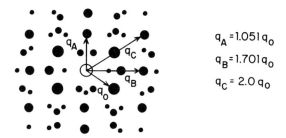

$q_A = 1.051 q_0$

$q_B = 1.701 q_0$

$q_C = 2.0 q_0$

FIG. 53. Most intense spots in a typical quasicrystal in reciprocal space normal to a twofold symmetry axis.

enough to accommodate ordering at both the fundamental wave, vector q_0 and $q_A = 1.052q_0$. The $n = 20$ and $n = 24$ peaks are in registry with the remaining second-generation peaks at $q_B = 1.701q_0$ and $q_C = 2.0q_0$. These observations make it plausible that an icosahedral quasicrystal could, under certain conditions, condense out of an undercooled liquid with structural correlations similar to those in metallic glass.

This hypothesis can be tested by using the density functional mean field theory of Ramakrishnan and Yussouff.[136] This theory examines at a mean field level the energetic and entropic differences between the uniform liquid and ordered crystals at a given temperature. The entropy is assumed to be the entropy associated with rearranging cells of particles with dimensions of order the translational correlation length. The structural energy is estimated to quadratic order in a density expansion by using the structure factor of the liquid. Starting with, say, the liquid structure factor at the triple point of argon or sodium, one can make accurate, quantitative predictions about the volume, entropy, and peak intensities in the coexisting crystalline phase.[136]

Recently, Sachdev and Nelson used model amorphous metal structure factors as input into the density functional theory.[134] These are assumed to be characteristic of undercooled liquids in metastable equilibrium just above the glass transition temperature T_g. For simplicity, the calculations have focused on one-component amorphous systems. It is not hard to generalize the density functional approach to alloys; partial structure factors are required as input.

The density-functional approach allows us to expand the density $\rho(\mathbf{r})$ in a set of trial reciprocal lattice vectors,

$$\rho(\mathbf{r})/\rho_0 = 1 + \sum_{\mathbf{G}} \mu_{\mathbf{G}} e^{i\mathbf{G} \cdot \mathbf{r}}, \tag{13.1}$$

[136]T. V. Ramarkrishnan and M. Yussouff, *Phys. Rev.* **B19**, 2775 (1979).

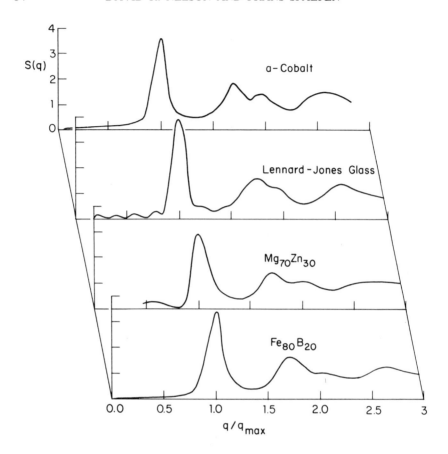

FIG. 54. Structure factors of vapor-deposited amorphous cobalt (P. K. Leung and J. C. Wright, *Philos. Mag.* **30**, 185 (1974)); a computer-cooled Lennard–Jones glass, (Ref. 106) the bimetallic glass Mg_{70}, Zn_{30} (H. Rudin, S. Jost and J.-J. Güntherodt, *J. Non-Cryst. Solids* **62/62**, 291 (1984)); and the metallic glass $Fe_{80}B_{20}$ (A. Defrain, L. Bosio, R. Cortes, and G. De Costa, *Non-Cryst. Solids* **61/62**, 439 (1984)). The structure factors have been rescaled to make their primary peaks coincide.

and search for states with a lower free energy than the isotropic liquid. Here ρ_0 is the mean density, and the $\{\mu_G\}$ may be viewed as dimensionless variational parameters. In addition to reciproval lattice sets corresponding to bcc and fcc crystals, we also use sets with the symmetry of an icosahedral crystal, generated by stars of reciprocal lattice vectors pointing to the vertices, edges, or faces of an icosahedron.[135] Although icosahedral crystals based on the edge or face models are always unstable in the calculation in Ref. 134, liquids near T_g are metastable with respect to a vertex model icosahedral crystal. Fcc crystals have the lowest free energy, in accord with one's expectations for single-component hard spheres. The free energy of bcc crystals is intermediate between an fcc lattice and a vertex model icosahedral crystal. The density-functional approach allows the prediction of the Bragg peak intensities in the metastable icosahedral crystal.

In Fig. 55 we show the intensities of the Bragg peaks in a metastable icosahedral crystal obtained by using the structure factor of a relaxed dense random packing model[64] as input. The pattern of peak positions

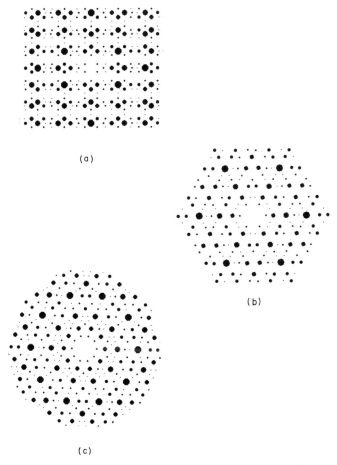

FIG. 55. Spot pattern perpendicular to the (a) twofold, (b) threefold and (c) fivefold axes of an icosahedron. The areas of the spots represent actual intensities predicted by the density-functional theory of Ref. 134.

and intensities normal to the three-, five-, and twofold icosahedral symmetry axes is quite similar to that observed experimentally by Shechtman *et al.*[3] The atomic positions in real space correspond to the decorated Penrose "bricks" shown in perspective in Fig. 56. Figure 57 shows a cross section of the density, normal to the fivefold symmetry axis, together with the corresponding cross section of the Penrose bricks.

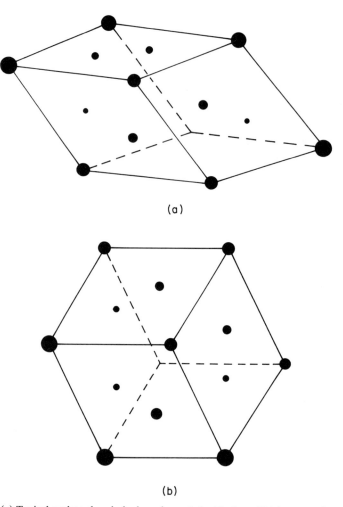

(a)

(b)

Fig. 56. (a) Typical prolate rhombohedron decorated with sites of high occupation number. The areas of the circles are proportional to the mean occupation number at the sites. Only the sites on the three faces of the rhombohedron nearest the viewer are shown. (b) The same as (a) for the oblate rhombohedron.

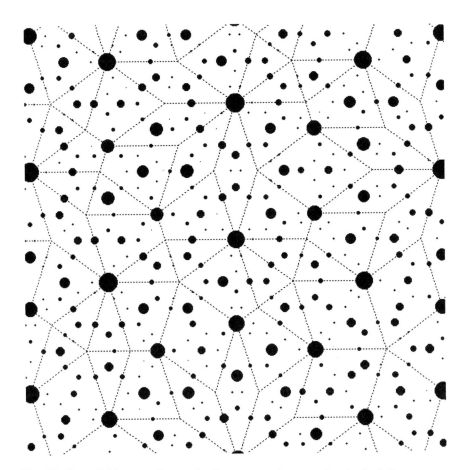

FIG. 57. Sites of high occupation number in a cross section normal to the fivefold symmetry axis of the quasicrystal solution of the density functional theory of Ref. 134.

Note that similar cross sections in Fig. 57 are decorated in approximately the same way.

The density-functional mean-field theory shows that one-component undercooled liquids are metastable with respect to a vertex-model icosahedral crystal, similar to icosahedral Al-Mn. The calculations assume that the structure of an undercooled liquid is similar to relaxed dense-random packing. The closely related structure factor of vapor-deposited amorphous cobalt was also used as input, because it appears to exhibit an exceptional degree of short-range icosahedral order, with similar results. The stability of the icosahedral crystal is related to the short-range icosahedral order already present in an undercooled liquid.

As shown in Fig. 57, the density-functional formalism allows us to visualize how the atoms of an icosahedral crystal are distributed in real space. If we use the particle spacing in amorphous cobalt as an effective hard-sphere diameter, the edges of the Penrose rhombahedra shown in Fig. 56 turn out to be about 4.1 Å. This is simialr to the edge length of 4.6 Å, which appears to be appropriate to icosahedral Al-Mn.[127] In their discussion of microscopic models for icosahedral Al-Mn, Elser and Henley place managanese atoms at the vertices of the Penrose rhombahedra and put aluminum atoms at various positions on the faces and edges.[127] We too find two different classes of sites, those with occupation numbers of order unity (on the vertices) and those with smaller occupation numbers (primarily on the faces). About 30% of the atoms sit at vertex sites, suggesting that these are indeed the relevant sites for the minority atoms (manganese) in Al-Mn alloys. The many small occupation numbers we find away from the vertices of the Penrose lattice suggest that, if we associate aluminum atoms with these sites, they are disordered, like the mobile phase in a superionic conductor.

As mentioned above, the structure function of polytetrahedral one-component dense-random packing was used as input in these calculations. This structure function, of course, is only a rough approximation to the melt of the Al-Mn quasicrystalline alloy discovered by Shechtman et al.[3] Henley and Elser have shown, however, that the quasicrystalline phase of $Mg_{32} (Al, Zn)^{49}$ is almost certainly polytetrahedral and have described in detail how the closely related Frank–Kasper phase that forms at slower coolings can be decomposed into distorted Penrose bricks, each decorated with a set of $-72°$ disclination lines.[129] There seems little doubt that quasicrystals provide yet another intriguing way in which nature copes with the frustration embodied in polytetrahedral particle packings.

Appendix:
The Statistical Honeycomb Model

Coxeter[6-8] has proposed a number of polytopes $\{3, 3, q\}$ that will formally fit into 3-D flat space provided one allows q, the number of tetrahedra per bond, to be nonintegral. One model, with $q = 5.115$, was used to model the Voronoi statistics of soap froth. Soap froth tends to be irregular and "glassy," in part because there is no regular packing of cells that can accommodate the constraint of 120° angles between walls and 109.5° angles at the vertices of the Voronoi cells. The soap froth problem is in some ways dual to the problem of packing spheres to form a metallic glass. Polytope $\{5, 3, 3\}$, the dual of polytope $\{3, 3, 5\}$, is a regular lattice of dodecahedra inscribed on S^3 (see Fig. 2a) and might serve as useful template for discussions of soap froth.

The statistical honeycomb model value of q for metallic glasses follows from noting that four particles in isolation will pack to form a perfect tetrahedron. Since the dihedral angle of a tetrahedron is $\cos^{-1}(1/3)$, exactly

$$q = 2\pi/\cos^{-1}(1/3) = 5.1043 \ldots \tag{A.1}$$

of these are necessary to fill the 2π of angle around a bond. Assuming all bonds have tetrahedra arranged in this way, we can construct Voronoi cells with Z identical faces formed by regular q-sided polygons. The Euler–Descartes theorem tells us that $Z - E + V = 2$, where E and V are the number of edges and vertices of each Voronoi cell. Since $E = (1/2)Zq$ and $V = (1/3)Zq$, the number of faces of each cell is

$$Z = \frac{12}{6 - q} \approx 13.4 \ldots . \tag{A.2}$$

Note that this nearest-neighbor coordination number is higher than that of an fcc crystal ($Z_{\text{fcc}} = 12$), so that this structure should have in some sense a lower energy than a conventional crystal for most pair potentials.

Continuing with the notion of Voronoi cells with nonintegral numbers of faces, each of which has a nonintegral number of edges, we can calculate the packing fraction. If the particle spacing is d, the packing fraction is

$$f = \frac{4}{3} \pi \left(\frac{d}{2}\right)^3 \Big/ V_{\text{cell}}, \tag{A.3}$$

where V_{cell} is the volume of the Voronoi cell. But

$$V_{\text{cell}} = Z\left(\frac{1}{3}\frac{d}{2}A_q\right), \tag{A.4}$$

where $1/3 \; d/2 \; A_q$ is the volume of a conical prism of height $d/2$ and a base of area A_q formed by a q-sided regular polygon. Upon dividing a q-sided regular polygon into q identical triangles, each with a vertex at the center, it is straightforward to show that

$$A_q = \tfrac{1}{4}qs^2 \cot(\pi/q), \tag{A.5}$$

where s is the length of a side. Now, s is the distance between the centers of two adjacent perfect tetrahedra with edge d, i.e.,

$$s = d/\sqrt{6}. \tag{A.6}$$

Combining Eqs. (A.3)–(A.6), we have

$$f = 24\pi/Zq \cot(\pi/q). \tag{A.7}$$

Using Eqs. (A.1) and (A.2) leads, after some algebraic manipulations, to the desired result,

$$f_{\text{ideal}} = \frac{1}{\sqrt{2}} [6 \cos^{-1}(1/3) - 2\pi]. \tag{A.8}$$

ACKNOWLEDGMENTS

It is a pleasure to acknowledge the advice and inspiration of David Turnbull. We also benefited from interactions with H. Andersen, L. C. Chen, S. Ebalard, V. Elser, R. B. Meyer, D. Minton, R. Mosseri, W. Ohashi, M. Rubinstein, S. Sachdev, J. F. Sadoc, J. Selinger, P. Steinhardt, J. P. Straley, and M. Widom. We are grateful to the Smithsonian Institution for providing the oolite sample for Figure 12. This work was supported by the National Science Foundation, through the Harvard Materials Research Laboratory (Grant No. DMR-86-14003) and through Grant No. DMR-85-14638.

SOLID STATE PHYSICS, VOLUME 42

Physical Properties of the New Superconductors

M. Tinkham and C. J. Lobb

Physics Department and Division of Applied Sciences
Harvard University
Cambridge, Massachusetts

1. Introduction

The discovery by Bednorz and Müller[1] of superconductivity above 30 K in LaBaCuO triggered a flood of research on high-temperature superconductivity that has been reported in hundreds of papers and at innumerable conferences. A sequence of new materials has subsequently been discovered, with the highest T_c being $\sim$125 K at the time of writing.[2] The crystal structures of many of these compounds feature Cu-O planes, as described in detail elsewhere in this volume, leading to unusually anisotropic superconducting properties. Although each of these new materials has its own individual characteristics, they share a largely common phenomenology. The superconducting state still appears to be based on electron pairs, but the presence of anistropy, granular structure, and an unusual range of parameter values leads to novel features relative to the classic superconductors.

[1] J. G. Bednorz and K. A. Müller, *Z. Phys. B* **64**, 189 (1986).
[2] S. S. P. Parkin, V. Y. Lee, E. M. Engler, A. I. Nazzal, T. C. Huang, G. Gorman, R. Savoy, and R. Beyers, *Phys. Rev. Lett.* **60**, 2539 (1988).

The key direct consequence of the high T_c is that the Bardeen–Cooper–Schrieffer (BCS) coherence length[3] $\xi_0 \sim 0.2\hbar v_F/kT_c$ is very small. (These materials have low carrier concentrations and Fermi velocities as well as the high T_c.) A representative value for ξ_0 is 10 Å (averaged over crystal axes), which is of the order of the unit cell dimension. Since the coherence length sets the length scale for variation of the superconducting order parameter, this small value implies that the superconductivity will be much more sensitive to small scale structural or chemical imperfections than in a classic superconductor, where $\xi_0 \sim 10^2$–10^4 Å. The small coherence length also gives a coherence volume so small as to contain only a few Cooper pairs, which implies that fluctuations will play a much larger role in these materials than in the classic superconductors. Of course the small coherence length also leads to extremely high values of the upper critical field H_{c2}, with potential benefits for applications.

In this review, we will emphasize the generic aspects of the new superconductors, taking YBCO($Y_1Ba_2Cu_3O_7$) as a typical example because it was the first of the 90-K superconductors to be discovered,[4] and it has been studied most thoroughly. Our discussion will emphasize those features where a reasonable degree of consensus has emerged, but some comments will also be made on the status of features where consensus is still lacking.

In the interest of compactness in this review, we assume that the reader is already familiar with the phenomenology of the classic superconductors, as presented, for example, in the text[5] written by one of the present authors. For the convenience of readers who lack familiarity with this material, we have often cited page references in that text, rather than referring back to the original literature of the earlier period that is summarized therein.

I. Magnetic and Resistive Properties

2. The Importance of Granularity

The properties of granular superconductors have been studied in two primary contexts: (a) in materials such as granular aluminum or niobium

[3]J. Bardeen, L. N. Cooper, and J. R. Schrieffer, *Phys. Rev.* **108,** 1175 (1957).

[4]M. K. Wu, J. R. Ashburn, C. J. Tong, P. H. Hor, R. L. Meng, L. Gao, Z. J. Huang, Y. Q. Wang, and C. W. Chu, *Phys. Rev. Lett.* **58,** 908 (1987).

[5]M. Tinkham, "Introduction to Superconductivity." McGraw-Hill, New York, 1975; reprinted by Robert E. Krieger Publishing Co., Malabar, Florida, in 1980 and 1985, cited hereafter as MT.

nitride, and (b) in fabricated two-dimensional arrays of superconducting thin film islands weakly connected to their neighbors by tunnel junctions or by normal metal proximity-effect weak links. Such systems typically show a two-stage transition to superconductivity upon cooling. First, at the transition temperature T_{c0} of the superconducting grain or island, the resistance drops, reflecting the loss of the resistance of the grains or islands. Upon further cooling, the resistance drops continuously as the Josephson coupling energy in the weak links begins to overcome thermal fluctuations and introduce correlations between the phases of the superconducting wave functions on the various islands. Finally, at a lower temperature, called simply T_c, the phases lock together to give long-range phase coherence and zero resistance. On physical grounds, one anticipates that $kT_c \sim E_J(T_c) = \hbar I_c(T_c)/2e$, where E_J is the Josephson coupling energy between adjacent grains, and I_c is the associated intergranular critical current. In fact, in the mean field approximation, $kT_c = (z/2)E_J(T_c)$, where z is the number of nearest-neighbor grains.

This sort of transition has been studied most thoroughly in two-dimensional systems, where it is called the Kosterlitz–Thouless[6] transition. Here a renormalization group calculation shows that $kT_c = (\pi/2)E_J(T_c)$, where the numerical coefficient (equivalent to $\sim$27 nA/Kelvin) reflects the two-dimensional geometry of a square array. Precisely speaking, T_c separates two regimes: Above T_c, there are free vortices (each containing one flux quantum) that move under the influence of a transport current, giving resistance. Below T_c, (in the absence of an externally applied magnetic field) all vortices are bound into pairs of opposite sign, forming entities that feel no net force from a transport current and, hence, give no resistance. The number of free vortices (and hence the resistance) goes to zero continuously at T_c, with a characteristic exponential dependence on $(T - T_c)^{-1/2}$.

In three dimensions, a qualitatively similar phase-locking transition takes place, but for a somewhat higher ratio of kT_c to E_J because of the larger number of nearest neighbors. However, the detailed nature of the transition in terms of vortex dynamics is much less clearly understood.

For this two-stage transition to be broad enough to be observable, the intergranular coupling must be weak enough that T_c is significantly below T_{c0}. Applying the criterion $kT_c \sim E_J(T)$ stated above, and noting that E_J typically increases linearly with $(T_{c0} - T)$, one can estimate this depression of T_c below T_{c0} for any given system. For the Kosterlitz–Thouless

[6]J. M. Kosterlitz and D. J. Thouless, *J. Phys.* **C6,** 1181 (1973). For the application to superconductivity, see B. I. Halperin and D. R. Nelson, *J. Low Temp. Phys.* **36,** 599 (1979); see also S. Doniach and B. A. Huberman, *Phys. Rev. Lett.* **42,** 1169 (1979).

case in two dimensions, it was shown by Beasley *et al.*[7] that the fractional transition width $(T_{c0} - T_c)/T_{c0}$ was of order $R_\square/R_q$, where $R_\square$ is the normal resistance per square of the film and $R_q \sim h/4e^2 \sim 6k\Omega$ is the characteristic quantum resistance. However this result is only valid in certain cases: a very thin homogeneous film or an array of islands coupled by ideal Josephson *tunnel* junctions. If instead the coupling is by SNS proximity-effect links, E_J decreases exponentially with the island separation distance (in units of the appropriate ξ) through the normal metal, while the normal resistance varies only linearly with separation. Hence one can have broad transitions, with deeply depressed T_c, even in systems with resistance much below R_q. For this reason, fabricated thin film systems of this type have been studied in detail by a number of groups.[8]

Now let us apply these ideas to the high-temperature superconductors. As noted above, because of the short coherence length associated with the high T_c, these materials are unusually sensitive to structural imperfections. Even so-called single crystal samples of the YBCO class of materials are in fact frequently interrupted (at intervals of ~1000 Å) by twin boundaries separating microcrystals with crystal axes rotated to interchange the *a* and *b* axes of the orthorhombic structure. Even if only a layer of one unit cell thickness is distorted by this twin boundary, that could be sufficient to introduce a significant weakening of the superconductivity, because this thickness would be comparable with the coherence length. Electron micrographs also show other defect structures in such crystals; moreover, variations in local stoichiometry might cause weakened superconductivity without showing up distinctly on micrographs. In the less perfectly aligned "textured" thin films, significantly boundaries are more prominent. Finally, in the clearly inhomogeneous structures typical of the ceramic "pellet" material that is involved in all large scale experiments [since only very tiny (<1 mm) single crystals have been grown to date], there is an obvious granularity on a length scale of order 1 micron. In such samples, one assumes one has grains of relatively good stoichiometric crystalline material, separated by off-stoichiometry material that may be superconducting at a lower temperature, nonsuperconducting but metallic, or even semiconducting.

It seems likely that the intergranular coupling is better modeled as an SNS-type weak link than as a tunnel junction directly connecting two bits

[7]M. R. Beasley, J. E. Mooij, and T. P. Orlando, *Phys. Rev. Lett.* **42,** 1165 (1979).

[8]J. Resnick, J. C. Garland, J. T. Boyd, S. Shoemaker, and R. S. Newrock, *Phys. Rev. Lett.* **47,** 1542 (1981); D. W. Abraham, C. J. Lobb, M. Tinkham, and T. M. Klapwijk, *Phys. Rev. B***26,** 5268 (1982).

of *ideal* material, because the short coherence length allows a continuous depression of the superconducting order parameter near the interface. Thus the Josephson coupling energy E_J is expected to be *less* than would be implied for the same normal resistance by the ideal Ambegaokar–Baratoff (AB) relation[9]

$$E_J = hI_c/2e = (\hbar/8e^2R_n)\Delta \tanh(\Delta/2kT), \qquad (2.1)$$

where the BCS energy gap $\Delta(T)$ approaches $1.76kT_c$ for $T \ll T_c$. To get a feel for the numbers, if the integranular resistance adds $\sim 10^{-4}$ Ω-cm to the normal resistivity, and if the grain size is $\sim 1\ \mu$m, the resistance of a typical intergranular link is ~ 1 ohm. For ideal AB junctions of YBCO, this would imply a link critical current (at $T = 0$) of $I_c(0) \sim 10^{-2}$ A, or $J_c(0) \sim 10^6$ A/cm^2. This value is found in good single crystals, but ceramic samples usually have J_c of only 10^3–10^4 A/cm^2. The latter values correspond to a depression of T_c by 1–10 K below T_{c0} in YBCO.

Note that the sort of transition width discussed here does *not* arise from the presence of macroscopic regions with different T_{c0} values but rather from the weakness of the phase-locking between grains with the *same* T_{c0}, which allows a sort of flux–flow resistance. Of course the *observed* transition widths will include contributions from macroscopic inhomogeneities as well, to an extent determined by the percolative paths for conduction through material of various T_{c0} values. On the other hand, a percolative path of particularly tightly coupled grains could give a much smaller depression of the resistive T_c than is estimated for *average* grains. Thus, the interpretation of data on individual samples is bound to involve much uncertainty unless they are particularly well characterized. In addition, as will be discussed below, the presence of an external magnetic field introduces substantial additional complications, as does the use of finite currents as in the measurement of critical current values.

Motivated by the above considerations, we shall begin our discussion of the properties of the high-temperature superconductors in terms of a model of grains of ideal crystalline superconductor connected by Josephson weak links. Because of the importance of this granularity, we shall first explore in some detail its effect on the electromagnetic properties of the materials. Then we explore the flux flow and creep phenomena, which are interesting in their own right and important in interpreting experiments that are used to infer material parameters. Finally, we discuss the current status of knowledge of other physical properties and parameters thought to characterize ideal material.

[9]V. Ambegaokar and A. Baratoff, *Phys. Rev. Lett.* **10,** 486 (1963).

3. Effective Medium Parameters

a. *Definition of the Model*

To facilitate a semiquantitative discussion of the effects of granularity on the properties of the high temperature superconductors, we adopt a highly simplified model of grains of ideal crystalline superconductor connected in a three-dimensional cubic array by weak links, each having critical current I_c. We take the lattice spacing of the array to be a and the grains to have volume V_g. In interpreting experimental results, it will be important to bear in mind that in real samples these parameters will have a *distribution* of values, and perhaps a *hierarchy* of grains *within* grains at different length scales may be needed to model the full range of behaviors observed. For simplicity, we initially ignore the large anisotropy of the grains and use angular averages of parameters for making estimates. When illustrative numerical values are quoted, they will refer to T well below T_c.

Two limiting regimes occur in granular superconductors, depending on how tightly coupled the grains are. A useful measure of this coupling strength is the ratio of the macroscopic critical current density set by the Josephson effect coupling, namely $J_{cJ} = I_c/a^2$, to that set by the critical current density J_{cg} inside the grains (which we presume are large compared to $\xi_g \sim 10$ Å, and fill the majority of the sample volume). Since we know that J_c values in excess of 10^6 A/cm^2 have been observed experimentally in crystalline material, we may presume that granular material with typical measured values of $J_c \sim 10^3$–10^4 A/cm^2 will correspond to the loosely coupled limit, in which the measured J_c is limited by J_{cJ}. If we also assume that J_{cg} has the classic form[10] of the depairing critical current $\propto H_{cg}/\lambda_g \sim (T_{c0} - T)^{3/2}$, and that I_{cJ} follows the classic[11] proportionality to $(T_{c0} - T)$, J_{cg} would eventually be the limiting factor *sufficiently close* to T_{c0}. However, this crossover would only occur for $(1 - T/T_{c0}) \sim [J_{cJ}(0)/J_{cg}(0)]^2 \sim 10^{-5}$ for typical parameters, and other complications would dominate so near to T_{c0}. A similar qualitative conclusion was reached by Clem[12] by an argument based on a comparison of the Josephson coupling energy and the condensation energy of the grains. For simplicity, we shall henceforth exclude this regime very near to T_c from our analysis and consider only loosely coupled grains. We emphasize that this regime is totally different from that found in, for

---

[10]See MT, Eqn. (4-36), p. 117.
[11]See MT, Eq. (6-5), p. 194, noting that $\Delta \sim (T_c - T)^{1/2}$.
[12]J. R. Clem, *Proc. Int. Conf. on High-Temp. Supercond. and Materials and Mech. of Supercond.*, March, 1988, Interlaken, Switzerland; *Physica* C**153–155**, 50 (1988).

example, granular aluminum, where the grain size is *small* compared to ξ_g. In that case, Clem[12] has shown that the effect of the weak links is simply to modify the parameter values in standard dirty-limit Ginzburg–Landau theory.

In analyzing the response of the granular composite medium, it is important to separate two regimes depending on the length scale a of the granular structure relative to the length scale $(\Phi_0/H)^{1/2}$ set by the size of an area containing one flux quantum. For fields low enough that $a < (\Phi_0/H)^{1/2}$ the electromagnetic response of the material can be treated as that of a homogeneous effective medium with suitably chosen parameters; the following analysis is devoted to finding these parameters. For higher fields, the inhomogeneous structure plays an explicit role, with intrinsic parameters of the grains themselves determining critical behavior.

b. *Critical Current*

We have already noted that the (zero-field) macroscopic critical current density will be set by the intergranular I_c through the relation

$$J_{cJ} = I_c/a^2. \tag{3.1}$$

In the following analysis, we shall take J_{cJ} as a given, empirical parameter, which, with the grain size a, characterizes the material.

c. *Penetration Depth*

In response to a magnetic field, a three-dimensional array of junctions sets up screening currents, analogous to those in a bulk sample, which prevent the field from penetrating deeply into the sample. To determine the effective penetration depth λ_J, we start with the expression for the current in a single Josephson junction. A junction which connects grains i and j in the presence of a vector potential A carries a supercurrent given by[13]

$$I = I_c \sin\left(\phi_i - \phi_j - \frac{2\pi}{\Phi_0} \int_i^j \mathbf{A} \, d\mathbf{l}\right), \tag{3.2}$$

where the expression in parentheses is called the gauge-invariant phase difference γ_{ij} between the grains i and j. [$\Phi_0 = hc/2e = 2.07 \times 10^{-7}$ G-cm^2 is the superconducting flux quantum.] Following the discussion of the

---

[13]See MT, Eq. (6-4), p. 194.

London equations in Tinkham,[5] we can choose a gauge such that all of the ϕ_i's are zero, the so-called London gauge. Furthermore, since fields and currents are assumed to be small, we approximate the sine by its argument. We also replace the integral by $\mathbf{A}a$, where a is the spacing between nearest neighbors, neglecting any phase gradient within the grain since it will be small if the grains are weakly coupled. Defining the macroscopic current density by $J = i/a^2$, leads to

$$\mathbf{J} = -2\pi \frac{J_{cJ}a}{\Phi_0} \mathbf{A},\tag{3.3}$$

where $J_{cJ} = I_c/a^2$. Taking the curl of this equation, and combining it with the curl of Ampere's law, gives

$$\nabla^2 \mathbf{h} = \frac{8\pi^2 J_{cJ}a}{c\Phi_0} \mathbf{h},\tag{3.4}$$

where $\mathbf{h}$ is the local value of the magnetic flux density. From this, one sees that the characteristic exponential screening length for the magnetic field is given by

$$\lambda_J = (c\Phi_0/8\pi^2 a J_{cJ})^{1/2}.\tag{3.5}$$

Comparing this with the usual formula[14] for the London penetration depth

$$\lambda_L = (mc^2/4\pi n_s e^2)^{1/2},\tag{3.6}$$

we see that we can define an effective density of superconducting electrons n_s [or $|\psi|^2$ in Ginzburg–Landau (GL) theory] that is proportional to aJ_{cJ}.

Note that (3.5) reduces to the usual expression[15] for the Josephson penetration depth in a classic tunnel junction if a is replaced by $(d + 2\lambda)$, the thickness of the region penetrated by flux if the barrier thickness is d, and λ is the Londong penetration depth into the electrodes. It also reduces essentially to the usual penetration depth if one replaces J_{cJ} by the GL critical current and replaces a by 2ξ; this allows easy conversion of our granular results to continuum ones.

Taking the representative values $J_{cJ} = 10^3$ A/cm^2 and $a = 10^{-4}$ cm, we

[14] See MT, Eq. (1-9), p. 6; see also Eq. (4-8), p. 107.
[15] See MT, Eq. (6-17), p. 200.

find $\lambda_J = 5 \times 10^{-4}$ cm. The physical significance λ_J is that a sufficiently weak magnetic field will be screened exponentially over this length. (Sufficiently weak means that the induced intergranular currents remain well below I_{cJ}, so that our linear approximation holds.) Since λ_J is five times the assumed grain size, this result is consistent with our assumption of slow spatial variations of the field on the scale of a.

d. Coherence Length

The other characteristic length of GL theory is the GL coherence length[16] $\xi(T)$. In conventional superconductors, $\xi(T)$ reduces, for $T \ll T_c$, essentially to the BCS ξ_0 if the metal is "clean" and to $(\xi_0 \ell)^{1/2}$ if it is "dirty", i.e., if the mean free path $\ell < \xi_0$; in both cases it diverges as $(1 - T/T_c)^{-1/2}$ near T_c. In the YBCO class materials, ξ_0 is so small that it is thought that the crystalline materials are in, or at least near, the clean limit. The question is what the appropriate value of $\xi(T)$ is to use in the effective medium treatment of the granular material.

The conventional interpretation of ξ as the distance scale over which $|\psi|$ varies is not applicable here, because in this loosely coupled grain regime the value of $|\psi|^2$ is determined by (3.5) and (3.6) and is not dependent upon applied fields or macroscopic boundary conditions. An alternative manifestation of ξ in standard GL theory[17] is that the maximum *phase* gradient $\nabla \gamma$ (i.e., that at J_c) is $1/\sqrt{3}\,\xi$. For the granular superconductor, the corresponding maximum phase gradient at J_{cJ} is $\nabla \gamma = \pi/2a$. Equating these two expressions, we obtain

$$\xi_J = 2a/\sqrt{3}\,\pi \sim 0.4a \tag{3.7}$$

for T well below T_c, but nominally independent of T.

e. Thermodynamic Critical Field

The thermodynamic critical field H_c is defined[18] by equating $H_c^2/8\pi$ with the condensation energy per unit volume. In conventional superconductors, H_c varies as $(1 - T/T_c)$ near T_c, and it reflects the free energy gain of the phase-coherent BCS ground state relative to the normal state. In the weak-field loosely coupled granular case, the most parallel energy is that resulting from the phase-locking of the Josephson coupling between grains. In our cubic array model, each grain has 6 neighbors.

---

[16] See MT, pp. 111–113.
[17] See MT, Eq. (4-37), p. 117.
[18] See MT, p. 38.

Summing over these 6 links, dividing by 2 to cancel double counting, and normalizing to a unit cell volume of a^3, we have

$$H_{cJ}^2/8\pi = 3E_J/a^3,$$

or

$$H_{cJ} = (12\Phi_0 J_{cJ}/ca)^{1/2}. \tag{3.8}$$

Inserting our standard set of representative parameters yields a numerical value of ~1.6 Oe.

In GL theory, H_c, ξ and λ are related by the formula[19] $\Phi_0 = 2\sqrt{2}\,\pi H_c\lambda\xi$. The physics underlying this relation is that the linear current response characterized by λ must break down at a maximum phase gradient defined by ξ^{-1} at which the energy increase in the linear approximation reaches the condensation energy defined by H_c. If we use this formula to check the consistency of the estimates (3.6), (3.7), and (3.8), we find that the product relation holds apart from a factor of $4/\pi$, which presumably stems from our rather crude definitions of ξ and H_c.

f. Ginzburg–Landau Parameter κ_J

Using the values of λ_J and ξ_J found above, we find for the GL parameter[20] κ_J the value

$$\kappa_J = \lambda_J/\xi_J = (3\Phi_0 c/32 J_{cJ} a^3)^{1/2}. \tag{3.9}$$

For our standard numerical values, this yields $\kappa_J \sim 10$.

g. Lower Critical Field H_{c1J}

According to GL theory, the minimum external field for which it is energetically favorable for a fluxon vortex to enter a high κ superconductor is[21]

$$H_{c1} = (\Phi_0/4\pi\lambda^2)\ln\kappa. \tag{3.10}$$

Inserting the values of λ_J and κ_J from above, we find

$$H_{c1J} = (2\pi a J_{cJ}/c)\ln(\lambda_J/\xi_J), \tag{3.11}$$

which has the value ~0.5 Oe for our representative parameters. Although

[19]See MT, Eq. (4-20), p. 112.
[20]See MT, Eq. (4-27), p. 113.
[21]See MT, Eq. (5-18), p. 149.

Clem[12] does not formally define ξ_J as we do, he obtains a result equivalent to (3.11) by noting on physical grounds that the inner cutoff for the energy integration (which gives the ln κ factor) must be taken to be $\sim a/2$, which is essentially the same as our ξ_J.

The physical significance of H_{c1J} is that it sets the limit for the strength of external fields that are screened exponentially in distance λ_J, by reversible surface screening currents. To screen fields $H > H_{c1J}$ in λ_J, the surface current density would need to exceed J_{cJ}, causing a breakdown of the Josephson weak links, allowing flux to penetrate further. In this way, a sort of "Bean-model"[22] penetration occurs for $H > H_{c1J}$, in which the flux penetrates *between* the grains, with field gradient $4\pi J_{cJ}/c \sim$ 1000 Oe/cm for our representative parameters.

(Throughout this discussion, the exclusion of flux by the diamagnetism of the grains should be taken into account by introduction of an effective permeability of the medium, as done by Clem.[12] We continue to omit this factor for maximum conceptual simplicity, but it *will* significantly affect quantitative results.)

h. *The Upper Critical Field H_{c2J}*

In conventional type-II superconductors, $H_{c2} = \Phi_0/2\pi\xi^2$ is the field at which the material makes a second-order phase transition into the normal state, or alternatively, it is the field at which *bulk* superconductivity can first nucleate in decreasing field.[23] It is a temperature-dependent quantity, which conventionally goes linearly to zero as T approaches T_c. If we apply the same formula to the granular system, using ξ_J, we obtain

$$H_{c2J} = 3\pi\Phi_0/8a^2, \tag{3.12}$$

which is dependent only on grain size and has a value ~ 24 Oe for $a = 1\,\mu$m. Clearly, apart from a numerical factor of $3\pi/8$ from our approximations, this is the field at which one flux quantum Φ_0 fits in each unit cell of area a^2.

Since the superconductivity of the grains is essentially unaffected by such small fields, this field does *not* have the conventional significance of marking the extinction of all superconductivity. Rather, it marks the point at which the flux enclosed in each unit cell is sufficient to change by 2π the sum of the four gauge-invariant phase differences γ_i around the perimeter of a square plaquette. Since fluxoid quantization[24] requires that

[22]C. P. Bean, *Phys. Rev. Lett.* **8**, 250 (1962); see also *Revs. Mod. Phys.* **36** 31 (1964), or see MT, p. 173.
[23]See MT, p. 128.
[24]See MT, pp. 121, 202.

this sum

$$\sum \gamma_i = 2\pi\Phi/\Phi_0 (\mathrm{mod}\ 2\pi), \tag{3.13}$$

where Φ is the magnetic flux enclosed, a flux change by Φ_0 is sufficient to complete a cycle. Simulations by Benz et al.[25] have shown that the *oscillatory* dependence of array average quantities washes out in a geometrically disordered array as soon as $\Phi\delta A_{rms}/a^2 = \delta\Phi_{rms} \sim \Phi_0/3$. Hence, these ensemble averages are straightforward in strongly disordered granular materials, so long as thermal energies are great enough to allow the system to explore freely its phase space of γ_i values; this is the so-called "ergodic" regime. (On the other hand, if phase slips by 2π are kinetically forbidden on the appropriate time scale, the system is in the "irreversible" regime, as discussed below in Section 5.) In the absence of any flux, all γ_i are zero in the ground state of the system to obtain the full binding energy E_J from each junction; accordingly, each can carry its full critical current I_c by increasing γ_i to $\pi/2$. The effect of the magnetic field in introducing essentially random initial values of γ_i is to reduce the capacity of the ("frustrated") network to carry a net macroscopic supercurrent by suitable subsequent readjustment of the γ_i. Thus, H_{c2J} gives a measure of the field at which the macroscopic critical current will be substantially reduced. A useful analogy is to a uniform Josephson junction in a magnetic field. The field introduces relative phase modulation in the two junction electrodes, which causes the overall I_c to vary as $(\sin x)/x$, where $x = |\pi\Phi/\Phi_0|$, Φ being the flux enclosed in the junction.[26] Allowing for randomness in the effective areas of the various junctions, one expects a nonoscillatory fall-off of the macroscopic critical current $\sim 1/\Phi \sim 1/H$, with H_{c2J} serving to set the scale for the roll-off to begin.

i. *Summary of Behavior in Magnetic Field*

We now recapitulate our results so far for the behavior of a virgin granular sample as an external field is applied and increased from zero. In doing so, we emphasize that the quoted numerical results are based on assumed parameter values of $J_{cJ} = 10^3\ \mathrm{A/cm^2}$ and $a = 10^{-4}\ \mathrm{cm}$; these parameters are structure sensitive and *illustrative only*.

(*i*) For $H < H_{c1J}$ (~ 0.5 Oe), the field is screened exponentially over a distance $\lambda_J \sim 5\ \mu\mathrm{m}$.

[25]S. P. Benz, M. G. Forrester, M. Tinkham, and C. J. Lobb, *Phys. Rev. B* **38**, 2869 (1988).
[26]See MT, Eq. (6-14), p. 199.

(*ii*) For $H_{c1J} < H < H_{c2J}$ (~25 Oe), the field penetrates (between the grains) a distance ~$cH/4\pi J_{cJ}$ in a Bean-type[22] critical state, leaving the grains as partially diamagnetic inclusions.

(*iii*) As H increases toward and above H_{c2J}, the effective J_{cJ} is reduced by a factor ~H_{c2}/H by the phase randomization, allowing even deeper penetration, to a distance ~$H/J_{cJ}(h) \sim H^2$, where h is the local value of the screened magnetic field. This is analogous to the generalization from the Bean model to that of Anderson and Kim.[27] In their model, the critical state condition is based on a field-dependent critical current density, obtained by equating the Lorentz force density to the maximum pinning force density, so this model also gives $J_c \sim 1/H$ for high fields. Presumably this similar field dependence is not coincidental but occurs because the random network of weak links acts to pin vortices in specific places, in a way analogous to the pinning effect of random inhomogeneities in a continuous superconducting medium.

(*iv*) So long as $H < H_{c1g}$ (~500 Oe), the field penetrates into each grain only to a depth $\lambda_g \sim 1500$ Å. When H reaches H_{c1g}, fluxons enter the grains, setting up another "Bean-model" screening *within* each grain but with penetration depth determined by J_{cg} instead of by J_{cJ}. Note that when $\lambda_g > a$, as will occur near enough T_c, $H_{c1g} < H_{c2J}$, and fluxons will start to enter the grains before the intergranular coupling is severely weakened by the field. Finally, at H_{c2g}, all superconductivity is finally extinguished.

4. RELATIONSHIP BETWEEN GRANULAR AND CONTINUUM MODELS

a. *Principles of Equivalence*

Because of its clear structural basis in the ceramic materials, the model of Josephson coupled grains has been emphasized in our discussion. However, in crystalline material, it seems more natural to consider a continuum model with inhomogeneities that pin vortices in place to allow nonzero resistanceless volume transport currents. A dichotomy is often made between these two models in terms of "spin glass" vs. "flux pinning." It seems more useful to consider these two models as limiting cases of the same picture, and we will now indicate how to "translate" the language and concepts between the two.

[27]P. W. Anderson, *Phys. Rev. Lett.* **9**, 309 (1962); P. W. Anderson and Y. B. Kim, *Rev. Mod. Phys.* **36**, 39 (1964); see also MT, pp. 175–181.

First, it is useful to note that the continuum model can be described in the granular framework by effectively discretizing the continuum using a length $\sim\xi$ as the mesh interval. The physical basis for this choice is simply that ξ sets the distance scale between points in the continuum at which ψ can take on independent values. We then identify the two parameters of the granular model with continuum variables as follows: we set $J_{cJ} = J_c = cH_c/3\sqrt{6}\,\pi\lambda$, the usual GL critical current density,[10] and we invert (3.7) and set $a = \sqrt{3}\,\pi\xi/2$. Within this framework, the critical current of a link is $I_c = J_c a^2$, and the associated Josephson coupling energy is $E_J = \hbar I_c/2e = (H_c^2/8\pi)(4\pi\xi^3)(\pi/2\sqrt{3})$, using the standard GL identity that $\Phi_0 = 2\sqrt{2}\,\pi\lambda\xi H_c$. It is immediately obvious that this energy is, within a small numerical factor, the characteristic energy[28] $(H_c^2/8\pi)(4\pi\xi^2)(\ln\kappa)$ associated with a length ξ of a flux line vortex and, hence, the characteristic energy from which the usual pinning energies are derived.

In the conventional continuum pinning view, the vortex pinning energy results from fractional modulation of this characteristic energy by modulation of physical parameters of the material over a length scale ξ. In the granular view, the pinning arises because of variations of E_J from link to link, which modulates the increase in Josephson energy associated with the presence of a vortex center within a plaquette of junctions. Since we have just seen that for the continuum the energy scale is the same in both points of view, and in both the depth of modulation comes from the degree of disorder in the material, it is clear that any difference is mostly semantic. It is convenient to continue to use concepts from the discrete granular model for *all* cases, with the understanding that E_J, J_{cJ}, and a are parameters that can describe physical grains, twin domains, or a modulated continuum, depending on the values that are inserted.

b. *Vortex Energy in the Discrete Model*

Tinkham et al.[29] have studied the energies of vortices in the case of a two-dimensional *regular* square lattice. For an isolated vortex in nominally zero field, they approximated the energy increment due to the vortex by

$$E = E_J \sum_p 4p(1 - \cos\gamma_p) \tag{4.1}$$

where the index p runs over concentric $p \times p$ contours ($p = 1, 3, 5 \cdots$)

[28]See MT, Eqn. (5-17a), p. 148.
[29]M. Tinkham, D. W. Abraham, and C. J. Lobb, *Phys. Rev.* **B28**, 6578 (1983).

surrounding the square containing the vortex center. Since the sum of the gauge-invariant phase differences around any contour enclosing a vortex line must be 2π, by symmetry we see that $\gamma_p = 2\pi/4p$, which decreases from $\pi/2$ in the core ring, to $\pi/6$ in the next, etc. By making the small angle approximation on the cosine, one sees that successive terms vary as $1/p$, so the sum is logarithmic, yielding $E = (\pi^2/4)E_J \ln p_{max}$, since $p_{min} = 1$. For an isolated vortex, this is held to a finite value by cutting off the sum at $p_{max} \sim \lambda_J/a \sim \kappa_J$. (Of course, in the continuum limit, $a \sim \xi$, and κ_J becomes the usual κ.)

When a finite flux density is present, such that the average flux per unit cell is the frustration index f, the sum over p must be cut off when $fp^2 = 1$, which defines the region containing one flux quantum. The contribution of successive contours p falls off so rapidly that the core ring ($p = 1$) dominates the total sum. If one combines this fact with the observation that the parameter variations which give pinning will tend to average out over larger areas (volumes), one sees that it should be a reasonable approximation to assume that *all* the energy *variations* that determine pinning come from the *core plaquette alone* that encloses the core of the vortex. For our cubic array model and $f \ll 1$, this energy is

$$E = 4E_J. \tag{4.2}$$

c. *Vortex Pinning in the Discrete Model*

We now use the discrete model to discuss the relation between pinning and critical currents. In doing so, we must distinguish the pinning caused by random inhomogeneities from that which arises from the mere existence of discreteness in the model.

We first examine the latter, whose only physical counterpart in ideal homogeneous continuum material stems from the unit cell structure of the crystal. Lobb et al.[30] investigated this issue numerically for a square lattice and found that this barrier height was only about $0.2E_J$. The key point to note is that this is only 5% of the basic core energy $4E_J$. Hence only a small amount of randomness in E_J will suffice to make the variations in these local equilibrium vortex core energies dominate the pinning. Next we consider the pinning due to these *random* parameter variations from plaquette to plaquette.

These larger scale inhomogeneities stem from compositional variations, second phases, grain boundaries, twin boundaries, dislocations, etc. Within our discrete model, these features are reflected by having E_J take

[30]C. J. Lobb, D. W. Abraham, and M. Tinkham, *Phys. Rev.* **B27**, 150 (1983).

on a range of values in the different links, by irregularities in the value of the lattice distance separating grains, and also by geometrical irregularities in the coordination number and number of grains in an elementary plaquette. For example, the core energies of a vortex in 3- and 5-member plaquettes are $4.5E_J$ and $3.5E_J$, respectively, instead of $4E_J$. Much larger variations are to be expected from differences in tightness of contact between "grains" or "coherence volumes" and, hence, in the E_J values themselves. These diverse variations introduce an inequivalence of the different plaquettes within which vortex cores may be found; this provides the pinning energy ΔE, or U.

Now, what is the effect of adding a transport current? When themal activation is taken into a ccount, some vortices are always present unless $E_J \gg kT$. On one side of the vortex, the transport current will add to the circulating current, tending to exceed I_c in the links; on the other side it will subtract. This contrast will cause a preferential vortex motion in one direction normal to the current direction, and consequently a $\mathbf{v} \times \mathbf{B}$ (resistive) electric field along the current direction,[31] unless the vortex motion is prevented by a sufficient pinning energy. In the presence of an *external magnetic field* or the *self-field* of the current, there will of necessity be unpaired vortices and hence resistance, unless the pinning is strong enough. In three dimensions, the vortex *lines* presumably move through a pinning structure by successive motions of short segments, but the same qualitative picture holds.

d. *Conclusions*

To obtain an explicit estimate of pinning effectiveness from variation of core energies from cell to cell, we note that $\mathbf{J} \cdot \mathbf{E}$ work done by the imposed transport current as a consequence of moving the vortex one cell period a is $Ja\Phi_0/c$ per unit length of vortex, or $Ja^2\Phi_0/c$ for a segment one cell long. If we equate this to an energy difference $U \sim \delta(4E_J)$, we find a critical current density

$$J_c = cU/a^2\Phi_0 = (U/2\pi E_J)J_{cJ}. \tag{4.3}$$

Thus, the expected bulk critical current density is reduced from the intrinsic one by a factor of the order of the fractional modulation of the core energy from cell to cell.

Returning to the language of the continuum model, the maximum pinning energy per length ξ of vortex line scales with $H_c^2\xi^3$. If this

[31]See MT, pp. 157–171.

maximal pinning were available everywhere, the critical current density would be the full depairing value $\sim H_c/\lambda$. If the pinning is weaker, the observed J_c will be scaled down by a partial pinning factor parallel to the factor $(U/2\pi E_J)$ in (4.3).

From this discussion we conclude that, from either the granular or the continuum point of view, it is appropriate to discuss thermally activated flux motion in terms of a characteristic pinning energy, barrier, or activation energy U. In the granular model, this U scales with $4E_J$, the core vortex energy per plaquette; in the continuum model, U scales with $(H_c^2/8\pi)(4\pi\xi^3)$. In both cases, these energies are reduced in proportion to the fractional modulation of the characteristic energies from point to point, or they can be increased if collective motions are involved. For continuum material, the two expressions are essentially equivalent, provided one associates E_J with the Josephson coupling energy of a critical current $J_c\xi^2$, where J_c is the depairing critical current density.

5. FLUX CREEP AND IRREVERSIBILITY

Because any diamagnetic property is sometimes carelessly referred to in the literature as a "Meissner effect" signal, it is important to make clear distinctions. The Meissner–Ochsenfeld discovery in 1933 was this: Not only was flux prevented from entering a superconductor already below T_c when an external field was applied (as expected from classical electrodynamics and "perfect conductivity"), *but also* if the sample was cooled down through T_c while in a magnetic field, the flux already inside was *expelled,* instead of being "frozen in," as would be predicted by the same classical electrodynamics. In other words, the perfect diamagnetism in fields below $H_c(T)$ reflects a thermodynamically stable state of lowest free energy that can be reached by lowering T and applying H, *in either order,* since the transformation of state is reversible. Observation of this ideal behavior is dependent on sample homogeneity, shape, and surface conditions, but it is a useful idealization.

With type-II superconductors, the reversible behavior is more difficult to achieve for fields above H_{c1} because the thermodynamically most stable state[32] then involves filling the sample with quantum fluxons, giving only a "partial Meissner effect" with $0 < -4\pi M < H$. Even a small degree of inhomogeneity is then sufficient to "pin" some fluxons at local free energy minima, so that some flux remains trapped in place when the external field is removed. Moreover, because this pinning of fluxons is necessary for the material to be able to carry a volume transport current

---

[32]See MT, pp. 150–157.

(as opposed to a surface current in the penetration depth), "practical" conductors are *designed* to be as irreversible as possible. In fact, it is customary to use measurements of the hysteresis loop of $M(H)$ to make a contactless determination of this nonequilibrium critical current density. Unfortunately the granular character of the YBCO ceramic materials complicates the interpretation of data from this method, because useful macroscopic transport current must flow from grain to grain, while the irreversible magnetic moment can be dominated by larger currents circulating *within* grains, or even within "subgrains" inside the visible granular structure. This situation has been noted for example, by Daeumling *et al.*[33] in interpreting their experimental data on flux penetration profiles in powdered $DyBa_2Cu_3O_7$.

In experiments designed to explore these irreversible magnetic effects, it is customary to make a three-step cycle of measurements, yielding zero-field cooled (ZFC), field-cooled (FC), and remanent (REM) magnetization data. In the ZFC case, the sample is cooled into the superconducting state in zero field, after which a finite field is applied, and the decreasing magnitude of the *excluded* flux is measured as the sample is warmed back up to T_c. In the FC part of the test cycle, the finite field is left on, and the magnetic moment that develops during the cooling from the normal state is a measure of the increasing flux *expulsion* associated with a true Meissner effect. The third measured characteristic is that of the *remanent moment* left when the external field is removed at the end of the FC cool-down. Recently, Malozemoff *et al.*[34] have shown experimentally that, in a variety of samples, these three measured quantities are related by $M_{rem} = M_{fc} - M_{zfc}$; they have argued that this can be understood in physical terms as a consequence of a flux pinning model.

An interesting feature noted by Müller *et al.*[35] in their early measurements of these magnetic properties in high T_c superconductors was the existence of a field-dependent range of temperature below T_c in which the ZFC and FC values of $M(T, H)$ were the same. (Similar results were obtained more recently[36] on single crystal samples.) At lower temperatures, the two values of $M(T)$ diverged from each other. Further, they noted that the line in the H, T plane separating the reversible from the irreversible regime had the form

$$1 - t \propto H^{2/3}, \tag{5.1}$$

[33]M. Daeumling, J. Seuntjens, and D. C. Larbalestier, *Appl. Phys. Lett.* **52,** 590 (1988).

[34]A. P. Malozemoff, L. Krusin-Elbaum, D. C. Cronemeyer, Y. Yeshurun, and F. Holtzberg, *Phys. Rev. B*, submitted.

[35]K. A. Müller, M. Takashige, and J. G. Bednorz, *Phys. Rev. Lett.* **58,** 1143 (1987).

[36]Y. Yeshurun and A. P. Malozemoff, *Phys. Rev. Lett.* **60,** 2202 (1988).

where $t = T/T_c$. In analogy with the spin-glass literature, they referred to this "irreversibility line" also as a "quasi de Almeida-Thouless line".

Recently, Yeshuran and Malozemoff[36] have proposed an attractive alternative interpretation, in which the disappearance of irreversibility is essentially attributed to the disappearance of the ability to carry a *persistent* macroscopic transport supercurrent. This occurs when thermal activation overcomes pinning within the time-scale of measurement, allowing the system to reach the *unique* minimum free energy state for given H and T, independent of past history. The speed of thermally activated processes of this sort also accounts[37] for the prominence of flux creep, logarithmic in time, which is much more readily observed in the high T_c materials than in conventional superconductors. We now consider these phenomena in more detail.

a. *Resistive Transition in Magnetic Field*

As a first example of the importance of these thermally activated dissipative processes, we discuss the interpretation of the universally observed broadening of the resistive transition of YBCO class materials in a magnetic field. Some particularly clear data on high quality crystals of YBCO have been reported by Y. Iye et al.,[38] and we take them as representative. In zero field, the transition width is less than 1 K at 91.5 K. In a 90 kOe field, this width increases to about 10 K, extending down to ~78 K, making the identification of H_{c2} depend on the fractional resistance criterion chosen. Iye et al. plot $H_{2c}(T)$ values for both the zero-resistance and half-resistance criteria, but they emphasize the zero-resistance criterion. In this way they obtain $-dH_{c2}/dT = 33$ kOe/K and 8.8 kOe/K for fields perpendicular and parallel to the c-axis, respectively. From these values, they infer extrapolated zero-temperature coherence lengths $\xi(0) = 24$ Å in the ab-plane and 6 Å along the c-axis.

Extending the work of Malozemoff and collaborators, Tinkham recently proposed[39] that the *shape* as well as the width of these transition curves can be interpreted by assuming that the observed ratio R/R_n of the bulk sample is equal that of one of the typical pinning-determined "weak links" forming a complex series-parallel array in the crystal. This allows one to apply the result of Ambegaokar and Halperin[40] (AH) on the $I-V$ curves in the presence of thermal fluctuations of highly damped

[37]M. Tinkham, *Helv. Physica Acta* **61**, 443 (1988).

[38]Y. Iye, T. Tamegai, H. Takeya, and H. Takei, *in* "Superconducting Materials" (S. Nakajima and H. Fukuyama, eds.), *Jap. J. Appl. Phys. Series* 1, 1988, p. 46.

[39]M. Tinkham, *Phys. Rev. Lett.* **61**, 1658 (1988).

[40]V. Ambegaokar and B. I. Halperin, *Phys. Rev. Lett.* **22**, 1364 (1969).

Josephson junctions, such as the presumed metallic weak links between "grains."

According to this interpretation, it is the *upper* part of the resistive transition that measures H_{c2} (i.e., the extinction of superconductivity) for the material; the zero-resistance point reflects the point where resistive phase slippage within the superconducting state becomes too slow to give an observable voltage. A consequence of this reanalysis is that the *inferred values of* H_{c2} *are increased by a factor of* ~4 from those mentioned above, *and correspondingly the inferred* $\xi(0)$ *values are reduced by a factor of* ~2, to something like 13 Å in the *ab*-plane and 2.5 Å along the *c*-axis. The other side of the coin is that these higher H_{c2} values have no practical implications, since they only mark the point at which the resistance *starts to decrease* from its normal value. Because of the importance and surprising nature of these conclusions, we will now summarize the argument leading to them.

First, we recall that AH derived the I–V relation of a highly damped junction, with the normalized barrier height $\gamma_0 = 2E_J/kT = \hbar I_{c0}/ekT$ as a parameter. Since resistive transition measurements are made at low current levels, we may apply their analytic result for the zero-current limit, namely,

$$R/R_n = [I_0(\gamma_0/2)]^{-2} \qquad (5.2)$$

where I_0 is the tabulated modified Bessel function. For large values of γ_0, R/R_n falls as $\gamma_0 e^{-\gamma_0}$, but for $\gamma_0 < 1$, it only falls quadratically in γ_0.

Second, we take account of the conceptual equivalence developed in Section 4 between the weak-link and continuum flux pinning models. For the continuum case, we identify the Josephson barrier energy $2E_J$ with the activation energy U_0, so that the AH parameter $\gamma_0 = U_0/kT$. The pinning energy U_0 has been shown above to scale with the characteristic energy $(H_c^2/8\pi)(4\pi\xi^3/3)$. However, in the presence of a magnetic field, Yeshurun and Malozemoff[36] have argued that this coupling energy should be scaled by a factor of order $(\Phi_0/\pi\xi^2 B)$, which amounts to taking the effective cross-sectional area of the link to be that of a unit cell of the Abrikosov flux line lattice instead of $\sim\xi^2$. Inserting this factor, we obtain

$$U_0(B, T) = \beta H_c^2 \xi \Phi_0/B, \qquad (5.3)$$

where β, expected to be ~1, absorbs all numerical factors.

In the spirit of our discussion of the equivalence of continuum pinning and a Josephson coupling viewpoint, we transform (5.3) by substitution of the Ginzburg–Landau relation $\Phi_0 = 2\sqrt{2}\,\pi H_c\xi\lambda$ and the intrinsic GL critical current density $J_c = cH_c/3\sqrt{6}\,\pi\lambda$, obtaining

$$U_0 = (3\sqrt{3}\,\Phi_0^2\beta/2c)(J_c/B). \qquad (5.4)$$

This can be interpreted physically as the coupling energy per vortex line tending to maintain phase continuity between points $\sim \xi$ apart along the line. It has the advantage over (5.3) of focusing on a *single* material parameter J_c, which has a relatively clear operational significance even in the presence of granularity. Near T_c, $J_c \propto (1 - t)^{3/2}$, so we can write

$$\gamma_0 = U_0/kT \approx A(1 - t)^{3/2}/B. \tag{5.5}$$

By inserting this in (5.2), we obtain

$$R/R_n = \{I_0[A(1 - t)^{3/2}/2B]\}^{-2}. \tag{5.6}$$

Normally one can insert T_c for T in the denominator of (5.5), since the temperature dependence is only good near T_c, so A can be treated as constant, except for a dependence on crystal orientation through J_c.

Equation (5.6) is the key to our interpretation of the shape of the resistive transitions in finite field. One sees immediately that the temperature width of the transition should scale (at *any* R/R_n level) as

$$\Delta T \propto H^{2/3}, \tag{5.7}$$

which is in excellent agreement with the data of Iye *et al.*[38] and also that of Oh *et al.*[41] In fact, (5.7) can also be viewed as a generalization of the expression (5.1) for the irreversibility line, which would be viewed as defining a level of R/R_n sufficiently low to enable apparently persistent nonequilibrium currents to flow.

Another scaling consequence of (5.6) is that, with H_{c2} defined at any chosen R/R_n level, the *resistively measured* "upper critical field" near T_c should vary as $(T_c - T)^{3/2}$, *not* linearly as in the classic superconductors where these thermally activated processes are usually negligible. Again the data of Iye *et al.* and of Oh *et al.* are well fitted by this dependence, as was pointed out by Oh *et al.*, but interpreted by them in terms of critical fluctuations.

The specific functional form of (5.6) also well approximates the temperature dependence of $R(T; H)$ found in the experimental data of Iye *et al.*, as shown in Tinkham.[39] The value of the parameter A is found to be about 5 times as large for the field in the *ab* plane as when it is along the *c*-axis, which is at least qualitatively consistent with the ratio of critical currents in these directions, as expected from (5.4) and (5.5).

[41]B. Oh, K. Char, A. D. Kent, M. Naito, M. R. Beasley, T. H. Geballe, R. H. Hammond, and A. Kapitulnik, *Phys. Rev.* **B37**, 7861 (1988).

There is also reasonably good agreement with the data of Dubson *et al.*[42] on ceramic material, but A is about 100 times smaller, again consistent with the known lower critical currents of ceramic samples.*

b. *Flux Creep*

With the insight gained from analysis of the resistive transition, we now address the magnetization measurements. We can convert the roughly exponential resistance variation (5.6) to a similar dependence for die-away time of induced transient currents by simply introducing an appropriate inductance L reflecting the geometry of the circulating currents. In this way, one can define a time constant

$$\tau = L/R = (L/R_n)(R/R_n)^{-1} = \tau_n(R/R_n)^{-1}. \qquad (5.8)$$

Given the exponential dependence of R/R_n, this factor can readily vary over extreme numerical ranges, leading to time constants exceeding the duration of laboratory experiments: in other words, to persistent currents and "dc" magnetization.

This highly simplified point of view is valid, however, only when the induced currents that are decaying are sufficiently small so that they do not significantly affect the barrier energy that is impeding the phase slip. The more realistic case is that in which the current induced by the changing applied field initially exceeds the critical current density in the surface screening layer of thickness λ, allowing flux to enter further, establishing a Bean-model critical state regime, as discussed in Section 3.i. (This will occur as soon as H_{c1} is exceeded.) By construction, then, the initial state is one in which $I = I_c$ in each current-carrying link. In the weak link model, the barrier to phase motion is reduced by the tilting of the "washboard potential" by the current, and $I = I_c$ is the condition for the barrier to go to zero. This current-dependent barrier height can be expressed to good approximation by replacing γ_0 by

$$\gamma(I) = \gamma_0(1 - I/I_c)^{3/2}. \qquad (5.9)$$

In any case, $\gamma(I_c) = 0$, so that at I_c the resistance has essentially the normal value R_n, leading to a very rapid initial decay of current. But as soon as some decay occurs, the barrier energy is not longer zero, the resistance is decreased, and the continuing decay of current is slower.

[42]M. A. Dubson, S. T. Herbert, J. J. Calabrese, D. C. Harris, B. R. Patton, and J. C. Garland, *Phys. Rev. Lett.* **60**, 1061 (1988).
* See notes added in proof, p. 134.

This process leads to the familiar decay of currents not exponentially, but *logarithmically*, with time. We now offer a simple derivation that illustrates the primary features of this phenomenon.

We consider a closed current loop of inductance L containing a large number of weak links in series, with total normal resistance R_n. We assume that in the superconducting state the resistance is reduced by the ratio found by AH, taking their finite current results, not the zero-current limit (5.2). These results have no simple analytic form, but the key feature for the present purpose is that R varies *exponentially* with current, roughly as $\exp(hI/4ekT)$. For analytic convenience we make the reasonable approximation that

$$R/R_n = e^{-\gamma_0(1-i)}, \qquad (5.10)$$

where $i = I/I_c$. This reduces to R_n at I_c and builds in the correct asymptotic exponential decrease with γ_0 and the exponential dependence on I. When the induced and resistive voltages are balanced, the decay of current is governed by the equation

$$-L\, dI/dt = IR = IR_n e^{-\gamma_0(1-i)} \qquad (5.11)$$

by using the approximation (5.10). Since the regime with I very near I_c occupies little time in the decay, we have made no major error by simplifying the current dependence of γ to a simple linear one instead of the more accurate 3/2 power form given in (5.9). For analytic simplicity, we also assume that the decrease in I is sufficiently small so that it can be treated as constant in the prefactor and as a variable only in the exponent. With these simplifying approximations one can integrate (5.11) exactly. Taking $i = 1$ $(I = I_c)$ at $t = 0$ as the initial condition, one obtains

$$i = 1 - \gamma_0^{-1} \ln(1 + \gamma_0 t/\tau_n). \qquad (5.12)$$

As expected, this reduces to *exponential* decay with time constant τ_n for *very* short times. However, since $\gamma_0 \gg 1$, this very rapidly shifts to the *logarithmic* asymptotic form

$$i = 1 - \gamma_0^{-1} \ln(\gamma_0 t/\tau_n). \qquad (5.12a)$$

An immediate consequence of this form is that the decay rate is

$$di/d \ln t = -\gamma_0^{-1} = -kT/U_0. \qquad (5.13)$$

Another consequence of (5.12a) is that the current decays completely to zero at a finite time, namely

$$t_{i \to 0} = (\tau_n / \gamma_0) \exp \gamma_0. \tag{5.14}$$

Although this can be astronomically long if γ_0 is large, it will be of the order of seconds for γ_0 of the order of 10 to 20, for reasonable values of τ_n. For $I = I_c/2$ to remain after some seconds, one would need γ_0 to be twice these values. In traditional superconductors, γ_0 is typically much larger than these values, and hence the amount of logarithmic decay of current is small and is not a serious problem in magnets for example. However, as pointed out earlier,[36,37] the small coherence lengths of the high-temperature superconductors make for small pinning energies, and the higher operating temperatures naturally increase thermal activation. Thus, these logarithmic decays are ubiquitous in them.

Clearly, if a sample is inhomogeneous on a macroscopic scale, so that different values of γ_0 are represented, the currents in weaker link (smaller γ_0) regions will die out proportionally faster. So long as all regions are decaying in the logarithmic range, the decay of the total magnetic moment is still simply linear in $\ln t$. However, as soon as currents drop to zero in some regions, as estimated by (5.14), the measured logarithmic slope of the magnetic moment from the sum of the remaining currents will decrease in magnitude. Thus, inhomogeneous samples might be expected to show departures from simple logarithmic behavior, with an initial logarithmic slope higher than that found after relaxation has proceeded for a period of time. In fact, examples have been seen, particularly in low field experiments,[43] in which a rather clear break in logarithmic slope occurs after a finite time, presumably because of the exhaustion of one class of trapped flux.

Although this model of the decay of current in independent rings is convenient for analysis, it omits the effects of mutual inductance. For example, essentially the same flux links two adjacent rings, so that their mutual inductance and self inductance are similar in magnitude. The current decays are then coupled; τ_n will be doubled because combining the two rings effectively halves the resistance and leaves the inductance the same. To take account of this type of effect, one should take τ_n in the formulas to characterize the whole structure, not a single ring of grains.

To relate this simplified model to the usual experimental regime, we note that the critical-state model implies that an applied field change of

[43]M. A.-K. Mohamed, W. A. Miner, J. Jung, J. P. Franck, and S. B. Woods, *Phys. Rev.* **B37**, 5834 (1988); also, F. de la Cruz, private communication.

ΔH induces a nonequilibrium current density J_c extending from the surface to a depth $d = c\Delta H / 4\pi J_c$; subsequently, this current decays logarithmically in time. For a sufficiently large ΔH, d would exceed the thickness D of the sample, and one reaches a limiting case in which the entire specimen is in the critical state. The total current is then proportional to $J_c D$, the magnetic moment to $J_c D^2$, and the magnetization to $J_c D$. By applying (5.13) and noting that normalization to I_c is equivalent to normalization to J_c macroscopically, we have

$$-dM/d \ln t = (J_c D/c)(kT/U_0) \qquad (5.15)$$

apart from small numerical factors depending on the specific geometry. This is the basic relation used to interpret experimental data.

Experimentally, this logarithmic time derivative is found[36] to increase approximately linearly with T for $T < T_c/3$, as expected from the explicit temperature dependence in (5.15). At higher temperatures, the logarithmic derivative reaches a maximum and then falls to zero at T_c. This behavior is determined by the temperature dependence of the ratio of J_c to U_0, both of which go to zero at T_c and, hence, is dependent on the details of the model considered.[36]

c. *Impact on Determination of $H_{c2}(T)$*

An important qualitative implication of these prominent flux creep effects is that nominally simple experiments will yield results that depend on the time-scale of the measurement. Attention has been drawn forcefully to this fact by Malozemoff et al.[44] They note that measurements of $H_{c2}(T)$ on samples of YBCO from the same source, but with different techniques (zero resistance, ac susceptibility, zero pinning force), obtained systematically different results. This discrepancy could be accounted for naturally in terms of the effective frequency of the measurements. To test this hypothesis, they performed ac susceptibility measurements and found that the apparent T_c in a field of 6 kOe increased systematically from 89.5 K to 91.5 K as the measurement frequency was increased from 10^4 to 10^8 Hz. The increase was approximately logarithmic in the measuring frequency, as would be expected from the flux creep model. From more detailed fitting of their data to a simple model, they inferred that the characterisitc "attempt frequency" for the creep process was of the order of 10^9 Hz. If so, presumably the data at 10^8 Hz are close

[44]A. P. Malozemoff, T. K. Worthington, Y. Yeshurun, F. Holtzberg, and P. H. Kes, *Phys. Rev. B*, in press.

to a natural limit that should reflect the actual $H_{c2}(T)$, undistorted by the dynamics of thermally activated flux motion.

Several other important pieces of the experimental situation concerning these $H_{c2}(T)$ determinations should be mentioned: (1) Measurements of Oh et al.[41] of the shift of the *upper* end of the resistive transition imply a very steep, possibly linear, increase of H_{c2} with decreasing temperature. According to the view presented in Section 5.a, this measured quantity should reflect $H_{c2}(T)$ fairly well, whereas the lower end of the transition (i.e., the $R = 0$ point) is strongly affected by dissipative flux motion in an already superconducting medium. (2) Measurements of Fang et al.[45] that use the onset of *reversible* magnetization as a measure of $H_{c2}(T)$ indicate a steep initial rise of this quantity, apparently as $(1 - t)^{1/2}$. This unexpected power law is attributed by the authors to a two-dimensional layer effect, but it also might result from a fluctuation effect near T_c. (3) Further magnetization measurements have been carried out by Athreya et al.,[46] who explored the entire reversible regime within 5 K of T_c (i.e., inside the irreversibility line). This allowed them to determine free energy surfaces, which again indicate that the true H_{c2} curve must be very steep, i.e., that $T_c(H)$ is depressed much more slowly by a magnetic field than had been thought previously.

The results of the resistance and magnetization measurements on YBCO are schematically shown in Fig. 1, which also provides a graphic recapitulation of the viewpoint that has been developed here. The analysis of the broadening of the resistive transition given in Section 5.a showed that it is caused by a regime of thermally activated vortex motion within a superconducting state, so that the *upper* rather than the *lower* end of the transition should be more indicative of the thermodynamic $H_{c2}(T)$ at which superconductivity first nucleates. Unfortunately, measurements of the upper end of the transition are ambiguous because of rounding and T-dependent normal resistance. Shown for the resistive transition in Fig. 1 are the results of Iye et al. for the criteria $R/R_n = 0.5$ and "zero," the latter of which we take to represent $R/R_n \approx 10^{-2}$. We have added a third resistive curve, using (5.6) to extrapolate to a criterion of $R/R_n \approx 10^{-7}$, which is taken to illustrate qualitatively the criterion for the onset of magnetic irreversibility as reported by Müller et al.[35] (In comparing resistive data on different samples, it is important to recognize from the form of (5.6) that the H scale really represents H/J_{c0}, so the

[45]M. M. Fang, V. G. Kogan, D. K. Finnemore, J. R. Clem, L. S. Chumbley, and D. E. Farrell, *Phys. Rev.* **B37,** 2334 (1988).

[46]K. Athreya, O. B. Hyun, J. E. Ostenson, J. R. Clem, and D. K. Finnemore, to be published.

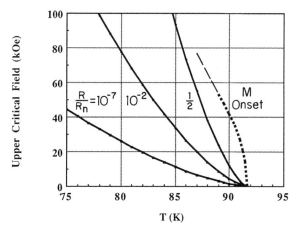

FIG. 1. Various indicated values of the upper critical field of crystalline YBCO for H parallel c-axis. Curve labelled Magnetization Onset represents the data of Fang *et al.* [*Phys. Rev. Lett.* B**37**, 2334 (1988)] for the field at which the linear rise in reversible magnetization begins, which should mark the thermodynamic transition field. The data end at 89 K, and a linear extrapolation is sketched in to lead the eye. Curves labelled $R/R_n = 1/2$ and 10^{-2} represent the data of Iye *et al.*, described by them as midpoint and zero-resistance point, respectively ["Superconducting materials" (S. Nakajima and H. Fukuyama, eds.), *Jap. J. Appl. Phys.*, Series 1, 1988, p. 46]. Curve labelled $R/R_n = 10^{-7}$, like the previous two curves, is obtained using (5.6) with the same value of the parameter $A = 1.2 \times 10^7$ Oe; it is meant to represent schematically the magnetic irreversibility line (for which experimental data are not available) for the same crystal. The same fractional resistance curves describe the data of Iye *et al.* for H in the ab-plane, if the H values are scaled *up* by a factor ~6, and data of Dubson *et al.* on ceramic material if the H values scaled *down* by a factor ~100 [*Phys. Rev. Lett.* **60**, 1061 (1988)]. According to (5.4), these factors reflect the variations of J_{c0} among these cases. [For simplicity, in this figure we have suppressed the transition width in zero field; in more quantitative comparisons, the origin of the $(T_c - T)^{3/2}$ curve for each R/R_n value should be at the T_c appropriate to that R/R_n level in zero field.]

absolute scale in Oe in general depends on sample quality, and, for example, would be at least 100 times lower for a ceramic sample.) Also shown is the $H_{c2}(T)$ determined by Fang *et al.* from measurements of onset of *reversible* magnetization near T_c. A *linear* extrapolation of their data further from T_c has been sketched in on the assumption that the $(1 - t)^{1/2}$ dependence found in their data is confined to T near T_c. (Their data extend only down to 89 K.) While this diagram is only tentative, it graphically illustrates the great width of the region in the (H, T) plane, in which the material is resistive despite being, thermodynamically speaking, in the "superconducting" phase. The contrast is obvious with classic superconductors, in which (because of longer coherence lengths and lower temperatures) thermally activated processes are negligible except within a fraction of a degree of T_c, so that the resistive and thermodynamic $H_{c2}(T)$ are essentially the same, both rising linearly with $(T_c - T)$.

For completeness, we should also point out that, because of the small coherence volume and high T_c, physical properties in the vicinity of T_c, where most of the data we have been discussing were taken, may be strongly influenced by *critical* fluctuation effects,[47,48] which are completely negligible in classic superconductors. In the critical region, physical quantities are expected to vary with critical exponents different from those in the mean field regime of classic BCS and GL theory. Since the unusual fractional exponents of power laws such as $H^{2/3}$ in the flux creep picture discussed above are based only on rather heuristic models, the possibility remains that the ultimate understanding of these phenomena may involve aspects of the critical phenomena, in addition to the considerations presented at more length in this review.

II. Other Physical Properties

The discussion in the prceding sections has shown that the low-frequency electromagnetic properties of the new superconductors can apparently be accounted for quite satisfactorily in terms of the standard phenomenology of superconductivity, based on the Ginzburg–Landau theory. Given the high T_c and the consequent short coherence length, then the high critical fields, the tendency to granular behavior, and the prominence of flux creep phenomena all follow quite naturally. Given the anisotropic crystal structure, together with coherence lengths comparable with unit cell dimensions, then the macroscopic anisotropy is also quite natural. In the present part we review briefly the status of the measurement of other physical properties, which, though less directly relevant to applications, may give additional insight into the basic microscopic physics underlying the condensed state in these materials.

6. ISOTOPE EFFECT

In the BCS theory[3] of the classic superconductors, the basic interaction causing the establishment of the superconducting state is the electron–phonon interaction. In second-order perturbation theory, this leads to an

[47]C. J. Lobb, *Phys. Rev.* B**36,** 3930 (1987).
[48]A. Kapitulnik, M. R. Beasley, C. Castellani, and C. DiCastro, *Phys. Rev.* B**37,** 537 (1988).

attractive interaction between electrons within a characteristic phonon energy $\hbar\omega_c$ of the Fermi surface. According to the BCS theory, this leads to an instability of the Fermi sea with respect to establishment of a highly ordered state, in which pairs of electron states with equal and opposite momentum and spin are always either both occupied or both empty. These paired electrons are often referred to as Cooper pairs, because of the pioneering work of Cooper in identifying this instability, which leads to the BCS ground state.

One of the qualitative consequences of this model is that the binding energy of the electrons into this ground state, and also the transition temperature T_c at which the pair condensation begins, should, for a given material, scale in proportion to the characteristic phonon frequency $\hbar\omega_c$. More specifically, in the weak coupling regime, the prediction is that

$$kT_c = 1.13\hbar\omega_c e^{-1/N(0)V}, \tag{6.1}$$

where $N(0)$ is the density of states (for one spin) at the Fermi level, and V is the average matrix element of the attractive interaction for energies within $\hbar\omega_c$ of the Fermi level. The energy $\hbar\omega_c$, in turn, should scale with $M^{-1/2}$, where M is the isotopic mass, while $N(0)V$ is expected to be independent of nuclear mass. This theoretical dependence of T_c on M has been tested experimentally for a number of elemental superconductors.[49] Results are usually quoted in terms of the exponent α, in the empirical relation

$$T_c \propto M^{-\alpha}. \tag{6.2}$$

For the classical metallic superconductors Pb, Sn, Tl, Hg, and Cd, α was found to have the expected value of $1/2$, within experimental error. However, in d-band elements such as Zr, Mo, Ru, Os, and Re, α values were found that ranged from 0 to 0.39, with error limits that easily excluded the value $1/2$.

Qualitatively, these departures from $\alpha = 1/2$ can be accounted for by including the Coulomb repulsion between electrons in addition to the electron-phonon interaction; this modifies the simple relation of T_c to isotopic mass. Very sophisticated calculations were made,[49] attempting to account for the observed values of α. These had some success, but considerable quantitative uncertainty remained about parameters entering the calculations. In none of these cases was there any serious

[49]R. Meservey and B. B. Schwartz, in "Superconductivity" (R. D. Parks, ed.), M. Dekker, New York, 1969, p. 124–127.

suggestion that the BCS theory was conceptually incapable of explaining the observations if enough was known about the electronic properties of the material. In other words, departures of the isotope shift exponent from 1/2 only show that the simple phonon model must be generalized to account for the electron–electron terms. They do *not* show that BCS theory is inapplicable. In fact, the basic BCS model for the condensed state could be used just as well if there were *no* phonon coupling and only some sort of electronic coupling mechanism to provide the matrix element V in (6.1). Excitonic and spin-mediated couplings are among the purely electronic mechanisms that have been examined. In such cases, no isotope effect would be expected, and the characteristic energy $\hbar\omega_c$ in (6.1) would not be that of the phonon but that of the exciton or magnon leading to the interaction.

With this background, we now turn to the new superconductors. The extraordinarily high T_c, together with the novel oxide chemical structure, have led many investigators to presume that a new mechanism is involved. An obvious check is the isotope effect. Measurements on YBCO, in which a shift in T_c was sought after a major fraction (75 to 90%) of the O^{16} was diffusively replaced by O^{18}, were carried out simultaneously at AT & T Bell Labs[50] and in Berkeley.[51] No shift in T_c was found by either group; both quoted $\alpha = 0$, with estimated accuracies of ±0.02 and ±0.027, respectively. On the other hand, the effect of the isotope substitution on phonon frequencies was confirmed directly by Raman measurements;[50] the expected 4% frequency shift was seen in modes involving oxygen vibrations. Although it is possible for electronic effects to modify the value $\alpha = 1/2$ expected for a purely phonon mechanism, as noted above, a precise cancellation to zero would be quite a coincidence. Thus, this result is generally taken to imply that any phonon contribution to the coupling is minor. This conclusion is also supported by the fact that it would be very difficult to account for the high T_c with a phonon energy as the characteristic scale factor in (6.1). On the other hand, if $\hbar\omega_c$ reflected an electronic energy ~1000 K or more, a relatively small value of the coupling constant $N(0)V$ could still yield a T_c as high as 100 K.

Subsequent to these two early measurements, two more elaborate

[50]B. Batlogg, R. J. Cava, A. Jayaraman, R. B. van Dover, G. A. Kourouklis, S. Sunshine, D. W. Murphy, L. W. Rupp, H. S. Chen, A. White, K. T. Short, A. M. Mujsce, and E. A. Rietman, *Phys. Rev. Lett.* **58**, 2333 (1987).

[51]L. C. Bourne, M. F. Crommie, A. Zettl, H.-C. zur Loye, S. W. Keller, K. L. Leary, A. M. Stacy, K. J. Chang, M. L. Cohen, and D. E. Morris, *Phys. Rev. Lett.* **58**, 2337 (1987).

measurements were carried out in Berkeley.[52,53] Although these disagree in detail, they appear to indicate a small positive value of α, essentially at the limits of measurement accuracy, namely a few percent. Such a result seems consistent with a small, but nonzero, phonon contribution to the coupling mechanism. At about the same time, a Los Alamos group reported apparent observation of a giant isotope effect between samples made *entirely* with different isotopes of oxygen, rather than by diffusive substitution in the same physical crystal. Because of the extreme sensitivity of T_c to the oxygen concentration (reduction of the oxygen number from 7 to 6.5 per formula unit makes the material a *semiconductor!*), it seems almost certain[54] that a difference in oxygen *concentration* rather than oxygen isotopic *mass* was responsible for these surprising observations.

Although overshadowed by the YBCO studies at 90 K, isotope effect measurements were also carried out on the 30 K LBCO system originally discovered by Bednorz and Müller. In this case, an isotope shift corresponding to the intermediate value $\alpha \approx 0.2$ was found,[55,56] indicating that phonon coupling is playing a significant, if not necessarily dominant, role. Theoretical estimates also suggest that a T_c of 30 K is marginally within range of a purely phonon coupling mechanism. These considerations jointly suggest that the phonon coupling mechanism is playing a role in this system, but it is probably supplemented by another purely electronic mechanism.

Summing up, the most straightforward (but not proven) phenomenological interpretation of the isotope shift experiments is that the phonon coupling mechanism is supplemented by an electronic mechanism in LBCO and is almost completely dominated by such a mechanism in YBCO. These conclusions are, however, not inconsistent with the supposition that the actual superconducting *state* induced by the various coupling mechanisms remains a BCS-like paired electron state in all cases.

[52]K. J. Leary, H.-C. zur Loye, S. W. Keller, T. A. Faltens, W. K. Ham, J. N. Michaels, and A. M. Stacy, *Phys. Rev. Lett.* **59**, 1236 (1987).

[53]D. E. Morris, R. M. Kuroda, A. G. Markelz, J. H. Nickel, and J. Y. T. Wei, *Phys. Rev.* **B37**, 5936 (1988).

[54]B. Batlogg, private communication.

[55]B. Batlogg, G. Kourouklis, W. Weber, R. J. Cava, A. Jayaraman, A. E. White, K. T. Short, L. W. Rupp, and E. A. Rietman, *Phys. Rev. Lett.* **59**, 912 (1987).

[56]T. A. Faltens, W. K. Ham, S. W. Keller, J. J. Leary, J. N. Michaels, A. M. Stacy, H.-C. zur Loye, D. E. Morris, T. W. Barbee III, L. C. Bourne, M. L. Cohen, S. Hoen, and A. Zettl, *Phys. Rev. Lett.* **59**, 915 (1987).

7. ENERGY GAP

According to the BCS theory,[3] the quasiparticle spectrum in the superconducting state should be characterized by an energy gap $\Delta(T)$, which has the value $1.76kT_c$ at $T = 0$ and vanishes as $(1 - T/T_c)^{1/2}$ at T_c. This gap is a hallmark of that theory, and its existence in the traditional superconductors has been demonstrated by a variety of classic experimental techniques:[57] The electronic specific heat tends to zero at low temperatures as $e^{-\Delta/kT}$; at $T \ll T_c$, the absorption of electromagnetic radiation is essentially zero until the photon energy exceeds $2\Delta \approx 3.5kT_c$; the rate of nuclear relaxation and of ultrasonic attenuation by exchange of energy with quasiparticles for frequencies well below the gap fall off with decisively different temperature dependences, as predicted by the BCS "coherence factors," but both are dominated by the exponential $e^{-\Delta/kT}$ for $T \ll T_c$; and finally, with most precision, by electron tunneling.

It has been natural to ask whether the new oxide superconductors also display a BCS-like gap, and, if so, what its width is in relation to kT_c, because the ratio $2\Delta/kT_c$ is known to increase from the BCS weak-coupling limit of 3.5 to values of up to 4.2 to 4.6 in superconductors with strong electron–phonon coupling such as Pb and Hg. Thus, in the framework of a generalized BCS theory, this ratio is a diagnostic for the coupling strength. A complicating factor here is that the unusually great anisotropy of the electronic transport properties of the new materials might be reflected in unusually great gap anisotropy. In the classic superconductors, such anisotropy is usually intrinsically small because the metals have nearly isotropic properties. In addition, as shown by Anderson,[58] in the common case of "dirty" meals with mean free path less than ξ_0, the gap anisotropy tends to be averaged out. In the high T_c materials, ξ_0 is so short that this averaging may not take place, allowing the full intrinsic anisotropy to remain. A further consequence of the coherence length being essentially of atomic dimensions is that one can imagine a spatial variation of the gap from atom to atom within a single unit cell. The observation[59] of different temperature dependences below T_c of the NMR relaxation rates for nuclei at inequivalent Cu sites in YBCO appears to illustrate such an effect.

[57]See MT, pp. 33–59.
[58]P. W. Anderson, J. Phys. Chem. Sol. 11, 26 (1959).
[59]W. W. Warren, Jr., R. E. Walstedt, G. F. Brennert, G. P. Espinosa, and J. P. Remeika, Phys. Rev. Lett. 59, 1860 (1987); R. E. Walstedt, W. W. Warren, Jr., R. Tycho, R. F. Bell, G. F. Brennert, R. J. Cava, L. Schneemeyer, and J. Waszczak, Phys. Rev. B, submitted.

Another possibility of interest is that the electron pairing may not be the conventional s-state pairing of BCS, but a higher angular momentum pairing (p-state or d-state). If this is the case, the energy gap will not be essentially isotropic but instead will go to zero on nodal lines or planes in k-space.[60] The consequence of such an angular structure is that there is no absolute energy gap, only a reduced density of states as E approaches zero. Such pairings are known to exist in such systems as superfluid ^{3}He and the heavy-fermion superconductors, but not in the classic superconductors. If such higher angular momentum pairings were present, one would expect the exponential freeze-out of quasiparticles at low temperatures to be replaced by a weaker power-law dependence.

With these considerations in mind, we now turn to examine the experimental situation. To anticipate, it must be stated that, at this time, only a modest amount can be said with confidence and consensus about the interesting issues mentioned above. This is not for any lack of experimental work but because of difficulties inherent in these materials: absence of large single crystals, sensitivity of results to granularity and anisotropy, sensitivity to poor surface conditions because of short coherence lengths, etc. Despite this unsatisfactory situation, and to focus attention on the need for further work, it seems worth commenting briefly on the status of the two most active approaches: far-infrared absorption spectroscopy and electron tunneling.

a. Far-Infrared Absorption

The early experiments, in which ceramic samples were used, revealed a complex spectrum in both normal and superconducting states. By contrast, in a conventional metallic superconductor the normal metal absorption is structureless, so that a drop in absorption below the gap frequency in the superconducting state is a clear indication of the gap width. The reason for this dramatic difference is that the complex unit cell of the oxide superconductors leads to a rich spectrum of vibrational modes in the far infrared, which can be strongly excited optically because of poor screening by the few conduction electrons. In fact, for radiation polarized along the c-axis, these vibrational modes dominate the interaction with the electromagnetic field, while in the ab-plane, the stronger conductive response dominates. In a composite of randomly oriented crystalline grains, smaller than a wave length in size, the electromagnetic response is determined by an effective medium, averaged over crystal directions and affected by the intergranular coupling. In this average, the

[60]T. M. Rice, Z. Phys. B**67**, 141 (1987).

poorly-conducting directions tend to dominate, since the same current must flow in series through them all in screening out the high-frequency field. This accounts for the prominence of the vibrational structure in the response of unoriented ceramic samples, which undermined attempts at direct interpretation of the early reflection results on ceramic samples, leading to inferences of unusually small values of $2\Delta/kT_c$. However, Timusk et al.[61] made the *indirect* inference that in YBCO the minimum gap width is such that $2.98 < 2\Delta/kT_c < 4.80$, on the basis that vibrational modes below this energy range narrowed in the superconducting state (presumably by loss of the damping mechanism of coupling to quasiparticles in the gap), while those above it did not.

In the latter half of 1987, data on single crystals and oriented epitaxial films of YBCO began to appear. In these, the sharp phonon lines were essentially eliminated from the in-plane far-infrared spectrum, leaving a simpler Drude-like spectrum.[62] Closer study, however, showed that one could obtain a more physically satisfactory fit to the spectrum by augmenting the Drude term at low frequencies by a "mid-band" highly damped Lorentz oscillator.[61,63] Alternatively, one could introduce Drude parameters for effective mass and dampling constant that were dependent on frequency as well as temperature.[64]

Given all these uncertainties about how to interpret the *normal* state absorption spectrum, it is not surprising that different groups reached different tentative conclusions about the interpretation of the superconducting state data. Schlesinger et al.[62] interpreted their data on single crystal YBCO in terms of BCS-based Mattis–Bardeen[65] frequency-dependent complex conductivities corresponding to a gap with $2\Delta/kT_c \approx 8$ at low temperatures. Such a large ratio would imply very strong coupling superconductivity.

On the other hand, Timusk et al.,[61] working from generally similar data, attributed the reflectance feature at $500\,\text{cm}^{-1}$ not to a superconducting gap but to a plasma edge, which was assumed to result from the interplay of the Mattis–Bardeen complex conductivity of the conduction electrons with the complex dielectric function of the other electrons and lattice modes; consequently, Timusk et al. concluded that no gap, nor

[61]T. Timusk, D. A. Bonn, J. E. Greedan, C. V. Stager, J. D. Garrett, A. H. O'Reilly, M. Reedyk, K. Kamaras, C. D. Porter, S. L. Herr, and D. B. Tanner, *Proc. Interlaken Conf., Physica C* **153–155,** 1744 (1988).

[62]Z. Schlesinger, R. T. Collins, D. L. Kaiser, and F. Holtzberg, *Phys. Rev. Lett.* **59,** 1958 (1987).

[63]M. S. Sherwin, P. L. Richards, and A. Zettl, *Phys. Rev. B37,* 1587 (1988).

[64]G. A. Thomas, J. Orenstein, D. H. Rapkine, M. Capizzi, A. J. Millis, R. N. Bhatt, L. F. Schneemeyer, and J. V. Waszczak, *Phys. Rev. Lett.* **61,** 1313 (1988).

[65]D. C. Mattis and J. Bardeen, *Phys. Rev.* **111,** 412 (1958); see also MT, p. 57.

even the existence of a gap, could be demonstrated reliably from the data. A rather similar point of view was adopted by Sherwin et al.[63] in interpreting their data on a reflectance edge in polycrystalline LaSrCuO in terms of a temperature-dependent plasma edge. They also concluded that "neither the existence nor the value of the energy gap are obvious from the far-infrared reflectance data."

In all of these experiments, it was noted that the measured reflectance fell at least a few percent short of 100% even at low frequencies and $T \ll T_c$, contrary to expectations for a resistanceless superconductor. This discrepancy was attributed to surface imperfections, diffuse reflection, etc., but it hinders any simple interpretation of the data. Very recently, Thomas et al.,[64] reported data on two YBCO crystals that have lower T_c (50 K and 70 K, presumably because of less complete oxygenation), but which show a reflectance that reaches 100% within experimental error (estimated <1%). Presuming that this high reflectivity indicates near ideal surface conditions, they suggest that these samples may be characteristic of good material. In them, they find that the reflectivity stays at 100% until a corner is reached at a "gap frequency," above which the reflectance drops roughly linearly with increasing frequency. Remarkably, in both crystals, with quite different T_c values, the gap is found to have the BCS value of $3.5kT_c$, with uncertainty estimated at $\pm 0.5kT_c$.

All the above experiments have inferred the absorption from the difference of the measured reflectivity from 100%, which involves the problem of working with the *difference* of two much larger quantities. In an interesting alternative approach, Gershenzon et al.[66] directly determined the absorption of radiation by measuring the resulting temperature rise of a sample of ceramic YBCO. They found essentially zero absorption up to $40 \, \text{cm}^{-1}$, which was followed by a steep rise in absorption that reached a saturation value at about $100 \, \text{cm}^{-1}$. This onset frequency corresponds to $2\Delta/kT_c$ of only 0.6! As the authors note, their technique is sensitive to the *lowest* energy gap or other absorption mechanism; it is not clear that they can distinguish electronic from lattice absorption, which may dominate in an unoriented ceramic sample.

To summarize, it appears that most of the FIR data give inconclusive information about the existence and width of an energy gap. There seems to be growing evidence in support of a gap value of 3 to $4kT_c$, near the BCS weak-coupling value of $3.5kT_c$, at least for currents in the ab plane, but a value as large as $8kT_c$ is not excluded.

[66]E. M. Gershenzon, G. N. Gol'tsman, B. S. Karasik, and A. D. Semenov, *JETP Lett.* **46**, 237 (1988).

b. *Electron Tunneling*

Since the pioneering work of Giaever,[67] the electron tunneling technique has been recognized as probably the most powerful single technique for probing the superconducting state. Not only does it yield direct information about the width of the energy gap and the density of states peaks just outside the gap, but also subtle structure on the I–V curves at higher voltages yields information about the spectrum of the electron–phonon coupling that causes the superconductivity.[68]

In the case of SIN tunneling, the electron tunnels through an insulating layer (often a native oxide) between a normal electrode and the superconductor under study; in that case, at $T = 0$ the current I is ideally zero until the voltage difference V between electrodes satisfies the relation

$$eV = \Delta, \tag{7.1}$$

after which it rises following (for the BCS density of states) the form

$$IR_n = [V^2 - (\Delta/e)^2]^{1/2}, \tag{7.2}$$

where R_n is the resistance of the tunnel junction in the normal state. In the other case, namely SIS tunneling between two superconducting electrodes, at $T = 0$ the current ideally is zero until V satisfies

$$eV = (\Delta_1 + \Delta_2), \tag{7.3}$$

where Δ_1 and Δ_2 refer to the energy gaps in the two electrodes; at that point the current jumps sharply to a value near that in the normal state at the same voltage. At $T > 0$, there is also a feature in the I–V curve at $V = |\Delta_1 - \Delta_2|/e$, which allows Δ_1 and Δ_2 to be separately determined.

Although this conceptually simple and powerful technique has worked well on the classic superconducting metals, it has proved more difficult to apply to the earlier generation of high T_c materials, such as Nb_3Sn, and very difficult indeed for the YBCO class materials. The distinguishing feature is that the classic superconductors have long coherence lengths and clean surfaces, so that the tunneling process samples the density of states of representative interior material, even though the electrons inevitably must pass through the surface. With the shorter coherence lengths of the A15 materials like Nb_3Sn, one can no longer neglect the

[67] I. Giaever, *Phys. Rev. Lett.* **5,** 147 and 464 (1960).
[68] W. L. McMillan and J. M. Rowell, *Phys. Rev. Lett.* **14,** 108 (1965); see also review by same authors in "Superconductivity," (R. D. Parks, ed.), chap. 11, Marcel Dekker, New York, 1969.

modification of the surface layer by the presence of the tunneling barrier and counter electrode. Although useful preliminary results could be obtained by use of simple point contact and evaporated film counter electrodes, quantitative results could be obtained only with rather sophisticated procedures[69] that require unfolding of the proximity effects in the complex barrier interface. Presumably these problems are even more severe in the YBCO class materials, but they have not yet been addressed in a quantitative manner. Instead, we are still at the stage of rather crude experiments and simple interpretations of data.

An additional complication enters the interpretation of data on the new materials because of their typically granular structure. This allows the possibility that the measured voltage may not appear across a *single* junction, as is implicit in the preceding discussion, but may appear across a series of several intergranular junctions within the bulk of the superconductor, in addition to that at the surface. If that is the case, of course, there is the possibility of confusing a resulting high-voltage feature with a gap width that is a multiple of the true value. This has been suggested as a possible explanation of data[70] showing features at voltages in the ratio of $1:3:5$; this could arise with a normal counter electrode making the surface contact of the SIN type, showing Δ, and the other features from a series of SIS intergrain contacts, each contributing 2Δ.

Yet another source of confusion in the data is the range of phenomena associated with the "Coulomb blockade."[71] This term refers to the fact that electron tunneling between microscopic conductors is strongly affected by the charging energy $E_c = e^2/2C$ associated with a single electronic charge in a small capacitiance C. Simply stated, it is not energetically favorable for an electron to tunnel until the voltage difference exceeds $e/2C$; this is $\sim 10\,\mathrm{mV}$ for $C \sim 10^{-17}F$, which corresponds to a structure with length scale $\sim 10^{-5}$ cm. Moreover, if one of the electrodes is an isolated grain, one expects a "Coulomb staircase"[72]

[69]E. L. Wolf, J. Zasadzinski, G. B. Arnold, D. F. Moore, J. M. Rowell, and M. R. Beasley, *Phys. Rev. B22*, 1214 (1980); D. A. Rudman and M. R. Beasley, *Phys. Rev. B30*, 2590 (1984).

[70]M. D. Kirk, D. P. E. Smith, D. B. Mitzi, J. Z. Sun, D. J. Webb, K. Char, M. R. Hahn, M. Naito, B. Oh, M. R. Beasley, T. H. Geballe, R. H. Hammond, A. Kapitulnik, and C. F. Quate, *Phys. Rev. B35*, 8850 (1987).

[71]I. Giaever and H. R. Zeller, *Phys. Rev. Lett.* **20**, 1504 (1968); E. Ben Jacob and Y. Gefen, *Phys. Lett.* **108A**, 289 (1985); D. V. Averin and K. K. Likharev, *J. Low Temp. Phys.* **62**, 345 (1986).

[72]J. B. Barner and S. T. Ruggiero, *Phys. Rev. Lett.* **59**, 807 (1987); L. S. Kuz'min and K. K. Likharev, *JETP Letters* **45**, 495 (1987).

in which a step increase in current should occur each time the voltage increases by a unit of e/C. Note that this phenomenon also leads to a possible explanation of multiple tunneling features at voltages in the ratio of $1:3:5:\cdots$. In this case, however, the voltage unit has nothing to do with a superconducting energy gap; it is a measure of the capacitance of a conducting grain! An important case arises when the tunneling is observed between the superconductor and the controlled point in a scanning tunneling microscope (STM) configuration. Even without the complication of multiple peaks, one has to take account of the fact that the observed gap feature is essentially shifted to higher voltage by the amount of the Coulomb gap $e/2C$, which will be significant for $C < 10^{-15} F$.

After this enumeration of the difficulties of obtaining reliable inferences from tunneling data, it is hardly surprising that an extremely broad range of tunnel I–V curves on YBCO have been reported in the literature. In fact, a single recent paper[73] has reported features in tunneling microscope data corresponding to $2\Delta/kT_c$ ranging all the way from 0.7 up to 13! Given the possibility of substantial gap anisotropy, it is not clear that a single value (or narrow range) should be expected, even after possible false multiple signals of the sorts described above are stripped away. None the less, as in the case of the FIR spectra, there appears to be some developing consensus in favor of values fairly close to, but somewhat above, the BCS weak-coupling limit $2\Delta/kT_c \approx 3.5$. We now briefly review some representative results in this literature.

An important early paper was that of Kirtley et al.,[74] which reported SIN tunneling from a pointed STM electrode driven sufficiently deeply into a YBCO crystal to obtain measureable currents. They obtained dI/dV curves with multiple peaks, but treating the lowest voltage peaks on each side of zeros as SIN peaks at $V = \pm\Delta/e$, they estimated $\Delta = 16 \pm 3$ meV, or $3.4 < 2\Delta/kT_c < 5.3$, taking $T_c = 85$ K. Their results with the point perpendicular and parallel to the c-axis were not distinguishable, considering the range of curves obtained at different points on the surface. Given the need for the point to penetrate into the crystal, it is possible that tunnel currents from it sampled both crystal directions in all cases.

To give the flavor of the diversity of the early data, we note that the three papers following that of Kirtley et al. in the same issue of the Physical Review also reported tunneling data on YBCO. The Stanford

[73]M. C. Gallagher, J. G. Adler, J. Jung, and J. P. Franck, *Phys. Rev.* **B37,** 7846 (1988).

[74]J. R. Kirtley, R. T. Collins, Z. Schlesinger, W. J. Gallagher, R. L. Sandstrom, T. R. Dinger, and D. A. Chance, *Phys. Rev.* **B35,** 8846 (1987).

group reported data from a tungsten tip in sintered powder YBCO, which they interpreted in terms of a gap with $2\Delta/kT_c \sim 13$, a value much larger than ever seen in classic superconductors.[70] In the next paper, the Berkeley group reported data from a copper tip pressed on polycrystalline YBCO pellet material, from which they inferred $2\Delta/kT_c \approx 3.9$ at $T = 0$, with indications of a BCS-like temperature dependence.[75] In the fourth paper in this group, the NBS group reported[76] data on YBCO obtained by using the break junction technique of touching together two pieces of bulk material that had been fractured under liquid helium; from their data, they estimated $2\Delta/kT_c \approx 4.8$.

In more recent work, Tsai et al.[77] reported measurements of apparent gap anisotropy from experiments in which oriented films of YBCO were broken in the cryogenic environment, to yield a clear surface, which was then contacted by Pb electrode. At 4.2 K, with the Pb superconducting, tunnel I–V curves were obtained showing features interpreted as $\Delta_{YBCO} \pm \Delta_{Pb}$, while at 10 K, where the Pb is normal, these features merged into a single one identified as Δ_{YBCO}. By choice of preferred growth orientation of the crystallites relative to the break direction, Tsai et al. were able to study tunneling in directions either parallel or perpendicular to the c-axis. In addition, they measured the temperature dependence of these tunneling features, and found that the gaps reduced to zero at different temperatures in the range 42 to 77 K for different films, and that the gap widths appeared to scale with the differing T_c values inferred in this way. Combining their data on 8 distinct samples in 4 films, they concluded that the gaps in the 2 principal directions had the values

$$\Delta_{ab} = 5.9 \pm 0.2kT_c \quad \text{and} \quad \Delta_c = 3.6 \pm 0.2kT_c. \tag{7.4}$$

Although the low and diverse T_c values introduce some uncertainty concerning the general applicability of these results, they are basically consistent with the estimated range of 3.4 to $5.3kT_c$ (without angular resolution) found by Kirtley et al.[74] by a different technique on a crystal with T_c estimated to be 85 K.

Another interesting recent publication that combines tunneling data

[75]M. F. Crommie, L. C. Bourne, A. Zettl, M. L. Cohen, and A. Stacey, Phys. Rev. B35, 8853 (1987).

[76]J. Moreland, J. W. Ekin, L. F. Goodrich, T. E. Capobianco, A. F. Clark, J. Kwo, M. Hong, and S. H. Liou, Phys. Rev. B35, 8856 (1987).

[77]J. S. Tsai, I. Takeuchi, J. Fujita, T. Yoshitake, S. Miura, S. Tanaka, T. Terashima, Y. Bando, K. Iijima, and K. Yamamoto, Proc. Interlaken Conf., Physica C 153–155, 1385 (1988).

with FIR absorption and Andreev reflection data on similar thin film YBCO samples has been provided by the Nijmegen group of van Bentum *et al.*[78] Their tunnel data were taken with an STM configuration using a tungsten tip. They found that, depending on the position and separation of the contact, the I–V curve may be better fitted by using the Coulomb blockade model (with *no* superconducting gap) than by using the BCS gap model (without taking account of the Coulomb "gap"). Since the two gaps may be of similar magnitude, clearly this can introduce ambiguity in the interpretation when both are present. In some cases, a Coulomb staircase of several steps, presumably due to the presence of small isolated metallic grains, was observed. In other experiments, the Nijmegen group made metallic contact to one grain and observed SIS tunneling from that grain to a neighboring grain, apparently through an insulating oxide barrier. In this case, the I–V curves showed the characteristic sharp rise of an SIS junction at a voltage $(\Delta_1 + \Delta_2)/e$. Assuming the two gaps were equal, they typically found $\Delta = 14$ meV, but with a range of 11 to 16 meV from point to point. Taking $T_c = 90$ K in all cases, this corresponds to a ratio $2\Delta/kT_c = 3.6 \pm 0.6$. In these experiments the current at voltages well below the gap was very far below the Ohmic line, indicating quite a clear gap. It is not clear whether the data are good enough to exclude the low-lying density of states expected in a higher angular momentum state pairing model.

In the same paper, the Nijmegen group report the results of their novel use of Andreev reflection to obtain an independent measure of the gap. The sample contains a thin film of Ag deposited on the sputter-cleaned surface of a YBCO film. Current is injected into the Ag through a tiny Ag–Ag point contact and taken out through the YBCO film. An electron approaching the Ag–YBCO interface with energy below the gap undergoes Andreev reflection[79] as a hole, with transfer of a Cooper pair to the YBCO. This contributes *twice* the current that would flow if the YBCO were normal and the electron simply entered it. So long as the Ag film is thin compared with the inelastic scattering length, the maximum energy of the incoming electron is controlled by the injection voltage. Thus, the differential resistance of the contact makes a step change when $eV = \Delta$, allowing direct determination of the gap. In these experiments, the rise in resistance occurs as a ramp extending from ~10 to ~15 mV,

[78]P. J. M. van Bentum, H. F. C. Hoevers, H. van Kempen, L. E. C. van de Leemput, M. J. M. F. de Nivelle, L. W. M. Schreurs, R. T. M. Smokers, and P. A. A. Teunissen, *Proc. Interlaken Conf., Physica C***153–155,** 1718 (1988); also H. F. C. Hoevers, P. J. M. van Bentum, L. E. C. van de Leemput, H. van Kempen, A. J. G. Schellingerhout, and D. van der Marel, *Physica C***152,** 105 (1988).
[79]G. E. Blonder, M. Tinkham, and T. M. Klapwijk, *Phys. Rev. B***25,** 4515 (1982).

very comparable with the range found in the SIS tunnel experiments described above. If we take the result of the Andreev reflection technique to be 12.5 ± 2 mV, it corresponds to $2\Delta/kT_c = 3.2 \pm 0.6$. Incidentally, the very *existence* of the Andreev reflection effect confirms that the superconductivity in YBCO is based on a condensate of Cooper pairs.

c. Concluding Summary

Taking together all the results cited above from several techniques, we note that the majority of the data identify the range 3 to 6 for $2\Delta/kT_c$. This range is similar to, but extends higher than, that measured in the classic superconductors. It also includes the BCS weak-coupling limit value of 3.53; with stronger coupling, higher values of the ratio are expected. Thus, it appears likely that the "center of gravity" value for $2\Delta/kT_c$ is in the expected range.

The *range* of values found may reflect a highly anisotropic gap, with different values sampled with different crystal orientations; the data of Tsai *et al.*, appear to indicate this explicitly, but further confirming work would be desirable. The more extreme values found in some experiments pose open questions as to their origin. It is intriguing that similar and sharply defined values 0.7 and 0.6 are found independently by tunneling and FIR absorption experiments in different laboratories. Likewise, the large value 13 was found by tunneling experiments in two different laboratories. Both too large and too small values of the gap can be explained away as artifacts of the surface conditions, Coulomb gap, etc., but such explanations do not account readily for independent observations of similar values.

In both the tunneling work of Tsai *et al.* and the FIR absorption work of Thomas *et al.*,[64] evidence was found that the observed gap scales linearly with T_c down to roughly half the full value of $\sim$90 K, as might be expected on general theoretical grounds. There is evidence indicating that the density of states in the "gap" is at least much less than at higher energies, but it is always hard to exclude the possibility of a small density of low-lying states, such as might arise from non-s-state pairing.

8. ELECTRONIC SPECIFIC HEAT

In typical classic superconductors, a large fraction of the specific heat at T_c and below is electronic in origin, because the electronic specific heat $C_{el} = \gamma T \sim T/T_F$ falls off more slowly at low temperatures than the lattice specific heat $\sim (T/\Theta_D)^3$. In the high-temperature superconductors, however, the same temperature dependences cause the lattice specific heat to dominate; hence, changes in the electronic specific heat associated with the establishment of the superconducting state are more difficult to measure. Despite this difficulty, reasonably good data on the jump in specific heat at T_c have been obtained to compare with the jump observed in the classic superconductors and predicted theoretically. In the BCS theory, the jump in specific heat is proportional to the value of $d\Delta^2/dT$ at T_{c-}. It is predicted to have the value $\Delta C = 1.43\gamma T_c$, and experimental data on the classic superconductors are generally in good agreement with this prediction. Two back-to-back papers by the Argonne[80] and Urbana[81] groups reported values for $\Delta C/\gamma T_c$ in YBCO of 1.5 and 1.23 ± 0.08, respectively, which nicely bracket the weak-coupling BCS theoretical prediction.

Because of the extremely short coherence length in these materials, fluctuation effects should be unusually large, and might be expected to give observable corrections very near T_c to the simple jump in specific heat predicted by the mean-field theory. Evidence for such effects was reported by Inderhees et al.,[82] and interpreted by the authors as confirming the three-dimensional nature of the superconductivity, but as being inconsistent in detail with Ginzburg–Landau theory. Further work will be required to confirm this interpretation of these rather small effects.

Another important open question is the significance of the nonvanishing linear specific heat term γT that is often observed (in addition to the dominant lattice T^3 term) even in the superconducting state of the high-temperature superconductors. This contrasts with the exponentially vanishing value $\sim\exp(-\Delta/kT)$ expected if there is a clean energy gap of width Δ. Although such a linear contribution to the specific heat is often found in the high T_c materials, its *magnitude* is found to vary widely between samples. This leads to the supposition that it may be an extrinsic effect depending on details of the sample composition or structure that are not central to the superconductivity. For example, it is known that so-called two-level systems can contribute a linear specific heat term of the observed order of magnitude in glasses, disordered solids, and some ceramics. On the other hand, recent experiments[83]

[80]M. V. Nevitt, G. W. Crabtree, and T. E. Klippert, *Phys. Rev. B36*, 2398 (1988).
[81]S. E. Inderhees, M. B. Salamon, T. A. Friedmann, and D. M. Ginsberg, *Phys. Rev. B36*, 2401 (1988).
[82]S. E. Inderhees, M. B. Salamon, N. Goldenfield, J. P. Rice, B. G. Pazol, D. M. Ginsberg, J. Z. Liu, and G. W. Crabtree, *Phys. Rev. Lett.* **60**, 1178 (1988).
[83]K. Kumagai, Y. Nakamichi, I. Watanabe, Y. Nakamura, H. Nakajima, N. Wada, and P. Lederer, *Phys. Rev. Lett.* **60**, 724 (1988).

appear to show a systematic correlation between the magnitude of γ and the observation of antiferrromagnetism vs. superconductivity as a function of composition in the LaBaCuO system. The authors interpret this correlation as supporting the resonating valence bond (RVB) ground state proposed by Anderson,[84] with an exchange energy $J \sim 1000$ K. Again, further investigation will be required to test this conjecture by confirming the generality of the observed correlation and evaluating possible alternative explanations.

9. Concluding Remarks

As is inevitable given the rapid evolution of knowledge of the high temperature superconductors, this review can only provide a snap-shot view of the field, as seen from the perspective of the authors at the beginning of July, 1988. We have emphasized what has been learned from experimental studies, together with the conceptual models from which a framework of phonomenological interpretation can be constructed. This emphasis has allowed us to side-step almost completely the deeper question of the microscopic nature of the superconducting state in the new materials and the interactions that produce it. Despite the extraordinary new parameter values for T_c and ξ_0, and the dramatic anisotropy, the basic BCS-Ginzburg–Landau phenomenology seems to provide a basis for understanding essentially everything that is observed in response to magnetic fields and electric currents. Nonetheless, working out the implications of the new regimes that arise with these new parameter ranges to help guide experimental work and development of applications will present interesting challenges for the future.

Acknowledgements

The authors are pleased to acknowledge the benefits of innumerable discussions with colleagues both at Harvard and elsewhere, which have helped shape their perspective. They hope the perspective is both true and useful, but they accept responsibility for shortcomings that are probably inevitable given the pressure of a practical deadline for completion. In particular, we regret that valuable contributions of many workers could not be included because of limitations of space and time.

This work was supported in part by the National Science Foundation (Grants DMR-84-04489 and DMR-86-14003), the U.S. Office of Naval Research (Contract N00014-83-K-0383), and the Joint Services Electronics Program (Contract N00014-84-K-0465).

[84] P. W. Anderson, *Science* **235**, 1196 (1987).

Notes Added in Proof

PHYSICAL PROPERTIES OF THE NEW SUPERCONDUCTORS

Very recently, Palstra *et al.* (*Phys. Rev. Lett.*, in press) have reported data on the resistive transition of $Bi_{2.2}Sr_2Ca_{0.8}Cu_2O_{8+\delta}$, with emphasis on the low resistance end of the transition. Their data show particularly clearly a thermally activated resistivity $\sim \exp(-U_0/kT)$, as expected from (5.2), but with U_0 having a different and weaker dependence on B than would be expected from (5.3). This difference may result from the exceptionally strong anisotropy of this material, which has very weak interlayer coupling. This work indicates that the specific dependences of U_0 on H and T discussed in Section 5a are not universal, but it strongly confirms the critical and qualitative importance of thermally activated resistive processes far below T_c.

The Structure of $Y_1Ba_2Cu_3O_{7-\delta}$ and Its Derivatives

R. BEYERS

IBM Research Division
Almaden Research Center
San Jose, California

T. M. SHAW

IBM Research Division
Thomas J. Watson Research Center
Yorktown Heights, New York

I. Introduction

In 1986, Georg Bednorz and Alex Müller discovered superconductivity above 30 K in the La–Ba–Cu–O system.[1,2] Several groups quickly

[1] J. G. Bednorz and K. A. Müller, *Z. Phys. B* **64,** 189 (1986).
[2] J. G. Bednorz, M. Takashige, and K. A. Müller, *Europhys. Lett.* **3,** 379 (1987).

reproduced their results and identified the superconducting phase as $La_{2-x}Ba_xCuO_4$. Many researchers then began to determine the effects of elemental substitutions and different processing conditions on the structure and superconducting properties of this oxide. Following this approach, in 1987, Paul Chu, Maw-Kuen Wu, and coworkers discovered superconductivity above 90 K in the Y–Ba–Cu–O system.[3] As before, the discovery was quickly reproduced, the 90 K phase was identified as $Y_1Ba_2Cu_3O_7$, and studies of the effects of substitutions and processing conditions were initiated. This cycle has been repeated twice in 1988 with the independent discoveries of superconductivity above 100 K in the Bi–Ca–Sr–Cu–O system by Hiroshi Maeda and coworkers[4] and in the Tl–Ca–Ba–Cu–O system by Allen Hermann and Zhengzhi Sheng.[5,6]

The purpose of this article is to give an overview of the structural studies performed on these superconducting oxides. We focus primarily on $Y_1Ba_2Cu_3O_{7-\delta}$, nicknamed the 123 structure, because this structure has been the most extensively studied to date. In the first two sections, we describe the basic atomic arrangements in $Y_1Ba_2Cu_3O_{7-\delta}$ and how these arrangements change as a function of processing. We then discuss the defects that are commonly found in the 123 structure, some of which play a dominant role in controlling superconducting properties. Next, we summarize the studies of elemental substitutions for Y, Ba, Cu and/or O in 123, and we compare 123 with the other families of superconducting oxides. These studies have both theoretical and practical interest because they provide clues to the roles that various structural features play in the mechanism of high-temperature superconductivity, which, in turn, may aid in the discovery of even higher temperature superconductors. Finally, we draw some conclusions and make suggestions for future structural studies. In all but the final section, we try to emphasize what is known and generally agreed upon about these oxides. The last section, however, is largely our own opinion and more speculative in nature.

II. Basic Structure

Soon after the discovery and confirmation of superconductivity in the Y–Ba–Cu–O system,[3,7-10] the phase responsible for the 90 K transition

[3]W. K. Wu, J. R. Ashburn, C. J. Torng, P. H. Hor, R. L. Meng, L. Gao, Z. J. Huang, Y. Q. Wang, and C. W. Chu, *Phys. Rev. Lett.* **58,** 908 (1987).

[4]M. Maeda, Y. Tanaka, M. Fukutomi, and A. Asano, *Jpn. J. Appl. Phys.* **27,** L209 (1988).

[5]Z. Z. Sheng and A. M. Hermann, *Nature* **332,** 55 (1988).

[6]Z. Z. Sheng and A. M. Hermann, *Nature* **332,** 138 (1988).

was found to have a cation stoichiometry of $1Y:2Ba:3Cu$. The unit cell dimensions determined by electron and x-ray diffraction identified the structure as being related to a cubic perovskite with one of the cube axes tripled.[11-23] In the basic perovskite structure ABO_3, there are two cation sites.[24] The A site lies at the center of a cage formed by corner-sharing anion octahedra and accommodates the larger cations in the structure. The B site lies at the centers of the anion octahedra and accommodates the smaller cations. It was therefore natural to place the larger Y and Ba ions in the A sites and the smaller Cu ions in the B sites. The tripling of

[7]Z. Zhao, L. Chen, Q. Yang, Y. Huang, G. Chen, R. Tang, G. Liu, C. Cui, L. Chen, L. Wang, S. Guo, S. Li, and J. Bi, *Kexue Tongbao* **32**, 1098 (1987).

[8]P. Ganguly, R. A. Mohan Ram, K. Sreedhar, and C. N. R. Rao, *Pramana J. Phys.* **28**, L321 (1987).

[9]J. M. Tarascon, L. H. Greene, W. R. McKinnon, and G. W. Hull, *Phys. Rev. B.* **35**, 7115 (1987).

[10]S. J. Hwu, S. N. Song, J. Thiel, K. R. Poeppelmeier, J. B. Ketterson, and A. J. Freeman, *Phys. Rev. B* **35**, 7119 (1987).

[11]R. J. Cava, B. Batlogg, R. B. van Dover, D. W. Murphy, S. Sunshine, T. Siegrist, J. P. Remeika, E. A. Rietman, S. Zahurak, and G. P. Espinosa, *Phys. Rev. Lett.* **58**, 1676 (1987).

[12]P. M. Grant, R. B. Beyers, E. M. Engler, G. Lim, S. S. P. Parkin, M. L. Ramirez, V. Y. Lee, A. Nazzal, J. E. Vazquez, and R. J. Savoy, *Phys. Rev. B.* **35**, 7242 (1987).

[13]R. M. Hazen, L. W. Finger, R. J. Angel, C. T. Prewitt, N. L. Ross, H. K. Mao, C. G. Hadidiacos, P. H. Hor, R. L. Meng, and C. W. Chu, *Phys. Rev. B* **35**, 7238 (1987).

[14]W. R. McKinnon, J. M. Tarascon, L. H. Greene, G. W. Hull, and D. A. Hwang, *Phys. Rev. B* **35**, 7245 (1987).

[15]W. J. Gallagher, R. L. Sandstrom, T. R. Dinger, T. M. Shaw, and D. A. Chance, *Solid State Commun.* **63**, 147 (1987).

[16]D. G. Hinks, L. Soderholm, D. W. Capone II, J. D. Jorgensen, and Ivan K. Schuller, *Appl. Phys. Lett.* **50**, 1688 (1987).

[17]T. Hatano, A. Matsushita, K. Nakamura, K. Honda, T. Matsumoto, and K. Ogawa, *Jpn. J. Appl. Phys.* **26**, L374 (1987).

[18]Y. Kitano, K. Kifune, I. Mukouda, H. Kamimura, J. Sakurai, Y. Komura, K. Hoshino, M. Suzuki, A. Minami, Y. Maeno, M. Kato, and T. Fujita, *Jpn. J. Appl. Phys.* **26**, L394 (1987).

[19]M. Hirabayashi, H. Ihara, N. Tereda, K. Senzaki, K. Hayashi, S. Waki, K. Murata, M. Tomumoto, and Y. Kimura, *Jpn. J. Appl. Phys.* **26**, L454 (1987).

[20]E. Takayama-Muromachi, Y. Uchida, Y. Matsui, and K. Kato, *Jpn. J. Appl. Phys.* **26**, L476 (1987).

[21]Y. Syono, M. Kikuchi, K. Ohishi, K. Hiraga, H. Arai, Y. Matsui, N. Kobayashi, T. Sasaoka, and Y. Muto, *Jpn. J. Appl. Phys.* **26**, L498 (1987).

[22]K. Semba, S. Tsurumi, M. Hikita, T. Iwata, J. Noda, and S. Kurihara, *Jpn. J. Appl. Phys.* **26**, L429 (1987).

[23]S. B. Qadri, L. E. Toth, M. Osofsky, S. Lawrence, D. U. Gubser, and S. A. Wolf, *Phys. Rev. B* **35**, 7235 (1987).

[24]J. B. Goodenough and M. Longo, *in* "Landolt-Börnstein" (K.-H. Hellwege and A. M. Hellwege, eds.), Group II.Vol. 4a, Chapter 3a, Springer-Verlag, New York, 1970.

the perovskite unit cell could then be accounted for by ordering the Y and Ba ions in the A sites such that the top and bottom cells in a stack of three contained Ba ions, while the middle cell contained a Y ion. This basic cation arrangement was not only consistent with the 1Y:2Ba:3Cu stoichiometry but also provided a reasonable fit to x-ray powder diffraction data obtained from nearly single-phase materials[11–14] and accounted for contrast seen in high-resolution images of the structure.[25–29] These results confirmed that the alternative suggestion[19,23] that the structure was related to the $La_3Ba_3Cu_6O_{14}$ structure proposed by Er-Rakho[30] was incorrect.

There are 3 anions per unit cell in the ideal perovskite structure, corresponding to 9 possible oxygen sites in a tripled perovskite unit cell. Formal balancing of the charges on the cations requires a maximum of 8 oxygen ions per unit cell if all of the copper is assumed to be in a +3 state, and 6.5 and 5 oxygen ions per unit cell if charges of +2 and +1 are assumed for all the copper cations. It was therefore clear that the 123 structure was oxygen deficient relative to the ideal perovskite structure. X-ray diffraction from small single crystals extracted from sintered polycrystalline material confirmed and refined the basic positions of the cations and also identified where the oxygen deficiency was accommodated in the structure.[13,14,31,32] The x-ray data clearly showed that the Y ion was surrounded by only 8 oxygen ions rather than by 12 as in the ideal perovskite structure. Oxygen deficiency was also noted in the basal copper plane between the Ba ions. Anion sites in this plane, which lie along the cell edges, were found to be only half occupied.[14,31,32] Refinements of the cell parameters in several studies[11,12,14,20,21,31] indicated that the unit cell was orthorhombic with b slightly larger than a, even though some single crystal studies[13,14,31,32] observed a tetragonal distribution of diffracted intensities. It was noted later that this apparent

[25]R. Beyers, G. Lim, E. M. Engler, R. J. Savoy, T. M. Shaw, T. R. Dinger, W. J. Gallagher, and R. L. Sandstrom, *Appl. Phys. Lett.* **50**, 1918 (1987).

[26]A. Ourmazd, J. A. Rentschler, J. C. H. Spence, M. O'Keeffe, R. J. Graham, D. W. Johnson Jr., and W. W. Rhodes, *Nature* **327**, 308 (1987).

[27]E. A. Hewat, M. Dupuy, A. Bourret, J. J. Capponi, and M. Marezio, *Nature* **327**, 400 (1987).

[28]Y. Matsui, E. Takayama-Muromachi, A. Ono, S. Horiuchi, and K. Kato, *Jpn. J. Appl. Phys.* **26**, L777 (1987).

[29]K. Hiraga, D. Shindo, M. Hirabayashi, M. Kikuchi, K. Oh-Ishi, and Y. Syono, *Jpn. J. Appl. Phys.* **26**, L1071 (1987).

[30]L. Er-Rakho, C. Michel, J. Provost, and B. Raveau, *J. Solid State Chem.* **37**, 151 (1981).

[31]T. Siegrist, S. Sunshine, D. W. Murphy, R. J. Cava, and S. M. Zahurak, *Phys. Rev. B* **35**, 7137 (1987).

[32]F. P. Okamura, S. Sueno, I. Nakai, and A. Ono, *Mat. Res. Bull.* **22**, 1081 (1987).

discrepancy was probably caused by microdomain twinning which prevented oxygen ordering in the single crystals from being observed by x-ray diffraction techniques. The twinning was readily apparent in transmission electron microscope (TEM) images of the material.[21,25]

The source of the orthorhombic distortion in the 123 structure was finally revealed by neutron diffraction data refined using the Rietveld technique.[33–41] The advantage of Rietveld refinement[42] is that the structure can be deduced directly from neutron powder diffraction data by fitting the data with calculated diffraction scans. Thus the larger scattering cross-section for oxygen by neutrons relative to x-rays can be taken advantage of without the need for large single crystals. The neutron diffraction experiments showed unambiguously that the oxygens in the basal copper plane of the unit cell were ordered in the orthorhombic structure and that the 90 K material contained nearly seven oxygens per unit cell.

The essential features of the orthorhombic structure are shown in Fig. 1, and atomic coordinates and thermal parameters for each atom, which agree well with those originally determined by Seigrest et al.,[31] are given in Table I.[43] The effect of oxygen ordering in the basal plane of the structure is to occupy one of the oxygen sites along a cell edge and to leave the other site vacant. The cell edge with the occupied site is thus lengthened relative to the cell edge with the vacant oxygen site, producing an orthorhombic unit cell. The ordering puts the copper ions in the basal plane of the structure, at the center of square arrangements

[33]J. J. Capponi, C. Chaillout, A. W. Hewat, P. Lejay, M. Marezio, N. Nguyen, B. Raveau, J. L. Soubeyroux, J. L. Tholence, and R. Tournier, *Europhys. Lett.* **3**, 1301 (1987).

[34]M. A. Beno, L. Soderholm, D. W. Capone, D. G. Hinks, J. D. Jorgensen, I. K. Schuller, C. U. Segre, K. Zhan, and J. D. Grace, *Appl. Phys. Lett.* **51**, 57 (1987).

[35]F. Beech, S. Miraglia, A. Santoro, and R. S. Roth, *Phys. Rev. B* **35**, 8778 (1987).

[36]J. E. Greedan, A. O'Reilly, and C. V. Stager, *Phys. Rev. B* **35**, 8770 (1987).

[37]F. Izumi, H. Asano, T. Ishigaki, E. Takayama-Muromachi, Y. Uchida, N. Watanabe, and T. Nishikawa, *Jpn. J. Appl. Phys.* **26**, L649 (1987).

[38]M. Francois, E. Walker, J. L. Jorda, and K. Yvon, *Solid State Commun.* **63**, 1149 (1987).

[39]T. Kajitani, K. Oh-Ishi, M. Kikuchi, Y. Syono, and M. Hirabayashi, *Jpn. J. Appl. Phys.* **26**, L1144 (1987).

[40]Q. W. Yan, P. L. Zhang, Z. G. Shen, J. K. Zhao, Y. Ren, Y. N. Wei, T. D. Mao, C. X. Liu, T. S. Ning, K. Sun, and Q. S. Yang, *Phys. Rev. B* **36**, 5599 (1987).

[41]W. I. F. David, W. T. A. Harrison, J. M. F. Gunn, O. Moze, A. K. Soper, P. Day, J. D. Jorgensen, D. G. Hinks, M. A. Beno, L. Solderholm, D. W. Capone II, I. K. Schuller, C. U. Segres, K. Zhang, and J. D. Grace, *Nature* **327**, 310 (1987).

[42]L. M. Rietveld, *J. Appl. Cryst.* **2**, 65 (1969).

[43]A. Williams, G. H. Kwei, R. B. Von Dreele, A. C. Larson, I. D. Raistrick, and D. L. Bish, *Phys. Rev. B* **37**, 7960 (1988).

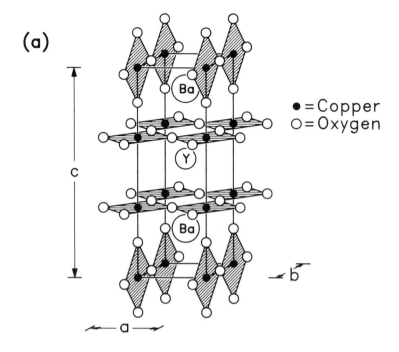

FIG. 1. The structure of $Y_1Ba_2Cu_3O_{7-\delta}$. In (a) the coordination of copper with oxygen is emphasized to show the location of copper-oxygen planes and chains in the structure. In (b) nearest-neighbor bonds are drawn in to show the puckering of the copper-oxygen planes. The notation used for the oxygen positions in this figure is used consistently throughout the remainder of this review. Note, however, that the oxygen site designations are often interchanged in the literature, especially the O(1) and O(4) sites.

of oxygen ions. The square planar arrangements of copper and oxygen ions are then linked together by sharing their corners to form "linear chains" along the b axis of the structure. A recent neutron diffraction study suggests that the chains may not be perfectly linear.[44] The effect of vacant oxygen sites around the Y ion can also be seen in Fig. 1. The absence of oxygen from the Y plane places the copper ions in five-fold coordinated square pyramidal sites. Linking of the square bases of the pyramids at their corners results in two-dimensional puckered sheets of copper-oxygen bonds that extend in the a-b plane of the structure. The chains and planes in the structure are linked through the oxygens that lie

[44]M. Francois, A. Junod, K. Yvon, A. W. Hewat, J. J. Capponi, P. Strobel, P. Marezio and P. Fischer, *Solid State Commun.* **66**, 1117 (1988).

(b)

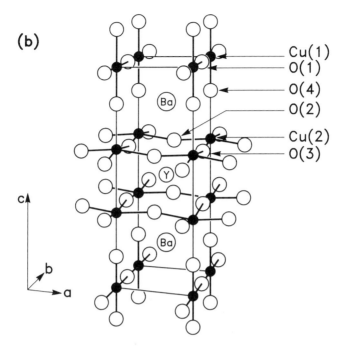

FIG. 1. (*Continued*)

at the apices of the square pyramids. The Y ion, which lies at the center of the cell, is coordinated by 8 oxygens that form a slightly distorted square prism. The Ba ion is 10-fold coordinated and is shifted slightly towards the Y ion relative to its position in the ideal perovskite structure.

Selected bond lengths and angles for the structure are given in Table II.[45] The Y–O and Ba–O distances are typical and compare well with distances observed in other structures. The Cu(2)–O(2) and Cu(2)–O(3) distances in the planes are 1.929 and 1.961, respectively, whereas the Cu(2)–O(4) bond, which joins the copper atom in the plane to the apical oxygen, is much longer, with a value of 2.341 Å. The Cu(1) atom in the chain has a bond length of 1.947 Å to the O(1) atom in the chain. The shortest Cu–O bond length, 1.834 Å, is between the Cu(1) atom in the chain and the apical O(4) atom. The long Cu(2)–O(4) distance suggests that the coppers in the planes are only weakly linked to those in the chains via the apical oxygen. It has been suggested that the short

[45]G. Calestani and C. Rizzoli, *Nature* **328**, 606 (1987).

TABLE I. STRUCTURAL PARAMETERS FOR ORTHORHOMBIC $Y_1Ba_2Cu_3O_{6.91}$ DETERMINED FROM A JOINT REFINEMENT OF X-RAY AND NEUTRON POWDER DIFFRACTION DATA TAKEN AT 297 K. THE NUMBERS IN PARENTHESES ARE THE ESTIMATED STANDARD DEVIATIONS IN THE LAST SIGNIFICANT DIGIT(S). (WILLIAMS et al.[43])

ATOM	SPACE GROUP x	Pmmm, y	z	$a = 3.82030(8)$ Å, $U_{11}(\text{Å}^2)$	$b = 3.88548(10)$ Å, $U_{22}(\text{Å}^2)$	$c = 11.68349(23)$ Å $U_{33}(\text{Å}^2)$	OCCUPANCY
Y	$\frac{1}{2}$	$\frac{1}{2}$	$\frac{1}{2}$	0.0085(8)	0.0106(8)	0.0085(6)	1
Ba	$\frac{1}{2}$	$\frac{1}{2}$	0.18393(6)	0.0078(6)	0.0096(7)	0.0198(5)	1
Cu(1)	0	0	0	0.0080(9)	0.0115(9)	0.0150(7)	1
Cu(2)	0	0	0.35501(8)	0.0033(5)	0.0036(5)	0.0207(5)	1
O(1)	0	$\frac{1}{2}$	0	0.0161(16)	0.0104(11)	0.0080(14)	0.910(8)
O(2)	$\frac{1}{2}$	0	0.37819(15)	0.0039(6)	0.0068(7)	0.0203(11)	1
O(3)	0	$\frac{1}{2}$	0.37693(16)	0.0109(8)	0.0084(7)	0.0056(11)	1
O(4)	0	0	0.15840(13)	0.0162(11)	0.0123(9)	0.0097(7)	1

TABLE II. SELECTED BOND LENGTHS (Å) AND ANGLES (°) IN ORTHORHOMBIC $Y_1Ba_2Cu_3O_7$. THE NUMBERS IN PARENTHESES ARE ESTIMATED STANDARD DEVIATIONS. (CALESTANI AND RIZZOLI[45])

Y–O(2)	$4 \times 2.418(15)$	Cu(1)–O(1)	$2 \times 1.947(5)$
Y–O(3)	$4 \times 2.399(15)$	Cu(1)–O(4)	$2 \times 1.834(27)$
Ba–O(1)	$2 \times 2.891(2)$	Cu(2)–O(2)	$2 \times 1.929(3)$
Ba–O(2)	$2 \times 2.980(19)$	Cu(2)–O(3)	$2 \times 1.961(3)$
Ba–O(3)	$2 \times 2.948(19)$	Cu(2)–O(4)	$2.341(28)$
Ba–O(4)	$4 \times 2.750(3)$		
O(1)–Cu(1)–O(4)		90.0(5)	
O(4)–Cu(1)–O(4)		180.0(7)	
O(1)–Cu(1)–O(1)		180.0(0)	
O(3)–Cu(2)–O(3)		166.3(1)	
O(2)–Cu(2)–O(2)		165.3(7)	
O(2)–Cu(2)–O(3)		89.1(0)	
O(4)–Cu(2)–O(3)		96.9(6)	
O(4)–Cu(2)–O(2)		97.4(6)	

Cu(1)–O(4) bond distance is a consequence of preferential location of Cu^{3+} ions in the chain sites.

In addition to the refinements done at room temperature, a number of refinements were conducted at lower temperatures in order to investigate any structural changes that occur on cooling.[33,35,36,40] As will be discussed in Section 4, these initial studies found no substantial changes in the basic orthorhombic structure down to temperatures as low as 5 K, but several anomalies were reported in subsequent studies.

The oxygen content in the 123 structure depends strongly on the processing conditions used to form the material. An O_7 stoichiometry is only reached if samples are slowly cooled in oxygen. Quenching from high temperatures or cooling in a reducing atmosphere results in a more oxygen deficient structure. If sufficient oxygen is removed, the structure undergoes an orthorhombic-to-tetragonal phase transformation. This transformation will be described in greater detail in Section 1. Variability in oxygen content due to different heat treatments can account for many of the differences in unit cell dimensions and atomic positions reported in early x-ray refinements. For example, the report of a tetragonal unit cell by Hazen et al.[13] is consistent with an oxygen deficient structure.

Rietveld refinements of neutron diffraction data taken from oxygen deficient materials show that oxygen is lost primarily from the O(1) site in

TABLE III. ATOM POSITIONS AND OCCUPANCIES FOR TETRAGONAL $Y_1Ba_2Cu_3O_{6.06}$. THE O(2) AND O(3) POSITIONS ARE EQUIVALENT IN THE TETRAGONAL STRUCTURE. (SANTORO et al.[48])

| ATOM | SPACE GROUP $P4/m\,mm$, $a = b = 3.8570(1)$ Å, $c = 11.8194(3)$ Å | | | | |
	x	y	z	B_{iso} (Å^2)	OCCUPANCY
Y	$\frac{1}{2}$	$\frac{1}{2}$	$\frac{1}{2}$	0.73(4)	1
Ba	$\frac{1}{2}$	$\frac{1}{2}$	0.1952(2)	0.50(4)	1
Cu(1)	0	0	0	1.00(4)	1
Cu(2)	0	0	0.3607(1)	0.49(3)	1
O(1)	0	$\frac{1}{2}$	0	0.9	0.028(4)
O(2)	0	$\frac{1}{2}$	0.3791(1)	0.73(4)	1
O(4)	0	0	0.1518(2)	1.25(6)	0.990(6)

the chains.[39,46-50] In reduced material with six oxygens per unit cell the chain site is completely empty, which reduces the coordination of the Cu(1) atom to two. This copper coordination is similar to that found in delafossites, which has led to the suggestion that reduced 123 contains Cu^{2+} ions in the planes and Cu^{1+} ions in the two-fold coordinated sites.[48,49] Refined cell parameters, atom coordinates, and selected bond lengths for the reduced 123 structure are given in Tables III and IV.[48] As can be seen by comparing the bond length data in Tables II and IV, the long bond length between the Cu(2) and O(4) atoms lengthens and the short bond length between the Cu(1) and O(4) atoms shortens in the reduced material. This indicates a further decrease in the coupling between Cu(2) planes and the Cu(1) site. The bond length changes are reflected in the unit cell dimensions as a lengthening of the c axis of the cell. Because the average of the a and b axes in the orthorhombic structure is about the same as the a axis of the tetragonal structure, the volume of the unit cell for the O_6 material is also increased. Two structural studies[39,49] reported cation disorder on the Ba and Y sites in

[46]A. Renault, G. J. McIntyre, G. Collin, J. P. Pouget, and R. Comes, *J. de Phys.* **48,** 1407 (1987).

[47]J. D. Jorgensen, M. A. Beno, D. G. Hinks, L. Soderholm, K. J. Volin, R. L. Hitterman, J. D. Grace, I. K. Schuller, C. U. Segre, K. Zhang, and M. S. Kleefisch, *Phys. Rev. B* **36,** 3608 (1987).

[48]A. Santoro, S. Miraglia, F. Beech, S. A. Sunshine, D. W. Murphy, L. F. Schneemeyer, and J. V. Waszczak, *Mat. Res. Bull.* **22,** 1007 (1987).

[49]P. Bordet, C. Chaillout, J. J. Capponi, J. Chenavas, and M. Marezio, *Nature* **327,** 687 (1987).

[50]C. C. Torardi, E. M. McCarron, P. E. Bierstedt, A. W. Sleight, and D. E. Cox, *Solid State Commun.* **64,** 497 (1987).

TABLE IV. SELECTED BOND LENGTHS (Å) IN
TETRAGONAL $Y_1Ba_2Cu_3O_{6.06}$. (SANTORO et al.[48])

Y–O(2)	$8 \times 2.4004(8)$
Ba–O(2)	$4 \times 2.905(1)$
Ba–O(4)	$4 \times 2.7751(5)$
Cu(1)–O(4)	$2 \times 1.795(2)$
Cu(2)–O(2)	$4 \times 1.9406(3)$
Cu(2)–O(4)	$2.469(2)$

reduced 123, but there have been no subsequent investigations of this possibility.

One final point regarding the basic structure of 123 that has not been fully resolved concerns the space group of the structure. The final refinements in x-ray and neutron diffraction studies[25,51–56] of the ortho-rhombic structure have all been conducted in the space group Pmmm. Attempts to refine in other space groups including the Pmm2 space group give either comparable or poorer fits to the data.[34,35,41] A number of studies[25,51–56] of the space group have been conducted using convergent beam electron diffraction,[57] which is sensitive to even small deviations from symmetry. These studies have led to conflicting results. Several authors have found evidence for the absence of one of the mirror planes parallel to the c axis of the structure, which would reduce the space group to Pmm2.[25,51,52,56] In one study this absence was attributed to the presence of defects in the crystals being studied.[52] However, in cases where care was taken to avoid such artifacts the Pmm2 space group was still observed in some crystals.[25,26] Moodie and Whitfield[54] suggest that the loss of a mirror plane normal to the a axis could be caused by the presence of oxygen vacancy strings along the b axis in oxygen deficient material. The distinction between the two space groups is of considerable interest as the Pmm2 space group is not centrosymmetric and therefore

[51]C. H. Chen, D. J. Werder, S. H. Liou, J. R. Kwo, and M. Hong, Phys. Rev. B **35**, 8767 (1987).

[52]D. J. Eaglesham, C. J. Humphreys, N. McN. Alford, W. J. Clegg, M. A. Harmer, and J. D. Birchall, Appl. Phys. Lett. **51**, 457 (1987).

[53]M. Tanaka, M. Terauchi, K. Tsuda, and A. Ono, Jpn. J. Appl. Phys. **26**, L1237 (1987).

[54]A. F. Moodie and H. J. Whitfield, Ultramicroscopy **24**, 329 (1988).

[55]X. D. Zou, C. Y. Yang and Y. Q. Zhou, J. Electron Microsc. Tech. **7**, 269 (1987).

[56]J. Zou, D. J. H. Cockayne, G. J. Auchterlonie, D. R. McKenzie, S. X. Dou, A. J. Bourdillon, C. C. Sorrell, K. E. Easterling, and A. W. S. Johnson, Phil. Mag. Lett. **57**, 157 (1988).

[57]B. F. Buxton, J. A. Eades, J. W. Steeds, and G. M. Rackham, Phil. Trans. Roy. Soc. A **281**, 171, (1976).

allows for phenomena such as ferroelectricity, which has been speculated to occur in the 123 structure.[58–60]

III. Phase Transformations and Oxygen Arrangements

The structure and properties of $Y_1Ba_2Cu_3O_{7-\delta}$ that are observed at and below room temperature depend critically on how the material is processed at higher temperatures. This situation arises because the superconducting properties are largely controlled by the oxygen content and arrangement in $Y_1Ba_2Cu_3O_{7-\delta}$, which, in turn, are controlled by the annealing times and temperatures, the oxygen partial pressures, and the quench rates used in preparing the material. In this section we review the many studies of the orthorhombic-to-tetragonal phase transition in 123 and summarize the oxygen arrangements that have been observed. We then use theoretical phase diagrams to try to tie these two areas together. Lastly, we briefly discuss some of the structural changes that have been reported to occur below room temperature.

1. THE ORTHORHOMBIC-TO-TETRAGONAL PHASE TRANSITION

The orthorhombic-to-tetragonal phase transition in 123 was initially identified by electron beam heating in a transmission electron microscope[25] (TEM) and was subsequently studied in situ by numerous techniques.[47,61–71] Hot-stage x-ray diffraction[61,62] showed that the lattice

[58]Z. Yang, J. Zhu, and Y. Xu, *J. Phys. C* **20**, L843 (1987).

[59]Z. J. Yang, J. Zhu, and Y. H. Xu, *Mater. Lett.* **6**, 19 (1987).

[60]S. K. Kurtz, L. E. Cross, N. Setter, D. Knight, A. Bhalla, W. W. Cao, and W. N. Lawless, *Mater. Lett.* **6**, 317 (1988).

[61]I. K. Schuller, D. G. Hinks, M. A. Beno, D. W. Capone II, L. Soderholm, J. P. Locquet, Y. Bruynserade, C. U. Segre, and K. Zhang, *Solid State Commun.* **63**, 385 (1987).

[62]R. Beyers, G. Lim, E. M. Engler, V. Y. Lee, M. L. Ramirez, R. J. Savoy, R. D. Jacowitz, T. M. Shaw, S. LaPlaca, R. Boehme, C. C. Tsuei, S. I. Park, M. W. Shafer, and W. J. Gallagher, *Appl. Phys. Lett.* **51**, 614 (1987).

[63]P. K. Gallagher, H. M. O'Bryan, S. A. Sunshine, and D. W. Murphy, *Mat. Res. Bull.* **22**, 995 (1987).

[64]J. M. Tarascon, W. R. McKinnon, L. H. Greene, G. W. Hull, B. G. Bagley, E. M. Vogel, and Y. LePage, *in* "Extended Abstracts on High Temperature Superconductors," (D. U. Gubser and M. Schluter, eds.), p. 65. Materials Research Society, Pittsburgh, PA, 1987.

[65]K. Kishio, J. Shimoyama, T. Hasegawa, K. Kitazawa, and K. Fueki, *Jpn. J. Appl. Phys.* **26**, L1228 (1987).

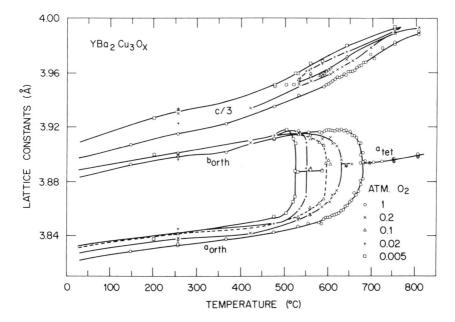

FIG. 2. Lattice constants versus temperature for various oxygen partial pressures. The orthorhombic-to-tetragonal transformation temperature decreases with decreasing oxygen pressure. The lines are to guide the eye. (Specht et al.[67])

parameters of 123 expanded linearly up to 500°C, with a thermal expansion along the c axis that is nearly twice that along the a and b axes. Above 500°C, however, the b axis contracts and the a axis expands at a supralinear rate. Finally, the a and b axes become equal and 123 is tetragonal. The temperatures at which the thermal expansion deviates from linearity and at which the orthorhombic-to-tetragonal transition occur are very dependent on the oxygen partial pressure[47,67] (Fig. 2). Thermogravimetric studies[62–67] have shown that O_7 starting material begins to lose oxygen reversibly above ~350–400°C. The

[66]S. Yamaguchi, K. Terabe, A. Saito, S. Yahagi, and Y. Iguchi, Jpn. J. Appl. Phys **27**, L179 (1988).

[67]E. D. Specht, C. J. Sparks, A. G. Dhere, J. Brynestad, O. B. Cavin, D. M. Kroeger, and H. A. Oye, Phys. Rev. B **37**, 7426 (1988).

[68]P. P. Freitas and T. S. Plaskett, Phys. Rev. B **36**, 5723 (1987).

[69]G. Van Tendeloo, H. W. Zandbergen, and S. Amelinckx, Solid State Commun. **63**, 389 (1987).

[70]G. Van Tendeloo, H. W. Zandbergen, and S. Amelinckx, Solid State Commun. **63**, 603 (1987).

[71]G. Van Tendeloo and S. Amelinckx, Phys. Stat. Sol. (a) **103**, K1 (1987).

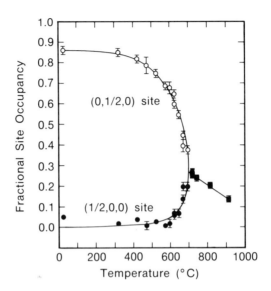

FIG. 3. Fractional oxygen site occupancies in the basal copper plane versus temperature for a sample heated in one atmosphere oxygen. (Jorgensen *et al.*[47])

orthorhombic-to-tetragonal transition is assumed to occur where there is a discontinuity in the curvature of weight loss versus temperature plots.[63]

In situ neutron diffraction[47] showed that the observed temperature-dependent changes in lattice parameters are caused by changes in oxygen content and order on the basal copper plane (between the barium layers) (Fig. 3). The oxygen that is lost above ~400°C comes primarily from the $O(1)$ $[= (0, 1/2, 0)]$ site, resulting in the negative thermal expansion of the b axis before the transition. Part of the oxygen that is removed from the $O(1)$ site goes on the normally vacant $(1/2, 0, 0)$ site, resulting in the enhanced expansion along the a axis just below the transition. These results show that the oxygen arrangement changes from fully ordered at room temperature, to partially ordered at elevated temperatures in the orthorhombic phase, to completely disordered at still higher temperatures in the tetragonal phase. Oxygen continues to be depleted from the basal copper plane above the transition. The first neutron diffraction study[47] found that the transition always occurs near an oxygen stoichiometry of 6.5. More recent studies[66,67] have concluded that the oxygen stoichiometry at the transition varies with oxygen partial pressure. An in situ x-ray diffraction and thermogravimetric study[67] found that the oxygen content at the transition decreases from 6.66 ± 0.01 to 6.59 ± 0.02 as the oxygen partial pressure is reduced from 1.0 atm to 5×10^{-3} atm

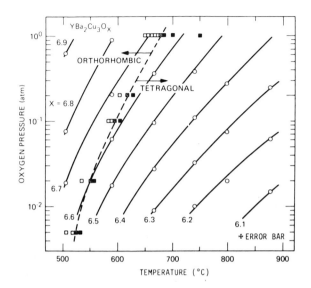

FIG. 4. Structural phase diagram for 123. Open squares indicate orthorhombic x-ray diffraction pattern, filled squares indicate tetragonal, and x-filled squares indicate the point of transformation. The oxygen content at the orthorhombic-to-tetragonal transition and the transition temperature both decrease with decreasing oxygen partial pressure. (Specht *et al.*[67])

(Fig. 4). Concurrently, the transition temperature decreases from 676 ± 5 to $621 \pm 10°C$. Lower oxygen pressures can reduce further the oxygen content at the transition. For example, samples prepared in 5×10^{-4} atm oxygen at 500°C by a solid-state ionic technique remained orthorhombic with an oxygen content of just 6.41 ± 0.02.[72]

There is no doubt that the orthorhombic-to-tetragonal transition is an order-disorder transformation. However, it is not yet clear whether the phase change is a classical Gibbsian first-order phase change or is a second-order transformation. Hysteresis observed in early hot-stage x-ray diffraction studies was interpreted as evidence for a first-order phase change.[61] Moreover, several x-ray diffraction studies found evidence for coexistence of orthorhombic and tetragonal 123, again indicative of a first-order phase change.[63] Conversely, others saw little or no hysteresis in their x-ray diffraction,[67] neutron diffraction,[47] or resistivity[68] studies of the transition and concluded that the phase change is second-order.

[72]R. Beyers, E. M. Engler, P. M. Grant, S. S. P. Parkin, G. Lim, M. L. Ramirez, K. P. Roche, J. E. Vazquez, V. Y. Lee, R. D. Jacowitz, B. T. Ahn, T. M. Gür, and R. A. Huggins, *Mater. Res. Soc. Symp. Proc.* **99**, 77 (1987).

Differential scanning calorimetry studies did not detect a latent heat at the transition. The difficulties encountered here in determining the order of the transition are analogous to those encountered in studies of order-disorder transformations in metal alloys. Rhines and Newkirk[73] provide an excellent discussion of the difficulties in making such a determination for metal alloys and conclude that order-disorder transformations in alloys are normal first-order phase changes if studied under true equilibrium conditions. More careful studies than those performed to date are required before the order of the transition can be definitively determined for 123.

2. ORDERED OXYGEN ARRANGEMENTS

Soon after the orthorhombic $Y_1Ba_2Cu_3O_7$ and tetragonal $Y_1Ba_2Cu_3O_6$ phases were identified, researchers began to investigate the structures and properties of $Y_1Ba_2Cu_3O_{7-\delta}$ samples with intermediate oxygen contents. Many studies[74–79] prepared these samples by rapidly quenching the high-temperature, low-oxygen-content tetragonal phase from 600–1000°C to room temperature or liquid nitrogen temperature. These studies usually found a smooth variation of T_c with oxygen content with the samples becoming insulators near $O_{6.5}$, corresponding to the orthorhombic-to-tetragonal phase transition.[74] Alternatively, intermediate oxygen content samples were prepared by removing oxygen at lower temperatures (typically 500°C) from O_7 starting material. These studies employed a variety of methods to remove the oxygen, including annealing in reducing atmospheres,[79,80] equilibrating O_6 and O_7 123 mixtures in sealed quartz tubes,[81] gettering with zirconium in a sealed

[73]F. N. Rhines and J. B. Newkirk, *Trans. ASME* **45,** 1029 (1953).

[74]J. D. Jorgensen, B. W. Veal, W. K. Kwok, G. W. Crabtree, A. Umezawa, L. J. Nowicki, and A. P. Paulikas, *Phys. Rev. B* **36,** 5731 (1987).

[75]E. Takayama-Muromachi, Y. Uchida, M. Ishi, T. Tanaka, and K. Kato, *Jpn. J. Appl. Phys.* **26,** L1156 (1987).

[76]H. Nozaki, Y. Ishizawa, O. Fukunaga, and H. Wada, *Jpn. J. Appl. Phys.* **26,** L1180 (1987).

[77]M. Tokumoto, H. Ihara, T. Matsubara, M. Hirabayashi, N. Terada, H. Oyanagi, K. Murata, and Y. Kimura, *Jpn. J. Appl. Phys.* **26,** L1565 (1987).

[78]S. Nakanishi, M. Kogachi, H. Sasakura, N. Fukuoka, S. Minamigawa, K. Nakahigashi, and A. Yanase, *Jpn. J. Appl. Phys.* **27,** L329 (1988).

[79]W. E. Farneth, R. K. Borida, E. M. McCarron III, M. K. Crawford, and R. B. Flippen, *Solid State Commun.* **66,** 953 (1988).

[80]P. Monod, M. Ribault, F. D'Yvoire, J. Jegoudez, G. Collin, and A. Revcolevschi, *J. de Physique* **48,** 1369 (1987).

[81]D. C. Johnston, A. J. Jacobson, J. M. Newsam, J. T. Lewandowski, D. P. Goshorn, D. Xie, and W. B. Yelon, *ACS Symposium Series* **351:** "Chemistry of High-Temperature Superconductors," 136 (1987).

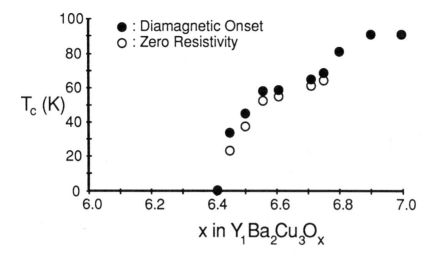

FIG. 5. Superconducting transition temperature vs oxygen content for samples prepared at 500°C by a solid-state ionic technique. (Beyers *et al.*[83])

tube,[82] and titrating electrochemically in a solid-state ionic cell.[72,83] The variation of T_c with oxygen content in samples prepared at lower temperatures is different from that observed in samples quenched from the tetragonal phase. The superconducting transition temperature does not vary smoothly with oxygen content in these samples (Fig. 5). Instead, T_c remains at 91 K between O_7 and $O_{6.9}$, then falls to a ~60 K "plateau" between $O_{6.7}$ and $O_{6.6}$, and finally drops to zero (i.e., not superconducting) between $O_{6.5}$ and $O_{6.4}$.

Complementary structural studies have tried to establish a link between the variations in T_c and variations in the oxygen structure. Numerous electron diffraction studies[70–72,84–88] found evidence for addi-

[82]R. J. Cava, B. Batlogg, C. H. Chen, E. A. Rietman, S. M. Zahurak, and D. Werder, *Phys. Rev. B* **36**, 5719 (1987).

[83]R. Beyers, G. Gorman, P. M. Grant, V. Y. Lee, R. M. Macfarlane, S. S. P. Parkin, S. J. LaPlaca, B. T. Ahn, T. M. Gür, and R. A. Huggins (unpublished).

[84]C. Chaillout, M. A. Alario-Franco, J. J. Capponi, J. Chenavas, J. L. Hodeau, and M. Marezio, *Phys. Rev. B* **36**, 7118 (1987).

[85]S. S. P. Parkin, E. M. Engler, V. Y. Lee, and R. B. Beyers, *Phys. Rev. B* **37**, 131 (1988).

[86]M. A. Alario-Franco, J. J. Capponi, C. Chaillout, J. Chenavas, and M. Marezio, *Mater. Res. Soc. Symp. Proc.* **99**, 41 (1988).

[87]Y. Kubo, T. Ichihashi, T. Manako, K. Baba, J. Tabuchi, and H. Igarashi, *Phys. Rev. B* **37**, 7858 (1988).

[88]C. H. Chen, D. J. Werder, L. F. Schneemeyer, P. K. Gallagher, and J. V. Waszczak, *Phys. Rev. B* **38**, 2888 (1988).

tional oxygen vacancy ordering near $O_{6.5}$. In [001] zone axis patterns, superlattice spots appear with wave vector $q = [1/2, 0, 0]$, corresponding to a doubling of the unit cell along the a axis in real space (Fig. 6(a)). The superlattice spots elongate into bands along the a^* direction in rapidly quenched samples. Tilting experiments revealed that the extra spots are in fact diffuse streaks running parallel to the c^* axis (Fig. 6(b)). The observed diffraction effects are consistent with an $O_{6.5}$ structure in

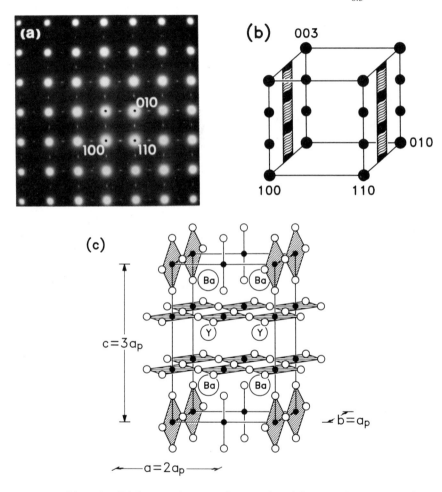

FIG. 6. (a) Typical [001] zone axis pattern from a twinned $O_{6.5}$ sample prepared at 500°C by a solid-state ionic technique. Sharp $q = [1/2, 0, 0]$ superlattice reflections are present along a^*. (Beyers et al.[83]) (b) Schematic of the reciprocal lattice obtained by electron diffraction tilting experiments and (c) proposed real space structure for the $O_{6.5}$ phase. (Alario-Franco et al.[86])

which oxygen is removed from the O(1) position in every other Cu(1)–O(1) linear chain in the 123 structure, thereby creating strings of oxygen vacancies along the b axis in the basal copper plane and doubling the unit cell along the a axis (Fig. 6(c)). Presumably the vacancy strings are not perfectly ordered in rapidly quenched material or in material that is off the $O_{6.5}$ stoichiometry, thus causing the superlattice spots to elongate into bands along the a^* direction in diffraction patterns from these samples. The diffuse streaking parallel to c^* indicates that the vacancy strings are uncorrelated between adjacent basal copper planes, which is not surprising given the layered nature of the 123 structure. These electron diffraction patterns have been observed in samples with overall oxygen contents ranging from $O_{6.4}$ to $O_{6.7}$, and it is widely believed that this structure is responsible for the "plateau" in T_c between $O_{6.6}$ and $O_{6.7}$. This speculation will be discussed in greater detail in the next section.

Three additional types of superlattices have been observed in electron diffraction studies. Alario-Franco et al.[86,89] observed continuous rods running parallel to c^* that give superlattice reflections with wave vector $q = [\pm 1/4, \pm 1/4, 0]$ in [001] zone axis patterns of orthorhombic $O_{6.85}$ and tetragonal $O_{6.15}$ samples. For the $O_{6.85}$ crystals, the superlattice was attributed to the ordered removal of every fourth O(1) atom from every other Cu(1)–O(1) linear chain in the basal copper plane, with the ordering being uncorrelated between the basal copper planes (Fig. 7). The ordering scheme is the same in the $O_{6.15}$ crystals, but the roles of oxygen and oxygen vacancies are reversed. Werder et al.[90] observed a superlattice with wave vector $q = [2/5, 0, 0]$ in $O_{6.72}$ material in addition to the more commonly reported $q = [1/2, 0, 0]$ superlattice, and Mitchell et al.[91] found a transient superlattice with wave vector $q = [1/3, 0, 0]$ in reduced samples. The electron diffraction patterns provide an existence proof for each of these superlattices. However, because they have not been studied as extensively as the $q = [1/2, 0, 0]$ superlattice, the precise conditions needed to form these superstructures reproducibly have yet to be determined. All of the superlattices observed indicate that intermediate oxygen stoichiometries in 123 are accommodated by ordered oxygen vacancy arrangements if the material is given time to equilibrate. A simplistic chemical explanation for this behavior is that the ordered

[89]M. A. Alario-Franco, C. Chaillout, J. J. Capponi, and J. Chenavas, *Mat. Res. Bull.* **22,** 1685 (1987).

[90]D. J. Werder, C. H. Chen, R. J. Cava, and B. Batlogg, *Phys. Rev. B* **37,** 2317 (1988).

[91]T. E. Mitchell, T. Roy, R. B. Schwarz, J. F. Smith, and D. Wohlleben, *J. Electron Microsc. Tech.* **8,** 317 (1988).

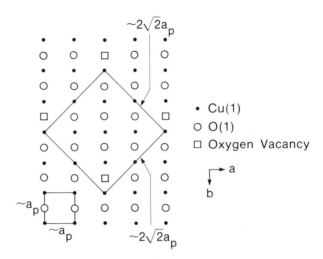

FIG. 7. Proposed oxygen arrangement in the basal copper plane of $O_{6.875}$ material. (Alario-Franco et al.[89])

arrangements minimize the number of copper atoms that are three-fold coordinated to oxygen, which is an energetically unfavorable state.[92]

When oxygen is removed from O_7 starting material at low temperatures, the loss of superconductivity does not coincide precisely with the orthorhombic-to-tetragonal phase transition. In these samples, T_c typically drops to zero between $O_{6.5}$ and $O_{6.4}$, but the orthorhombic-to-tetragonal transition does not occur until $O_{6.40-6.35}$. Several bond lengths change rapidly just before the transition, such as the Cu(1)–O(4) and Cu(2)–O(4) bonds. Consequently, it has been suggested that these bonds may play a role in superconductivity. However, because several distances are changing simultaneously, it is not clear at the present time which, if any, is critical for high T_c. A comparison with bond length changes in related superconducting structures may help to determine if there is more to these empirical observations. Finally, we note that a solid-state ionic study[72] concluded that oxygen inhomogeneity is essential for observing small amounts of superconductivity in samples with average oxygen contents below 6.5. It is currently speculated that these inhomogeneities arise from stress-induced oxygen diffusion, which in turn results from the anisotropic thermal expansion in 123.

[92]J. L. Hodeau, C. Chaillout, J. J. Capponi, and M. Marezio, *Solid State Commun.* **64,** 1349 (1987).

3. Phase Diagrams

The many structural changes that occur in 123 as a function of processing are most simply understood and discussed using phase diagrams. Pseudoternary[93-95] and quaternary[96,97] phase diagrams describing the equilibria between 123 and the other phases in the Y–Ba–Cu–O system have been determined by several groups. In this section, we restrict our discussion to the theoretical pseudobinary phase diagrams that have been used to explain variations in the oxygen substructure that occur with processing.[98-105] Most of these calculated diagrams use two-dimensional Ising models to explain the oxygen arrangements in the basal copper plane as a function of oxygen content and temperature. They neglect any changes in the remainder of the 123 structure.

Treating only nearest-neighbor interactions between oxygen atoms, Bakker et al.[98] derived an open-system order-disorder model of the orthorhombic-to-tetragonal phase transition that explained the observed variations in lattice constants with temperature. By incorporating a concentration-dependent heat of oxygen solution, Salomons et al.[99] extended Bakker's treatment to explain oxygen pressure-composition isotherms in 123. Wille and de Fontaine[102] analyzed the stability of a two-dimensional Ising model with one nearest-neighbor interaction, V_1, and two second-nearest-neighbor interactions, V_2 and V_3 (Fig. 8(a)). Stable oxygen arrangements for average oxygen stoichiometries of 7.0 and 6.5 were derived as a function of two phenomenological parameters, the ratios V_2/V_1 and V_3/V_1 (Fig. 8(b) and (c)). Assuming the effective pair

[93]K. G. Frase, E. G. Liniger, and D. R. Clarke, *J. Amer. Ceram. Soc.* **70**, C204 (1987).

[94]R. S. Roth, K. L. Davis, and J. R. Dennis, *Adv. Ceram. Mater.* **2**, 303 (1987).

[95]G. Wang, S. J. Hwu, S. N. Song, J. B. Ketterson, L. D. Marks, K. R. Poeppelmeier, and T. O. Mason, *Adv. Ceram. Mater.* **2**, 313 (1987).

[96]B. T. Ahn, T. M. Gür, R. A. Huggins, R. Beyers, and E. M. Engler, *Mater. Res. Soc. Symp. Proc.* **99**, 171 (1987).

[97]B. T. Ahn, T. M. Gür, R. A. Huggins, R. Beyers, E. M. Engler, P. M. Grant, S. S. P. Parkin, G. Lim, M. L. Ramirez, K. P. Roche, J. E. Vazquez, V. Y. Lee, and R. D. Jacowitz, *Physica C.* **153-155**, 590 (1988).

[98]H. Bakker, D. O. Welch, and O. W. Lazareth, Jr., *Solid State Commun.* **64**, 237 (1987).

[99]E. Salomons, N. Koeman, R. Brouwer, D. G. de Groot, and R. Griessen, *Solid State Commun.* **64**, 1141 (1987).

[100]J. M. Bell, *Phys. Rev. B.* **37**, 541 (1988).

[101]D. de Fontaine, L. T. Wille, and S. C. Moss, *Phys. Rev. B* **36**, 5709 (1987).

[102]L. T. Wille and D. de Fontaine, *Phys. Rev. B* **37**, 2227 (1988).

[103]L. T. Wille, A. Berera, and D. de Fontaine, *Phys. Rev. Lett.* **60**, 1065 (1988).

[104]A. G. Khachaturyan, S. V. Semenovskaya, and J. W. Morris, Jr., *Phys. Rev. B* **37**, 2243 (1988).

[105]A. G. Khachaturyan and J. W. Morris, Jr., *Phys. Rev. Lett.* **61**, 215 (1988).

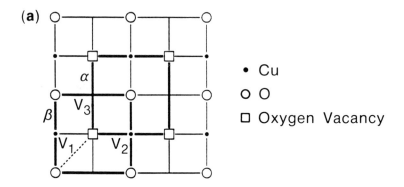

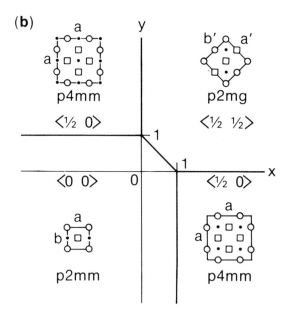

FIG. 8. (a) Ising model for the basal copper plane in $Y_1Ba_2Cu_3O_7$. V_2 and V_3 are second-nearest-neighbor interactions, with V_2 mediated by copper. (Willie et al.[103]) Stable ground states as a function of $x = V_2/V_1$ and $y = V_3/V_1$ for (b) O_7 and (c) $O_{6.5}$ (Wille and de Fontaine[102]).

(c)

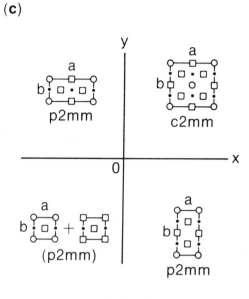

FIG. 8. (c)

interactions do not depend on oxygen content, the following restrictions must be placed on V_1, V_2, and V_3 to agree with the experimentally observed oxygen arrangements at $O_{6.5}$ and O_7: $V_1 > 0$, $V_2 < 0$, and $0 < V_3 < V_1$.

Wille et al.[103] arbitrarily set $V_1 = 1$, $V_2 = -0.5$, and $V_3 = 0.5$ to meet these restrictions and used the cluster variation method to calculate the phase diagram shown in Fig. 9. This diagram predicts that the orthorhombic-to-tetragonal transition is a second-order transition between points b and c, but is a first-order transition to the left of point b (i.e., for the transformation from the orthorhombic cell doubled along the a axis, *p2mm II*, to the disordered tetragonal phase, *p4mm*). Thus, the observation of a first-order or a second-order orthorhombic-to-tetragonal transition depends on how the phase diagram is traversed. The open circles in the diagram are the data reported by Specht et al.[67] One of these points was used to fit the estimated temperature scale on the right side of the drawing. The dotted phase boundaries below kT/V_1 ~0.8 are conjectured extrapolations. This diagram comes closest to matching the oxygen arrangements observed in 123 to date because it includes the orthorhombic phase near $O_{6.5}$ with the doubled unit cell.

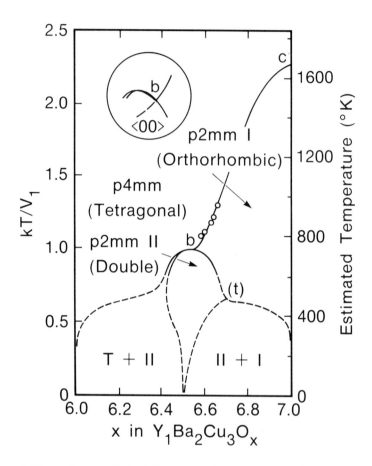

FIG. 9. Phase diagram obtained by the cluster variation method. Dotted lines are conjectured extensions. The diagram contains a bicritical point, b, and a tricritical point, t. Open circles are experimental data from Fig. 4. (Wille et al.[103])

Previous analyses equated the two second-nearest-neighbor interactions $(V_2 = V_3)$ and thus predicted a miscibility gap between tetragonal O_6 and orthorhombic O_7 phases at low temperatures and no distinct phase at $O_{6.5}$. Such a miscibility gap has never been observed, whereas there is widespread agreement on the existence of the $O_{6.5}$ phase. However, the phase diagram in Fig. 9 cannot be considered completely correct either, given the electron diffraction evidence for additional superstructures. It seems likely that some of these superstructures exist between $O_{6.5}$ and

O_7. Higher order interactions must be included in the Ising models to stabilize such structures. Alternatively, using a "maximum amplitude principle," Khachaturyan and Morris[105] explain all of the observed superstructures except that reported by Alario-Franco et al.[89] Khachaturyan and Morris[105] also propose that these superstructures are metastable transients that precede a spinodal decomposition into the O_6 and O_7 phases. As noted above, however, separation into O_6 and O_7 phases has never been observed experimentally.

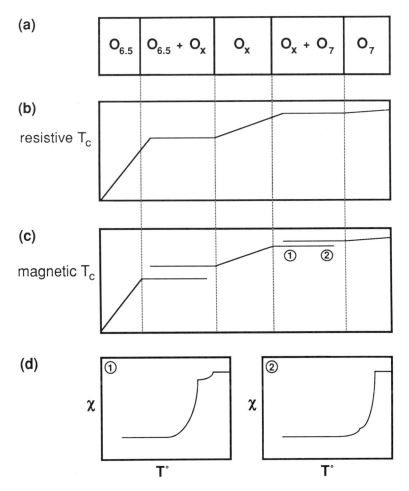

FIG. 10. For the pseudobinary diagram in (a), the expected variations in T_c measured by (b) resistivity and (c) susceptibility. In (d) susceptibility versus temperature plots are shown for points 1 and 2 in (c).

The variation of T_c with oxygen content in samples prepared at low temperatures may also point to the existence of additional phases between $O_{6.5}$ and O_7. If the "plateaus"[106] at 91 and ~60 K in the T_c-versus-O_x curve arise from crossing two-phase regions in the pseudobinary diagram, then there must be at least one additional ordered oxygen phase between $O_{6.5}$ and O_7. Figure 10 shows the expected variations in T_c measured by resistivity and susceptibility if an additional phase exists in the pseudobinary phase diagram at low temperature and the samples are prepared under equilibrium conditions. In the single-phase regions, T_c is allowed to vary as the oxygen content within the single phase varies. In the two-phase regions, however, the properties of the coexisting phases are fixed because the oxygen contents in the coexisting phases are fixed by the Gibbs phase rule (assuming that the pseudobinary approximation is valid). Consequently, the T_c measured resistively in the two-phase regions would be that of the higher oxygen content phase if there were connectivity between grains of this phase (Fig. 10(b)). Susceptibility data in the two-phase regions would show evidence for two phases with different T_c (Fig. 10(c)). The diamagnetic contributions of each phase would vary according to the lever rule as the two-phase region was crossed (Fig. 10(d)). If this pseudobinary interpretation is valid, then the "plateau" at ~60 K could be caused by either the additional phase between $O_{6.5}$ and O_7 or the $O_{6.5}$ phase. Moreover, it is also possible that the additional phase is in fact a series of phases with only small differences in oxygen content and superstructure.

4. LOW-TEMPERATURE STRUCTURE CHANGES

High superconducting transition temperatures are often associated with lattice instabilities brought on by strong electron-phonon coupling.[107] Consequently, many investigators have looked for structural changes in 123 below room temperature. While no substantial changes have been found, there have been a number of anomalies reported. A high-resolution x-ray scattering study by Horn et al.[108] detected an anomaly in the orthorhombic splitting, b-a, near the superconducting transition. They suggested that uniaxial anisotropy within the ab plane in the

[106]The word 'plateau' is put in quotes because the observed T_c varies somewhat in these regions, possibly due to insufficient equilibration time. Alternatively, the T_c variation in the "plateaus" could indicate that this simple phase rule explanation for the shape of the T_c versus oxygen curve is not correct.

[107]J. C. Phillips, Phys. Rev. Lett. 26, 543 (1971).

[108]P. M. Horn, D. T. Keane, G. A. Held, J. L. Jordan-Sweet, D. L. Kaiser, F. Holtzberg, and T. M. Rice, Phys. Rev. Lett. 59, 2772 (1987).

superconducting wave function might be the cause of this distortion. David et al.[109] also found an anomaly in the volume expansivity at T_c with neutron powder diffraction, but tentatively associated this anomaly with a change in the c axis, rather than with a change in the ab plane. Conversely, Francois et al.[44] verified the enhancement in the orthorhombic splitting at T_c with neutron powder diffraction and found evidence for a second anomaly at 240 K. Laegreid et al.[110] observed a specific heat anomaly at 90 K. Using ultrasound propagation, Bhattacharya et al.[111] found evidence for a phase transition at 120–130 K, probably occurring in the ab planes. In a related material with the 123 structure, $CaLaBaCu_3O_x$, both electron and neutron diffraction[112] found a phase transition between 160 and 230 K that increases the basic unit cell from a $1a_p \times 1a_p \times 3a_p$ cell to a $2a_p \times 2a_p \times 6a_p$ cell, where $a_p \sim 3.85$ Å (Fig. 11). The reproducibility of these effects, their root causes, and the implications for pairing mechanisms remain largely unresolved.

There is definitely a second type of phase transition occurring in 123 that may be important for pairing mechanisms not involving electron-phonon coupling, namely antiferromagnetic ordering of the magnetic moments on the copper ions. Antiferromagnetism was first detected by Nishida et al.[113] using muon spin relaxation and was subsequently studied by several groups using neutron diffraction.[114–118] There are differences in the magnetic structures derived by these groups. Tranquada et al.[114,115]

[109]W. I. F. David, P. P. Edwards, M. R. Harrison, R. Jones, and C. C. Wilson, *Nature* **331,** 245 (1988).

[110]T. Laegreid, K. Fossheim, E. Sandvold, and S. Julsrud, *Nature* **330,** 637 (1987).

[111]S. Bhattacharya, M. J. Higgins, D. C. Johnston, A. J. Jacobson, J. P. Stokes, D. P. Goshorn, and J. T. Lewandowski, *Phys. Rev. Lett.* **60,** 1181 (1988).

[112]Y. Tokura, S. J. La Placa, T. C. Huang, R. Beyers, R. F. Boehme, A. I. Nazzal, S. S. P. Parkin, and J. B. Torrance (unpublished).

[113]N. Nishida, H. Miyatake, D. Shimada, S. Okuma, M. Ishikawa, T. Takabatake, Y. Nakazawa, Y. Kuno, R. Keitel, J. H. Brewer, T. M. Riseman, D. L. Williams, Y. Watanabe, T. Yamazaki, K. Nishiyama, K. Nagamine, E. J. Ansaldo, and E. Torikai, *Jpn. J. Appl. Phys.* **26,** L1856 (1987).

[114]J. M. Tranquada, D. E. Cox, W. Kunnmann, H. Moudden, G. Shirane, M. Suenaga, P. Zolliker, D. Vaknin, S. K. Sinha, M. S. Alvarez, A. J. Jacobson, and D. C. Johnston, *Phys. Rev. Lett.* **60,** 156 (1988).

[115]J. M. Tranquada, A. H. Moudden, A. I. Goldman, P. Zolliker, D. E. Cox, G. Shirane, S. K. Sinha, D. Vaknin, D. C. Johnston, M. S. Alvarez, A. J. Jacobson, J. T. Kewandowski, and J. M. Newsam, *Phys. Rev. B* **38,** 2477 (1988).

[116]W. J. Li, J. W. Lynn, H. A. Mook, B. C. Sales, and Z. Fisk, *Phys. Rev. B* **37,** 9844 (1988).

[117]H. Kadowaki, M. Nishi, Y. Yamada, H. Takeya, H. Takei, S. M. Shapiro, and G. Shirane, *Phys. Rev. B* **37,** 7932 (1988).

[118]J. W. Lynn, W. H. Li, H. A. Mook, B. C. Sales, and Z. Fisk, *Phys. Rev. Lett.* **60,** 2781 (1988).

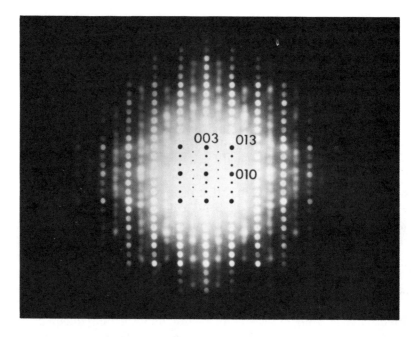

FIG. 11. [100] electron diffraction pattern from tetragonal $CaLaBaCu_3O_x$ at 77 K. The superlattice spots arise from a low-temperature phase transformation that doubles the unit cell along all three axes. (Tokura et al.[112]).

found that copper spins order antiferromagnetically in the CuO_2 planes, but not in the linear chains (i.e., not in the basal copper plane). More recent studies by Kadowaki et al.[117] and Lynn et al.[118] observed a second transition at lower temperature that they attributed to ordering of the spins in the basal copper plane, but the spin structures that these two groups deduced are not the same. All groups agree that the antiferromagnetic properties depend strongly on oxygen content. The Néel temperature for three-dimensional ordering in the CuO_2 planes varies from ~500 K at $O_{6.0}$ to ~0 K at $O_{6.4}$. Local antiferromagnetic ordering may exist in the superconducting state, but this has not been proven experimentally. A more detailed discussion of the magnetic properties of 123 and its derivatives is beyond the scope of this review.

IV. Defect Structures

X-ray and neutron diffraction are widely used to examine the average structure of 123 and its derivatives. Microscopic examination, however, shows that nearly all 123 materials contain defects. Transmission electron microscopy (TEM) has played a central role in determining the structure of these defects, but has some inherent limitations. High-resolution TEM has shown that while the location of the cations can be readily identified under the proper focus conditions,[25–29] the local oxygen structure is much more difficult to determine. Image simulations indicate that the images are only sensitive to the presence of oxygen in the basal copper plane over a narrow range of focus and are highly sensitive to the thickness of the crystal.[27,119–122] Moreover, the electron beam used to obtain the images can change the oxygen substructure.[25] In practice, these difficulties make analysis of the local oxygen distribution all but impossible. High-resolution studies have thus focussed on the analysis of defects in the cation sublattice. (However, as discussed in Section 2, ordered oxygen arrangements have been inferred from electron diffraction studies.) In this section we describe the structures of the various defects and briefly examine their effects on superconducting properties.

5. TWINS

The most commonly observed defects in 123 are the twins that arise from the orthorhombic-to-tetragonal phase transformation.[21,25,51–55,69,92,123–135] The twins form when oxygen ordering in the

[119]A. Ourmazd and J. C. H. Spence, *Nature* **329**, 425 (1987).

[120]A. Ourmazd, J. C. H. Spence, J. M. Zuo, and C. H. Li, *J. Electron Microsc. Tech.* **8**, 251 (1988).

[121]W. Krakow and T. M. Shaw, *J. Electron Microsc. Tech.* **8**, 273 (1988).

[122]N. P. Huxford, D. J. Eaglesham, and C. J. Humphreys, *Nature* **329**, 812 (1987).

[123]S. Nakahara, T. Boone, M. F. Yan, G. J. Fisanick, and D. W. Johnson, Jr., *J. Appl. Phys.* **63**, 451 (1988).

[124]E. A. Hewat, M. Dupuy, A. Bourret, J. J. Capponi, and M. Marezio, *Solid State Commun.* **64**, 517 (1987).

[125]M. Hervieu, B. Domenges, C. Michel, G. Heger, J. Provost, and B. Raveau, *Phys. Rev. B* **36**, 3920 (1987).

[126]A. I. Kingon, S. Chevacharoenkul, J. Mansfield, J. Brynestad, and D. G. Haase, *Adv. Ceram. Mater.* **2**, 678 (1987).

[127]M. Sarikaya, B. L. Thiel, I. A. Aksay, W. J. Weber, and W. S. Frydrych, *J. Mater. Res.* **2**, 736 (1987).

[128]S. Iijima, T. Ichihashi, Y. Kubo, and J. Tabuchi, *Jpn. J. Appl. Phys.* **26**, L1478 (1987).

[129]S. Iijima, T. Ichihashi, Y. Kubo, and J. Tabuchi, *Jpn. J. Appl. Phys.* **26**, L1790 (1987).

[130]J. C. Barry, *J. Electron Microsc. Tech.* **8**, 325 (1988).

[131]A. Brokman, *Solid State Commun.* **64**, 257 (1987).

[132]K. Lukaszewicz, J. Stepiendamm, R. Horyn, Z. Bukowski, and M. Kowalski, *J. Appl. Cryst.* **20**, 505 (1987).

[133]M. M. Fang, V. G. Kogan, D. K. Finnemore, J. R. Clem, L. S. Chumbley, and D. E. Farrell, *Phys. Rev. B* **37**, 2334 (1988).

[134]S. Takeda and S. Hikami, *Jpn. J. Appl. Phys.* **26**, L848 (1987).

[135]M. Sarikaya, R. Kikuchi, and I. A. Aksay, *Physica C* **152**, 161 (1988).

basal copper plane causes elongation of the b axis and contraction of the a axis. At the twin boundaries the cell edge with the enhanced oxygen content switches from one cell edge to the other, effectively rotating the unit cell through almost 90°. The two orientations share a common (110) plane in the twin boundary. Dense arrays of parallel twins with spacings from 20 nm to over 100 nm are observed. A typical example of parallel twins formed in 123 is shown in Fig. 12. Arrangements in which both {110} variants occur in the same grain are also frequently seen.[25,54,69,70,125,129,135] The misorientation across the twin boundary deviates from 90° by an amount ϕ, where $\phi = 2\arctan(b/a) - \pi/2$, which is on the order of 1° for the O_7 structure.

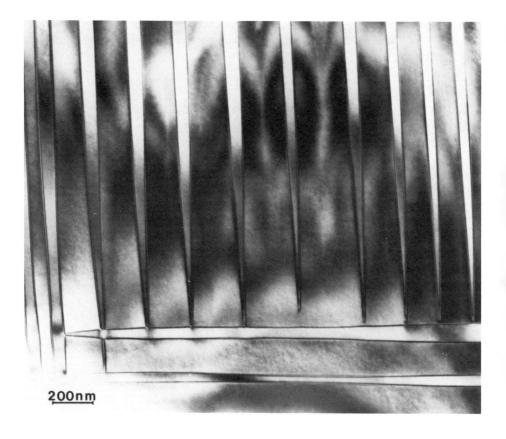

200nm

FIG. 12. Transmission electron micrograph of parallel twins formed on the {110} planes in $Y_1Ba_2Cu_3O_7$.

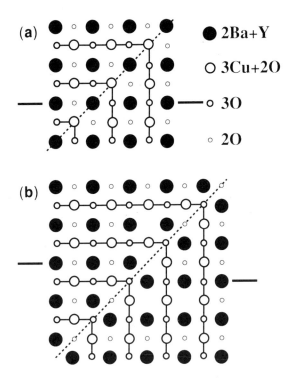

FIG. 13. Two models for twin boundaries in 123. In (a) both copper and oxygen ions are shared at the twin boundary whereas in (b) only oxygen ions are shared. Note the offset in the rows of (2Ba + Y) columns in model (b). (Hewat et al.[124]) Only model (a) is consistent with high-resolution TEM images of the twin boundaries.

Two models for the twin boundary structure have been proposed, one in which the twin plane contains both copper and oxygen and one in which it contains only oxygen.[54,124,125] Both models are shown schematically in Fig. 13. High-resolution images show that the planes containing columns of cations bend slightly on crossing the twin boundary but remain continuous.[54,124,130,136] The observation of continuous lattice planes is only consistent with the model in which the twin plane contains both copper and oxygen (Fig. 13(a)). Examination of the lattice plane bending in well-annealed material indicates that the lattice distortion associated with the twin boundary is confined to one or two unit cells at the boundary.[130,136] The high-resolution images do not reveal the location

[136]H. W. Zandbergen, G. Van Tendeloo, T. Okabe, and S. Amelinckx, Phys. Stat. Sol. (a) **103**, 45 (1987).

of the oxygen ions at the boundary. From the model in Fig. 13(a), however, it is evident that if all the oxygen sites in the two twin-related crystals are occupied, then the oxygens at the boundary share the same cell. Some repositioning of oxygens at the twin boundary plane therefore seems likely. It has also been suggested that the boundary is a likely site for peroxide formation.[131]

It is well known that the strain energy produced in transformations that cause small shear strains can be partially relieved by twinning.[137] The reduction in strain energy is balanced by the energy required to form the twin boundaries. This energy balance leads to an equilibrium spacing for the twins that depends on the size of the transformed region, the magnitude of the shear strain, and the twin boundary energy. For polycrystalline materials, when a complete phase transformation takes place, the relevant scale for the transformed region is the grain size.[138] The twin spacing, d, is then related to the grain size, G, by the relation

$$d = \sqrt{\frac{128\pi\gamma G}{ES_s^2}},$$

where E is the elastic modulus of the material, γ is the twin boundary energy, and S_s is a measure of the transformation strain and is related to the lattice parameters by $S_s = (a/b) - 1$. Measurements of twin spacings in different grain size materials are consistent with this relationship,[139] clearly demonstrating that the twin spacings in 123 are controlled by the transformation strains. This observation also explains why attempts to correlate twin spacings to the local oxygen content of a grain have failed, since the effect of grain size was not taken into consideration in these studies.[135,140]

While the cause of twinning in 123 is clear, the precise mechanism by which the transformation takes place and the twinning develops is not well-established. In samples rapidly quenched to below the transformation temperature, in situ observations found that a fine tweed-like microstructure develops that subsequently coarsens into a set of parallel twins.[70,129] On the other hand, in polycrystalline materials cooled slowly through the transformation temperature, dense patches of intersecting twin boundaries form in the final microstructure.[139] Similar twin struc-

[137]A. G. Khachaturyan, "Theory of Structural Transformations in Solids," John Wiley and Sons, New York, 1983.

[138]G. Arlt, D. Hennings, and G. de With, *J. Appl. Phys.* **58,** 1619 (1985).

[139]T. M. Shaw, S. L. Shinde, D. Dimos, R. F. Cook, P. R. Duncombe, and C. Kroll, *J. Mater. Res.* (1988) (in press).

[140]M. Sarikaya and E. A. Stern, *Phys. Rev. B* (1988) (submitted).

tures are seen in single crystals after oxygen treatment.[141–143] Convergent beam electron diffraction patterns and lattice images from twinned crystals also indicate that there are substantial local changes in the a and b lattice parameters in polycrystalline materials.[129,140,144] These variations have been interpreted as evidence that phase separation[103,104] may accompany the transformation.

Twins may affect a number of superconducting and normal state properties of 123. Several authors suggest that twinning could lead to an enhancement of T_c by either altering the ordering of oxygen in the twin plane,[145] altering the symmetry and periodicity of the crystal,[146] or changing the low-frequency phonon spectrum, which also alters the specific heat and normal state resistivity of the structure.[147] Indirect evidence for the enhancement of T_c by twins, similar to that reported in other superconducting materials,[148] has been obtained from magnetization experiments on field orientated samples[133] of $Y_1Ba_2Cu_3O_{7-\delta}$. The presence of a superconducting glass state, deduced from magnetization and microwave absorption experiments,[149,150] has also been attributed to the twins. In this case the twins are thought to act as weak links. Direct assessment of the effects of twins on superconducting properties, however, is hampered by the difficulty in controlling the twin structure without altering other features of the material, such as the grain size, doping, or oxygen content. In the absence of such controlled experiments, the effects of twins remain uncertain.

6. STACKING FAULTS

Several different fault structures occur on the (001) planes in 123. In high resolution images of the faults, an offset in the (100) planes of $a/2$ is

[141]H. A. Hoff, A. K. Singh, and C. S. Pande, *Appl. Phys. Lett.* **52,** 669 (1988).

[142]Y. Tajima, M. Hikita, A. Katsui, Y. Hidaka, T. Iwata, and S. Tsurumi, *J. Cryst. Growth* **85,** 665 (1987).

[143]K. Semba, H. Suzuki, A. Katsui, Y. Hidaka, M. Hikita, and S. Tsurumi, *Jpn. J. Appl. Phys.* **26,** L1645 (1987).

[144]Z. Hiroi, M. Takano, Y. Ikeda, Y. Takeda, and Y. Bando, *Jpn. J. Appl. Phys.* **27,** L141 (1988).

[145]A. Robledo and C. Varea, *Phys. Rev. B* **37,** 631 (1988).

[146]F. M. Mueller, S. P. Chen, M. L. Prueitt, J. F. Smith, J. L. Smith, and D. Wohlleben, *Phys. Rev. B* **37,** 5837 (1988).

[147]B. Horovitz, G. R. Barsch, and J. A. Krumhansl, *Phys. Rev. B* **36,** 8895 (1987).

[148]I. N. Khlustikov and A. I. Buzdin, *Adv. Phys.* **36,** 271 (1987).

[149]K. A. Müller, K. W. Blazey, J. G. Bednorz, and M. Takashige, *Physica B & C* **148,** 149 (1987).

[150]K. W. Blazey, K. A. Müller, J. G. Bednorz, W. Berlinger, G. A. Moretti, E. Buluggiu, A. Vera, and F. C. Matacotta, *Phys. Rev. B* **36,** 7241 (1987).

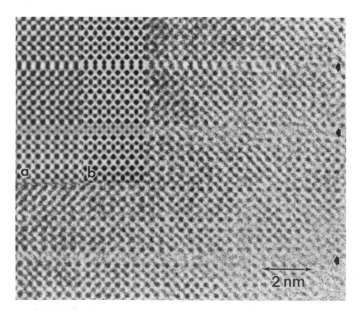

FIG. 14. High-resolution TEM image of $Y_1Ba_2Cu_3O_7$ in [010] orientation showing stacking faults (indicated by black arrows). The uppermost fault shows mirror symmetry while the lower two faults show glide symmetry. Note that the uppermost fault gradually assumes glide character near the right edge, suggesting that the length of the a axis is different on both sides of the defect. An averaged experimental image (a) and a calculated image (b) are inset. (Zandbergen et al.[152])

observed across some faults and not others.[26,28,151–153] Other faults appear to cause an offset in the {110} planes.[152,153] In each case the fault is accompanied by an increase in the c lattice parameter. Different displacement vectors have been assigned to the faults based on the high-resolution images. From diffraction contrast experiments only two faults with similar displacement vectors of $[a/2, 0, c/6]$ and $[0, b/2, c/6]$ were identified.[154] It was shown later that the displacements observed in high-resolution images are consistent with displacement vectors of this type.[152] High-resolution images of the faults (Fig. 14) matched the contrast calculated for a defect in which an extra Cu–O layer is inserted between the Ba–O layers.[152] Examples are also reported of faults where an extra Cu–O layer is inserted between the Y and Ba–O layers.[28,153]

[151]Y. Hirotsu, Y. Nakamura, Y. Murata, S. Nagakura, T. Nishihara, and M. Takata, Jpn. J. Appl. Phys. **26**, L1168 (1987).
[152]H. W. Zandbergen, R. Gronsky, K. Wang, and G. Thomas, Nature **331**, 596 (1988).
[153]Y. Matsui, E. Takayama-Muromachi, and K. Kato, Jpn. J. Appl. Phys. **27**, L350 (1988).
[154]J. Tafto, M. Suenaga, and R. L. Sabatini, Appl. Phys. Lett. **52**, 667 (1988).

In bulk material, the faults appear to be extrinsic defects that result from the reaction of 123 with the environment, rather than intrinsic growth defects. The faults can grow in from the surfaces of crystals if they are left exposed to air.[155] The process of ion milling, used to prepare specimens for TEM, can also introduce faults.[127,152] The models for the faults indicate that a local enrichment in copper accompanies fault formation. The copper enrichment might result from copper segregated at the surface diffusing into the crystals.[156] Alternatively, copper enrichment could result from the loss of yttrium or barium from the crystals. It is unclear what role reactive species such as water vapor or carbon dioxide play in fault formation, and it is possible that the defects contain other species such as carbon or OH^- groups. Extensive fault formation has also been observed in 123 crystals heated in a reducing atmosphere and then quenched.[62,157,158] In this case the faults appear to be part of the reaction mechanism through which 123 decomposes to $Y_2Ba_1Cu_1O_5$, $BaCuO_2$, and Cu_2O.

The periodic introduction of stacking faults in the 123 structure gives rise to two new phases, $Y_1Ba_2Cu_4O_8$ and $Y_2Ba_4Cu_7O_{14}$.[159–162] In $Y_1Ba_2Cu_4O_8$, extra Cu–O layers are inserted between the Ba–O layers in *every* 123 unit cell, while in $Y_2Ba_4Cu_7O_{14}$, extra Cu–O layers are inserted between the Ba–O layers in every *other* 123 unit cell. Both structures are orthorhombic with space group *Ammm*. The unit cell parameters for $Y_1Ba_2Cu_4O_8$ are $a \sim b = 3.86(1)$ and $c = 27.24(6)$ Å, and those for $Y_2Ba_4Cu_7O_{14}$ are $a = 3.851(1)$, $b = 3.869(1)$, and $c = 50.29(2)$Å. The thermodynamic stability of these new phases has yet to be determined; $Y_2Ba_4Cu_7O_{14}$ has been produced in bulk samples, but $Y_1Ba_2Cu_4O_8$ has only been made in thin films. The superconducting transition temperature does not change substantially with the periodic introduction of the extra Cu–O layers. $Y_1Ba_2Cu_4O_8$ has a T_c of 81 K and a ytterbium derivative of $Y_2Ba_4Cu_7O_{14}$ has a T_c of 86 K. These results

[155]G. Van Tendeloo and S. Amelinckx, *J. of Electron Microsc. Tech.* **8**, 285 (1988).

[156]H. W. Zandbergen, R. Gronsky, and G. Thomas, *Phys. Stat. Sol.* (*a*) **105**, 207 (1988).

[157]G. Van Tendeloo, H. W. Zandbergen, T. Okabe, and S. Amelinckx, *Solid State Commun.* **63**, 969 (1987).

[158]L. D. Marks, J. P. Zhang, S. J. Hwu, and K. R. Poeppelmeier, *J. Solid State Chem.* **69**, 189 (1987).

[159]A. F. Marshall, R. W. Barton, K. Char, A. Kapitulnik, B. Oh, R. H. Hammond, and S. S. Laderman, *Phys. Rev. B* **37**, 9353 (1988).

[160]P. Marsh, R. M. Fleming, M. L. Mandich, A. M. DeSantolo, J. Kwo, M. Hong, and L. J. Martinez-Miranda, *Nature* **334**, 141 (1988).

[161]T. Kogure, R. Kontra, G. J. Yurek, and J. B. Vander Sander, *Physica C* **156**, 45 (1988).

[162]P. Bordet, C. Chaillout, J. Chenavas, J. L. Hodeau, M. Marezio, J. Karpinski, and E. Kaldis, *Nature* **334**, 596 (1988).

support the view that it is the CuO_2 planes that are essential for obtaining high-temperature superconductivity. In addition, these new phases may aid in understanding the role of the layers that separate the CuO_2 planes. A similar situation occurs in the thallium-containing superconductors (see Section 14) where the CuO_2 planes are separated by either Tl–O monolayers or Tl–O bilayers.

7. GRAIN BOUNDARIES

Numerous investigations implicate grain boundaries as the cause of the low critical current densities in polycrystalline 123.[163-169] There have been relatively few investigations, however, of intrinsic grain boundary structures. Part of the reason for this is that boundaries in materials processed under typical conditions contain a number of extrinsic features. Photoemission, inverse photoemission, and Auger spectroscopy of fracture surfaces in polycrystalline materials indicate the presence of carbonate phases at the grain boundaries,[170-172] which TEM observations confirm.[173] In many cases a discrete second phase is observed at the grain boundary,[173,174] but even in "clean" materials that show no evidence of second phases in the TEM, evidence of carbon segregation and carbonate formation is still seen in Auger spectra from the boundaries.[172]

[163]J. W. Ekin, *Adv. Ceram. Mater.* **2,** 586 (1987).

[164]S. Jin, R. A. Fastnacht, T. H. Tiefel, and R. C. Sherwood, *Appl. Phys. Lett.* **51,** 203 (1987).

[165]S. Jin, R. C. Sherwood, T. H. Tiefel, R. B. van Dover, D. W. Johnson, and G. S. Grader, *Appl. Phys. Lett.* **51,** 855 (1987).

[166]R. A. Camps, J. E. Evetts, B. A. Glowwaki, S. B. Newcomb, R. E. Somejh, and W. M. Stobbs, *Nature* **329,** 229 (1987).

[167]S. Jin, R. A. Fastnacht, T. H. Tiefel, and R. C. Sherwood, *Phys. Rev. B* **37,** 5828 (1988).

[168]J. W. Ekin, A. I. Braginski, A. J. Panson, M. A. Janocko, D. W. Capone, N. J. Zaluzec, B. Flandermeyer, O. F. Delima, M. Hong, J. Kwo, and S. H. Liou, *J. Appl. Phys.* **62,** 4821 (1987).

[169]R. L. Peterson, and J. W. Ekin, *Phys. Rev. B* **37,** 9848 (1988).

[170]J. A. Yarmoff, D. R. Clarke, W. Drube, U. O. Karlsson, A. Taleb-Ibrahimi, and F. J. Himpsel, *Phys. Rev. B* **36,** 3967 (1987).

[171]A. G. Schrott, S. L. Cohen, T. R. Dinger, F. J. Himpsel, J. A. Yarmoff, K. G. Frase, S. I. Park, and R. Purtell, *in* "Thin Film Processing and Characterization of High Temperature Superconductors," Conference Proceedings No. 165, p. 349 American Institute of Physics, 1987.

[172]S. Nakahara, G. J. Fisanick, M. F. Yan, R. B. van Dover, T. Boone, and R. Moore, *J. Cryst. Growth* **85,** 639 (1987).

[173]M. F. Chisholm (1987) (unpublished work).

[174]Y. Ishida, Y. Takahashi, M. Mori, K. Kishio, K. Kitazawa, K. Fueki, and M. Kawasaki, *J. Electron Microsc.* **36,** 251 (1987).

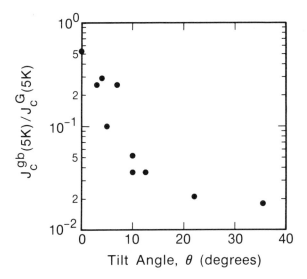

FIG. 15. Ratio of the grain boundary critical current density to the average critical current density in the two grains at 5 K versus the misorientation angle in the basal copper plane. (Dimos et al.[176])

Moreover, the thermal expansion anisotropy in the 123 structure leads to large internal stresses in polycrystalline materials that can cause cracking and extensive defect formation at grain boundaries,[172] especially in thermally cycled material.

The extrinsic grain boundary features described above certainly contribute to the low critical current densities, but recent experiments on individual boundaries in thin films graphically demonstrate that the critical current density (J_c) can be lowered by as much as a factor of 50 by *intrinsic* grain boundaries.[175,176] The ratio of the J_c supported by grain boundaries to the J_c of the grains is plotted versus grain boundary misorientation angle in Fig. 15. The rapid drop in the critical current density ratio with misorientation highlights the fact that the behavior comes from the intrinsic structure of the boundary and is therefore sensitive to the way the dislocation structure in the boundary changes with misorientation.

Characterization of the structures of intrinsic grain boundaries in 123

[175]P. Chaudhari, J. Mannhart, D. Dimos, C. C. Tsuei, J. Chi, M. M. Oprysko, and M. Scheuermann, *Phys. Rev. Lett.* **60,** 1653 (1988).
[176]D. Dimos, P. Chaudhari, J. Mannhart, and F. LeGoues, *Phys. Rev. Lett.* **61,** 219 (1988).

has been limited. Observations of the structure of low-angle tilt boundaries indicate that the misorientation between grains is accommodated by
walls of defects with closure failures of 0.39 nm and 1.167 nm.[177] The
0.39 nm closure failure is consistent with dislocations of Burgers vector
[010] as observed for perfect lattice dislocations.[178] The defect with the
1.167 nm closure failure indicates a dislocation with an unusually large
Burgers vector, corresponding to the c axis of the structure, and it is
proposed that the defect is actually a group of three closely spaced
$\frac{1}{3}$[001] type partials. Each defect disturbs the continuity of the lattice
planes across the boundary for several lattice planes. Given the short
coherence length in 123 materials, the disruption of the structure caused
by individual dislocations may be sufficient to lower the critical current
density that can flow across grain boundaries made of walls of such
defects. The misorientation dependence of J_c shown in Fig. 15, however,
cannot be completely accounted for by a model in which the dislocations
simply act as regions of "bad" material whose spacing decreases with
increasing misorientation. Dimos et al.[176] therefore propose that the
defects could also influence flux pinning in the boundary by providing an
easy path for flux penetration.

No direct observations of the structures of intrinsic high-angle grain
boundaries have been reported so far. Measurements of misorientations
across grain boundaries from electron diffraction patterns[179] and from the
external morphologies of clusters of faceted crystals[180] indicate that
high-angle grain boundaries with certain preferred misorientations form.
The observed misorientations are similar to those observed for special
low Σ boundaries in cubic materials. Because most of the observed
boundaries are twist type boundaries with the c axis as the misorientation
axis, this behavior may simply reflect the similarity of the lengths of the a
and b axes.

Two types of grain boundaries also occur in which the adjoining grains
are misoriented by exactly 90°.[179,181,182] In one type the c axis in one grain
interchanges with the a or b axis in the adjacent grain. In the other type

[177]M. F. Chisholm and D. A. Smith, *Phil. Mag.* (1989).

[178]S. Ikeda, T. Hatano, A. Matsushita, T. Matsumoto, and K. Ogawa, *Jpn. J. Appl. Phys.*
26, L729 (1987).

[179]L. A. Tietz, B. C. De Cooman, C. B. Carter, D. K. Lathrop, S. E. Russek, and R. A.
Buhrman, *J. Electron Microsc. Tech.* **8,** 263 (1988).

[180]D. A. Smith, M. F. Chisholm, and J. Clabes, *Appl. Phys. Lett.* (to appear December 1,
1988).

[181]H. You, J. D. Axe, X. B. Kan, S. C. Moss, J. Z. Liu, and D. J. Lam, *Phys. Rev. B* **37,**
2301 (1988).

[182]H. W. Zandbergen, R. Gronsky, M. Y. Chu, L. C. de Jonghe, G. F. Holland, and A.
M. Stacy (1987) (unpublished).

the a and b axes interchange upon crossing adjacent (001) planes. These boundaries, sometimes referred to as "90° twins," are most commonly seen in partially reacted materials, suggesting that they are growth defects that develop when the 123 structure first forms. Prolonged annealing at 900–950°C greatly reduces the density of these boundaries.

V. Structural Variants

Following the discovery of superconductivity in the Y–Ba–Cu–O system, numerous researchers began to explore the feasibility of substituting different elements into the $Y_1Ba_2Cu_3O_{7-\delta}$ structure. To date, some degree of substitution has been achieved on each site in the structure. Part of the initial enthusiasm for substitutions came from the possibility of further raising the superconducting transition temperature. More recent studies have focussed on the structural, valence, and magnetic effects of cation substitutions in an effort to understand the relevance of various structural features to superconductivity. Several clear issues have emerged from these studies. The aim of this section is to examine the most important of these issues rather than to provide an exhaustive survey of the vast literature that has grown in this area. Several of these topics are still very active areas of research. Thus, it is only possible to present a subjective assessment of the field. After discussion of the various cation substitutions, reports of anion substitutions and investigations of the isotope effect are described.

8. RARE EARTH SUBSTITUTIONS

Early studies[183–185] of substitutions concentrated on the replacement of Y by rare earth ions. These surveys found that all but a few of the lanthanide series could be incorporated in the $Y_1Ba_2Cu_3O_{7-\delta}$ structure. Variability in reported results suggested that some lanthanides form the 123 structure more readily than others. Subsequent studies have confirmed that all the lanthanides except Ce and Tb can fully replace Y in

[183]Z. Fisk, J. D. Thompson, E. Zirngiebl, J. L. Smith, and S. W. Cheong, *Solid State Commun.* **62,** 743 (1987).

[184]P. H. Hor, R. L. Meng, Y. Q. Wang, L. Gao, Z. J. Huang, J. Bechtold, K. Forster, and C. W. Chu, *Phys. Rev. Lett.* **58,** 1891 (1987).

[185]E. M. Engler, V. Y. Lee, A. I. Nazzal, R. B. Beyers, G. Lim, P. M. Grant, S. S. P. Parkin, M. L. Ramirez, J. E. Vazquez, and R. J. Savoy, *J. Am. Chem. Soc.* **109,** 2848 (1987).

the $Y_1Ba_2Cu_3O_{7-\delta}$ structure.[186-198] ($Pm_1Ba_2Cu_3O_{7-\delta}$ has not been synthesized due to its rapid radioactive decay, but is suspected to form.) The inability to form Ce and Tb 123 structures is explained by the ease with which Ce^{4+} and Tb^{4+} ions form. Of the structures that have been synthesized only $Pr_1Ba_2Cu_3O_{7-\delta}$ is not superconducting.[199-203] Similarly, this is attributed to the tendency of the +4 ion to form, but experimental evidence for the presence of Pr^{4+} is mixed. Magnetic susceptibility measurements[199,203] detect the presence of an effective moment of 2.9 μ_B on the Pr ions. By assuming contributions of 3.58 μ_B and 2.54 μ_B from the Pr^{+3} and Pr^{+4} ions, respectively, it is estimated that only 10% of the

[186]J. M. Tarascon, W. R. McKinnon, L. H. Greene, G. W. Hull, and E. M. Vogel, *Phys. Rev. B* **36**, 226 (1987).

[187]J. P. Golben, S. Lee, S. Y. Lee, Y. Song, T. W. Noh, X. Chen, and J. R. Gaines, *Phys. Rev. B* **35**, 8705 (1987).

[188]X. T. Xu, J. K. Liang, S. S. Xie, G. C. Che, X. Y. Shao, Z. G. Duan, and C. G. Cui, *Solid State Commun.* **63**, 649 (1987).

[189]Q. R. Zhang, Y. T. Qian, Z. Y. Chen, W. Y. Guan, Y. Zhao, H. Zhang, L. Z. Cao, J. S. Xia, G. Q. Pang, M. J. Zhang, Z. H. He, D. Q. Yu, S. F. Sun, T. Zhang, M. H. Fang, and Z. P. Yang, *Solid State Commun.* **63**, 497 (1987).

[190]K. N. Yang, Y. Dalichaouch, J. Ferreira, B. W. Lee, J. J. Neumeier, M. S. Torikachvili, H. Zhou, and M. B. Maple, *Solid State Commun.* **63**, 515 (1987).

[191]S. Lee, J. P. Golben, Y. Song, S. Y. Lee, T. W. Hoh, X. Chen, J. Testa, and J. R. Gaines, *Appl. Phys. Lett.* **51**, 282 (1987).

[192]S. Tsurumi, M. Hikita, T. Iwata, K. Semba, and S. Kurihara, *Jpn. J. Appl. Phys.* **26**, L856 (1987).

[193]S. Ohshima and T. Wakiyama, *Jpn. J. Appl. Phys.* **26**, L815 (1987).

[194]F. Garcia-Alvarado, E. Moran, M. Vallet, J. M. González-Calbet, M. A. Alario, M. T. Pérez-Frías, J. L. Vicent, S. Ferrer, E. García-Michel, and M. C. Asensio, *Solid State Commun.* **63**, 507 (1987).

[195]R. Escudero, T. Akachi, R. Barrio, L. E. Rendón-Diazmirón, C. Vázquez, L. Baños, G. Gonzáles, and F. Estrada, *Solid State Commun.* **64**, 235 (1987).

[196]E. A. Hayri, K. V. Ramanujachary, S. Li, M. Greenblatt, S. Simizu, and S. A. Friedberg, *Solid State Commun.* **64**, 217 (1987).

[197]J. J. Neumeier, Y. Dalichaouch, J. M. Ferreira, R. R. Hake, B. W. Lee, M. B. Maple, K. N. Torikachvili, K. N. Yang, and H. Zhou, *Appl. Phys. Lett.* **51**, 371 (1987).

[198]R. Liang, Y. Inaguma, Y. Takagi, and T. Nakamura, *Jpn. J. Appl. Phys.* **26**, L1150 (1987).

[199]B. Okai, K. Takahashi, H. Nozaki, M. Saeki, M. Kosuge, and M. Ohta, *Jpn. J. Appl. Phys.* **26**, L1648 (1987).

[200]L. Soderholm, K. Zhang, D. G. Hinks, M. A. Beno, J. D. Jorgensen, C. U. Segre, and I. K. Schuller, *Nature* **328**, 604 (1987).

[201]Y. Dalichaouch, M. S. Torikachvili, E. A. Early, B. W. Lee, C. L. Seaman, K. N. Yang, H. Zhou, and M. B. Maple, *Solid State Commun.* **65**, 1001 (1988).

[202]J. K. Liang, X. T. Xu, S. S. Xie, G. H. Rao, X. Y. Shao, and Z. G. Duan, *Z. Phys. B* **69**, 137 (1987).

[203]A. B. Okai, M. Kosuge, H. Nozaki, K. Takahashi, and M. Ohta, *Jpn. J. Appl. Phys.* **27**, L41 (1988).

Pr ions are in a Pr^{+3} state. This is inconsistent with lattice parameter and bond length measurements made on $Pr_1Ba_2Cu_3O_{7-\delta}$, which follow the trend with ionic radius expected if the Pr ions are predominately in a $3 +$ state.[204] Partial substitution of Pr for Y in the structure produces a monotonic decrease in the superconducting transition temperature, and substitution of more than half the Y with Pr produces a semiconducting material.[200-203] Again it is found that partial substitution of Pr for Y causes a change in lattice parameters that is consistent with Pr^{3+}, but magnetic susceptibility measurements indicate that Pr is mainly present as Pr^{4+}. Evidence for partial mixing of Pr on the Ba site is also observed,[203] which may contribute to the decrease in T_c. The anomalous behavior of the lattice parameters suggests that there may be structural differences between the Pr 123 structure and the other rare earth 123 structures. One possibility is that Pr^{4+} draws oxygen into the Y layer causing it to expand.[203] For the remaining $R_1Ba_2Cu_3O_{7-\delta}$ structures, transition temperatures on the order of 91–95 K are reported for the fully substituted structures. Remarkably, the presence of magnetic rare earth ions such as Gd and Ho has little effect on the superconducting transition temperature.

Many x-ray and neutron diffraction studies have been made of the rare earth substituted structures.[204-212] A summary of unit cell dimensions and atomic coordinates for the different orthorhombic structures is given in Table V. As expected, the unit cell volume decreases with increasing atomic number, due to the well-known variation of ionic radius with f electron count in the rare earths.[213] Based on data in an early study,[186] it was suggested that T_c for the rare earth compounds increases as the unit cell volume increases.[214] A more comprehensive survey of data from

[204]Y. Le Page, T. Siegrist, S. A. Sunshine, L. F. Schneemeyer, D. W. Murphy, S. M. Zahurak, J. V. Waszczak, W. R. McKinnon, J. M. Tarascon, G. W. Hull, and L. H. Greene, *Phys. Rev. B* **36**, 3617 (1987).

[205]T. Ishigaki, H. Asano, and K. Takita, *Jpn. J. Appl. Phys.* **26**, L987 (1987).

[206]T. Ishigaki, H. Asano, K. Takita, H. Katoh, H. Akinaga, F. Izumi, and N. Watanabe, *Jpn. J. Appl. Phys.* **26**, L1681 (1987).

[207]T. Ishigaki, H. Asano, and T. Takita, *J. Appl. Phys.* **26**, L1226 (1987).

[208]M. Onoda, S. Shamoto, M. Sato, and S. Hosoya, *Jpn. J. Appl. Phys.* **26**, L876 (1987).

[209]K. Takita, H. Akinaga, H. Katoh, T. Ishigaki, and H. Asano, *Jpn. J. Appl. Phys.* **26**, L1023 (1987).

[210]H. Asano, T. Ishigaki, and K. Takita, *Jpn. J. Appl. Phys.* **26**, L714 (1987).

[211]H. Asano, K. Takita, T. Ishigaki, H. Akinaga, H. Katoh, K. Masuda, F. Izumi, and N. Watanabe, *Jpn. J. Appl. Phys.* **26**, L1341 (1987).

[212]H. Asano, K. Takita, H. Katoh, H. Akinaga, T. Ishigaki, M. Nishino, M. Imai, and K. Masuda, *Jpn. J. Appl. Phys.* **26**, L1410 (1987).

[213]R. D. Shannon, *Acta Cryst. A* **32**, 751 (1976).

[214]T. J. Kistenmacher, *Solid State Commun.* **65**, 981 (1988).

TABLE V. UNIT CELL DIMENSIONS, R^{3+} RADII, z COORDINATES, AND O(1)
OCCUPANCY FOR $R_1Ba_2Cu_3O_{7-\delta}$, WHERE R = RARE EARTH. THE ESTIMATED
STANDARD DEVIATIONS (IN PARENTHESES) REFER TO THE LAST DIGIT
PRINTED. (LE PAGE et al.[204])

	$a(Å)$	$b(Å)$	$c(Å)$	RADIUS (Å)
Pr	3.905(2)	3.905(2)	11.660(10)	1.013
Sm	3.891(1)	3.894(1)	11.660(1)	0.964
Eu	3.869(2)	3.879(3)	11.693(6)	0.950
Gd	3.854(2)	3.896(2)	11.701(7)	0.938
Dy	3.830(3)	3.885(2)	11.709(3)	0.908
Y	3.827(1)	3.877(1)	11.708(6)	0.905
Ho	3.846(1)	3.881(1)	11.640(2)	0.894
Er	3.812(3)	3.851(4)	11.626(2)	0.881
Tm	3.829(3)	3.860(3)	11.715(2)	0.869

	$z(Ba)$	$z(Cu(2))$	$z(O(4))$	$z(O(2,3))$	occ(O(1))
Pr	0.1840(3)	0.3507(8)	0.157(5)	0.371(4)	1.0(2)
Sm	0.1853(2)	0.3554(4)	0.158(3)	0.374(2)	0.6(2)
Eu	0.1847(3)	0.3530(8)	0.159(5)	0.374(4)	0.4(2)
Gd	0.1855(2)	0.3546(4)	0.151(2)	0.377(2)	0.7(2)
Dy	0.1864(2)	0.3564(3)	0.157(2)	0.378(2)	0.7(2)
Y	0.1874(3)	0.3565(4)	0.157(3)	0.379(3)	0.8(2)
Ho	0.1858(2)	0.3577(4)	0.159(3)	0.378(2)	0.6(2)
Er	0.1866(2)	0.3579(4)	0.159(3)	0.381(2)	0.8(2)
Tm	0.1892(1)	0.3593(2)	0.155(2)	0.380(1)	0.5(1)

several sources,[184–186,190,192] however, suggests that the correlation of T_c
with unit cell volume is weak if it is present at all. Le Page et al.[204] noted
that the main effect of rare earth substitution is to change the bond
lengths that involve the rare earth ion, as can be seen in their plot of rare
earth-Cu(2) bond lengths in Fig. 16. By comparison, the average Ba–Cu
distance remains almost constant across the series (Fig. 16 also). These
trends, they suggest, indicate that rare earth substitutions have little
effect on the substructure formed by the copper-oxygen chains and planes
that surround the Ba ions. Subsequent structural studies support this
view.[205–212]

Rare earth ions towards the La end of the lanthanide series are large
enough to partially substitute on the Ba^{+2} site. The resulting compounds,
with the general formula $R_{1+x}Ba_{2-x}Cu_3O_{7\pm\delta}$, exhibit more complex
behavior than the other rare earth 123 structures because of their cation
solid solution range. Historically, $La_3Ba_3Cu_6O_{14+\delta}$ was identified before
$Y_1Ba_2Cu_3O_{7-\delta}$. Based on x-ray diffraction patterns it was suggested that
the La structure was an oxygen deficient perovskite in which ordering of

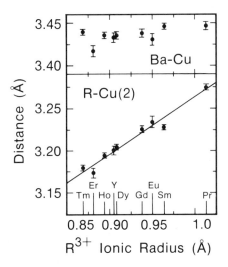

FIG. 16. Averaged Ba–Cu distance and R–Cu(2) distance versus R^{3+} radius, where R = rare earth. The Ba–Cu distance is independent of the R^{3+} radius. The R–Cu(2) distance is composed of an R–O and a Cu(2)–O distance at right angles (the latter distance is independent of the R^{3+} radius) and therefore varies approximately with a slope of $1/\sqrt{2}$. (Le Page et al.[204])

the oxygen vacancies tripled one of the cubic perovskite unit cell dimensions.[30] More recent neutron diffraction studies[215–218] found that $La_3Ba_3Cu_6O_{14+\delta}$ is essentially isomorphous with tetragonal $Y_1Ba_2Cu_3O_{7-\delta}$, with La fully replacing Y and the excess La partially replacing Ba in the structure; i.e., the structure is better described as $La_{1.5}Ba_{1.5}Cu_3O_{7+\delta}$. Studies of superconductivity in the La–Ba–Cu–O system have reported transition temperatures that vary widely. Systematic studies of $La_{1+x}Ba_{2-x}Cu_3O_{7+\delta}$, in which x was varied over the range 0 to 0.5, found that T_c decreased with increasing La substitution.[218,219]

[215]W. I. F. David, W. T. A. Harrison, R. M. Ibberson, M. T. Weller, J. R. Grasmeder, and P. Lanchester, *Nature* **328**, 328 (1987).

[216]R. Yoshizaki, H. Sawada, T. Iwazumi, Y. Saito, H. Ikeda, K. Imai, and I. Nakai, *Jpn. J. Appl. Phys.* **26**, L1703 (1987).

[217]F. Izumi, H. Asano, T. Ishigaki, E. Takayama-Muromachi, Y. Matsui, and Y. Uchida, *Jpn. J. Appl. Phys.* **26**, L1153 (1987).

[218]C. U. Segre, B. Dabrowski, D. G. Hinks, K. Zhang, J. D. Jorgensen, M. A. Beno, and I. K. Schuller, *Nature* **329**, 227 (1987).

[219]E. Takayama-Muromachi, Y. Uchida, A. Fujimori, and K. Kato, *Jpn. J. Appl. Phys.* **26**, L1546 (1987).

Single phase materials were obtained only if x was greater than 0.25. Neutron diffraction refinements[218] showed that the more barium-rich compositions are orthorhombic, but the orthorhombicity decreases with increasing La substitution, and the structure becomes tetragonal at $x = 0.5$.

Neutron diffraction was also used to determine the total oxygen content of $La_{1+x}Ba_{2-x}Cu_3O_{7+\delta}$ and the distribution of oxygen in the unit cell.[218] The oxygen content refined to a value greater than seven in all the structures and varied only slightly with the degree of La substitution. In the orthorhombic structures, both oxygen sites in the basal copper plane are partially occupied and show greater disordering of oxygens between the two sites than in orthorhombic $Y_1Ba_2Cu_3O_7$. Increasing La substitution decreased the oxygen order in the basal plane and produced a smooth monotonic decrease in T_c that correlated well with the order parameter. This observation, coupled with the fact that the formal charge on the copper remained constant at a value of about $+2.3$ across the series, was taken as evidence that ordering of the Cu–O chains was essential for superconductivity.

Studies of partial substitution of La for Ba in $Y_1Ba_{2-x}La_xCu_3O_{7+\delta}$[198,220–224] also found that increasing La substitution induces an orthorhombic-to-tetragonal transition and lowers T_c. The composition at which the $Y_1Ba_{2-x}La_xCu_3O_{7+\delta}$ structure becomes tetragonal is $x \simeq 0.3$–0.4.[198,221] In contrast to the results reported in the $La_{1+x}Ba_{2-x}Cu_3O_{7+\delta}$ system, the orthorhombic-to-tetragonal transition does not appear to be associated with a loss of superconductivity in $Y_1Ba_{2-x}La_xCu_3O_{7+\delta}$. Measurements of the oxygen content by wet iodometric titration indicated that the substitution produces intercalation of oxygen in excess of the O_7 stoichiometry. At low La concentrations $(0 < x < 0.05)$, oxygen appears to intercalate as a monomeric species. At high La concentrations, however, the oxygen may intercalate as dimeric O_2^{2-} peroxide ions.[221]

Substitution in the $Nd_{1+x}Ba_{2-x}Cu_3O_{7-\delta}$ system produces similar be-

[220]B. Chevalier, B. Buffat, G. Demazeau, B. Lloret, J. Etourneau, M. Hervieu, C. Michel, B. Raveau, and R. Tournier, *J. de Phys.* **48**, 1619 (1987).

[221]A. Manthiram, X. X. Tang, and J. B. Goodenough, *Phys. Rev. B* **37**, 3734 (1988).

[222]A. Maeda, T. Yabe, K. Uchinokura, M. Izumi, and S. Tanaka, *Jpn. J. Appl. Phys.* **26**, L1550 (1987).

[223]Y. Tokura, J. B. Torrance, T. C. Huang, and A. I. Nazzal, *Phys. Rev. B* **38**, 7156 (1988).

[224]A. Tokiwa, Y. Soyono, M. Kikuchi, R. Suzuki, T. Kajitani, N. Kobayashi, T. Sasaki, O. Nakatsu, and Y. Muto, *Jpn. J. Appl. Phys.* **27**, L1009 (1988).

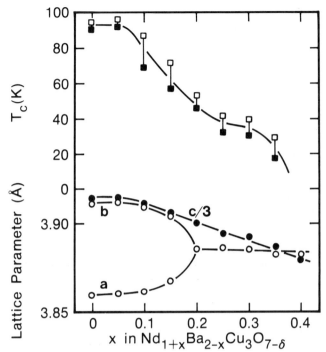

FIG. 17. Dependence of lattice parameters and T_c upon x in $Nd_{1+x}Ba_{2-x}Cu_3O_{7-\delta}$. Open and filled squares represent onset and zero-resistance temperatures, respectively. (Takita *et al.*[229])

havior to that in the La 123 compounds,[225–232] with a substitution-induced orthorhombic-to-tetragonal transition occurring at $x = 0.2$.[227,229] As in $Y_1Ba_{2-x}La_xCu_3O_{7+\delta}$, the orthorhombic-to-tetragonal transition has no influence on the decrease in T_c that occurs with increasing Nd substitution, and tetragonal material can be superconducting. Typical data for the variation of T_c and unit cell dimensions in $Nd_{1+x}Ba_{2-x}Cu_3O_{7-\delta}$ as a function of x are shown in Fig. 17. At higher substitution levels

[225]F. Izumi, S. Takekawa, Y. Matsui, N. Iyi, H. Asano, T. Ishigaki, and N. Watanabe, *Jpn. J. Appl. Phys.* **26**, L1616 (1987).

[226]R. J. De Angelis, J. W. Brill, M. Chung, W. D. Arnett, X. D. Xiang, G. Minton, L. A. Rice, and C. E. Hamrin, Jr., *Solid State Commun.* **64**, 1353 (1987).

[227]S. Takekawa, H. Nozaki, Y. Ishizawa, and N. Iyi, *Jpn. J. Appl. Phys.* **26**, L2076 (1987).

[228]H. Nozaki, S. Takekawa, and Y. Ishizawa, *Jpn. J. Appl. Phys.* **27**, L31 (1988).

[229]K. Takita, H. Katoh, H. Akinaga, M. Nishino, T. Ishigaki, and H. Asano, *Jpn. J. Appl. Phys.* **27**, L57 (1987).

[230]S. Tsurumi, T. Iwata, Y. Tajima, and M. Hikita, *Jpn. J. Appl. Phys.* **27**, L80 (1988).

[231]K. Takita, H. Akinaga, H. Katoh, H. Asano, and K. Masuda, *Jpn. J. Appl. Phys.* **27**, L67 (1988).

[232]K. Takita, H. Akinaga, H. Katoh, and K. Masuda, *Jpn. J. Appl. Phys.* **27**, L607 (1988).

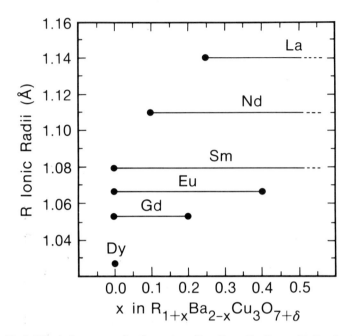

FIG. 18. Solid solution ranges for the systems $R_{1+x}Ba_{2-x}Cu_3O_{7+\delta}$ with R = La, Nd, Sm, Eu, Gd, and Dy. The full lines indicate the region in which a single phase is observed, while the broken lines indicate the possible extension of the single phase into regions not studied. (Zhang *et al.*[234])

($x = 0.25$—0.4) the Nd-substituted material is reportedly semi-conducting,[228] but treatments in high-pressure oxygen at 450°C have restored superconductivity in materials with x as high as 0.5.[230] Neutron diffraction refinements[225] and hydrogen reduction[227] of the tetragonal superconducting materials indicate that the oxygen content is nearly 7 and the two basal plane oxygen sites are approximately 50% occupied, as in the La system. At higher Nd concentrations ($x = 0.8$) the presence of a superlattice that doubles the a and c axes of the structure has been detected by electron diffraction.[233]

Solid solution ranges for several other large rare earth ions have now been explored.[234–236] A graph summarizing the data from a study by Zhang *et al.*[234] is shown in Fig. 18. Although the solubility ranges differ

[233]Y. Matsui, S. Takekawa, and N. Iyi, *Jpn. J. Appl. Phys.* **26,** L1693 (1987).

[234]K. Zhang, B. Dabrowski, C. U. Segre, D. G. Hinks, I. K. Schuller, J. D. Jorgensen, and M. Slaski, *J. Phys. C* **20,** L935 (1987).

[235]H. Akinaga, H. Katoh, K. Takita, H. Asano, and K. Masuda, *Jpn. J. Appl. Phys.* **27,** L610 (1988).

[236]T. Iwata, M. Hikita, Y. Tajima, and S. Tsurumi, *Jpn. J. Appl. Phys.* **26,** L2049 (1987).

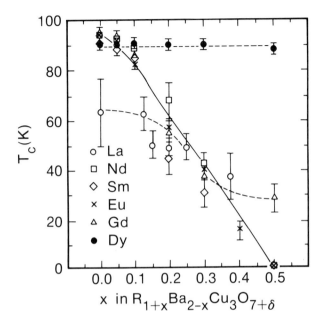

FIG. 19. Superconducting transition temperatures as a function of x for the systems $R_{1+x}Ba_{2-x}Cu_3O_{7+\delta}$ with R = La, Nd, Sm, Eu, Gd, and Dy. The error bars indicate the 90% and 10% points of the resistivity transition. Note the deviations from universal behavior for La, Gd, and Dy, indicating the ends of their solubility ranges. (Zhang et al.[234])

somewhat from studies that used different preparative conditions, overall solubility trends with ionic radius are evident. Figure 18 suggests that there may be a lower solubility limit for the larger rare earth ions determined by whether or not the ion is small enough to fit in the yttrium site. An upper solubility limit is determined by the size match to the larger barium site. Intermediate-sized ions such as samarium thus have extensive solubility. Zhang et al.[234] also suggest that the difficulty in forming the ytterbium and lutetium structures could be caused by their being too small to fit in even the yttrium site. The plot of T_c versus x shown in Fig. 19 indicates that over the composition ranges where solid solutions form, T_c decreases as a more or less universal function of x for each of the rare earths. As a result, dysprosium, which has almost no solid solution range, produces only a small variation in T_c, whereas samarium, which has extensive solubility, varies T_c over a wide composition range. Iqbal et al.[237] report that the solid solution

[237]Z. Iqbal, F. Reindinger, A. Bose, N. Cipollini, T. J. Taylor, H. Eckhardt, B. L. Ramakrishna, and E. W. Ong, Nature 331, 326 (1988).

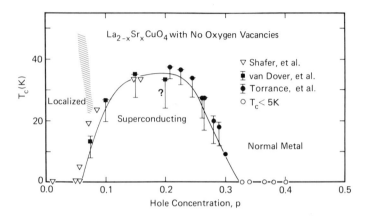

FIG. 20. The dependence of T_c on hole concentration in $La_{2-x}Sr_xCuO_4$ with no oxygen vacancies. The question mark refers to the possibility that this sample may have some oxyen vacancies. (Torrance et al.[238])

$Y_{1+x}Ba_{2-x}Cu_3O_{7-\delta}$, with $x = 0.5$ and $x = 0.375$, can form as a metastable phase. The yttrium ion, however, is smaller than Dy^{+3}, and therefore a stable Y–Ba solid solution is unlikely under normal preparative conditions. The fact that a wide range of compositions in the Y–Ba–Cu–O system have almost identical superconducting transition temperatures supports this conclusion.

Considerable speculation has arisen about the cause of the variations in T_c across the rare earth solid solution ranges. Initial attempts to correlate the decrease in T_c with the disruption of chains and the degree of order in the basal plane of the structure[218] appear to be incorrect, since T_c varies even in composition ranges where the tetragonal phase forms and hence the degree of order is constant. Hole concentration measurements (made using a wet chemical method) in the simpler $La_{2-x}Sr_xCuO_{4-\delta}$ ($x = 0-0.4$) system indicate that T_c is strongly correlated to the concentration of holes in the structure,[238] as can be seen from the data in Fig. 20. A similar correlation can be made for the 123 rare earth solid solutions if the hole concentration is estimated using a simple model in which it is directly controlled by the substitution of trivalent rare earth ions for the divalent Ba ions.[231] The model is based on the assumption that the oxygen concentration remains constant on substitution. Meas-

[238] M. W. Shafer, T. Penney, and B. Olsen, *Phys. Rev. B* **36**, 4047 (1987); J. B. Torrance, Y. Tokura, A. I. Nazzal, A. Bezinge, T. C. Huang, and S. S. P. Parkin, *Phys. Rev. Lett.* **61**, 1127 (1988); and R. B. van Dover, R. J. Cava, B. Batlogg, and E. A. Rietman, *Phys. Rev. B* **35**, 5337 (1987).

urements on $Y_1Ba_{2-x}La_xCu_3O_{7+\delta}$, however, indicate that the oxygen content of the structure increases with increasing substitution[221] and compensates for the introduction of the La^{3+} ion if the oxygen is introduced as O^{-2} ions. This observation led to the suggestion that oxygen may intercalate as O_2^{-2} peroxide ions, which would then require partial compensation for the charge on the substituted ions by a reduction in the number of holes.[221] Indirect evidence for peroxide formation in the

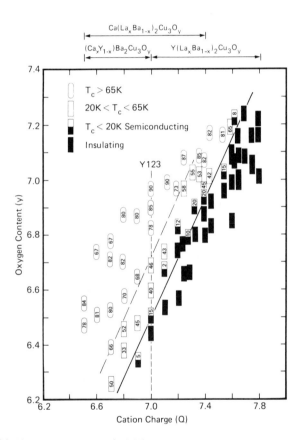

FIG. 21. (a) The oxygen content (± 0.03) measured for a series of isostructural samples made with different total charge, Q, on the non-copper metals. The short dashed line labelled Y123 is for $Y_1Ba_2Cu_3O_y$. The solid line is drawn to separate insulating and superconducting samples, while the long dashed line is drawn to separate the high T_c samples from those with $20\,K < T_c < 65\,K$, where T_c is the zero resistance temperature. In (b) the data are replotted using the relation $Q + 3(2 + p) = 2y$ to calculate the average $[Cu-O]^{+p}$ charge. (Tokura et al.[223])

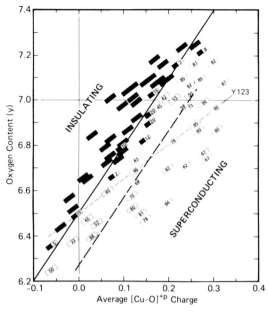

FIG. 21. (b)

$Y_1Ba_2Cu_3O_{7-\delta}$ structure has been obtained by Shafer *et al.*[239] by dissolving isotope-tagged materials. In spite of this possible difficulty with the simple compensation model of Takita *et al.*,[231] measurements of hole concentrations made using the Hall effect correlate with the trend predicted by the model for Nd-doped materials.[232] Further experiments, including measurements of charge on the copper ions and hole concentrations from Hall measurements, are needed to confirm these results and determine if the same type of behavior occurs for the other rare earth solid solutions.

Tokura *et al.*[223] looked at 123 materials in which La substitution on the Ba site, Ca substitution on the Y site, and annealing in different pressures of oxygen were used to vary independently the effective copper valence $(2 + p)$ and the oxygen content (y) over a wide range (see Fig. 21). They found that a high effective copper valence in the structure alone was not sufficient to produce superconductivity and suggested that the hole concentrations in the planes must be high for superconductivity. Nonsuperconducting behavior observed in materials with a high average

[239]M. W. Shafer, R. A. de Groot, M. M. Plechaty, and G. J. Scilla, *Physica C* **153–155**, 836 (1988).

hole concentration was then explained by the possibility that holes preferentially form on the chains rather than on the planes. By postulating a threshold concentration for hole formation in the planes that depends on the oxygen content of the chains, Tokura et al.[223] rationalized much of their experimental data and obtained a correlation between T_c and the hole concentration within the planes. Direct verification of these ideas, however, will be difficult, because it requires localized measurements of hole concentrations in the planes and in the chains.

9. TRANSITION ELEMENT SUBSTITUTIONS

Most of the work on copper substitution has focussed on elements in the series from Fe to Ga. Preliminary studies showed that each element strongly suppresses T_c, but differed as to which element has the largest effect.[240–242] From investigations of the variation in T_c with the degree of substitution,[241,243,244] it is now generally agreed that T_c is most strongly suppressed by zinc. Iron and cobalt, which act similarly, have a lesser effect. Nickel reportedly[241–243] has the least effect on T_c, but was found to suppress T_c as strongly as zinc in one study.[244] With nickel or zinc additions, the fully oxygenated phase remains orthorhombic.[242] In contrast, transition to a tetragonal structure occurs with small additions of iron, cobalt, or gallium. The retention of an orthorhombic structure in the Ni- and Zn-substituted materials has been rationalized on the basis that Ni and Zn substitute as $2+$ ions whereas Ga, Co, and Fe substitute with higher valence states ($+3$ or $+4$). The higher valence ions could then draw excess oxygen into the structure in order to increase their oxygen coordination, causing disordering of the oxygens in the basal copper plane. This argument is based on the assumption that the higher valence ions substitute into the Cu(1) chain sites. As shown in Fig. 22, the variation in T_c that occurs on substitution is not influenced by the orthorhombic-to-tetragonal transition.

[240]G. Xiao, F. H. Streitz, A. Gavrin, Y. W. Du, and C. L. Chien, *Phys. Rev. B* **35**, 8782 (1987).

[241]S. B. Oseroff, D. C. Vier, J. F. Smyth, C. T. Salling, S. Schultz, Y. Dalichaouch, B. W. Lee, M. B. Maple, Z. Fisk, J. D. Thompson, J. L. Smith, and E. Zirngiebl, *Solid State Commun.* **64**, 241 (1987).

[242]Y. Maeno, T. Tomita, M. Kyogoku, S. Awaji, Y. Aoki, K. Hoshino, A. Minami, and T. Fujita, *Nature* **328**, 512 (1987).

[243]E. Takayama-Muromachi, Y. Uchida, and K. Kato, *Jpn. J. Appl. Phys.* **26**, L2087 (1987).

[244]J. M. Tarascon, P. Barboux, P. F. Miceli, L. H. Greene, G. W. Hull, M. Eibschutz, and S. A. Sunshine, *Phys. Rev. B* **37**, 7458 (1988).

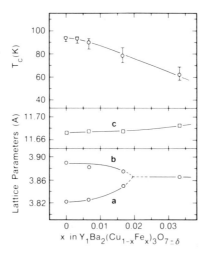

FIG. 22. Dependence of lattice parameters and T_c upon x in $Y_1Ba_2(Cu_{1-x}Fe_x)_3O_{7\pm\delta}$. Doping with Fe induces an orthorhombic-to-tetragonal phase transition at $\cong 0.02$ but the variation in T_c with x is smooth and insensitive to the structural transition. (Maeno et al.[242])

Analysis of substitutions for copper is complicated by the fact that two distinct copper sites exist in the $Y_1Ba_2Cu_3O_{7-\delta}$ structure. Much effort has therefore been directed towards identifying the way in which the substituents are distributed between the two sites. Conflicting conclusions have been reached in many cases. For example, neutron diffraction data from a $Y_1Ba_2Cu_{2.7}Zn_{0.3}O_{6.75}$ sample located 75% of the Zn^{+2} ions in the Cu(2) site,[245] but a second study on a sample with less zinc, $Y_1Ba_2Cu_{2.8}Zn_{0.2}O_{6.8}$, reached the opposite conclusion and placed all of the zinc in the Cu(1) site.[246] Results also disagree for the nickel substituted materials. Here, neutron diffraction indicates that the Ni ions are located entirely on the Cu(2) site in the CuO_2 planes,[245] but Raman scattering suggests that Ni^{+2} substitutes into the Cu(1) site in the chains.[247] For cobalt it is agreed[245,248–250] that substitution occurs prima-

[245]T. Kajitani, K. Kusaba, M. Kikuchi, Y. Syono, and M. Hirabayashi, *Jpn. J. Appl. Phys.* **27,** L354 (1988).

[246]G. Xiao, M. Z. Cieplak, D. Musser, A. Gavrin, F. H. Streitz, C. L. Chien, J. J. Rhyne, and J. A. Gotaas, *Nature* **332,** 238 (1988).

[247]Z. Iqbal, S. W. Steinhauser, A. Bose, N. Cipollini, and H. Echhardt, *Phys. Rev. B* **36,** 2283 (1987).

[248]T. Kajitani, K. Kusaba, M. Kikuchi, Y. Syono, and M. Hirabayashi, *Jpn. J. Appl. Phys.* **26,** L1727 (1987).

[249]P. F. Miceli, J. M. Tarascon, L. H. Greene, P. Barboux, F. J. Rotella, and J. D. Jorgensen, *Phys. Rev. B* **37,** 5932 (1988).

[250]Y. K. Tao, J. S. Swinnea, A. Manthiram, J. S. Kim, J. B. Goodenough, and H. Stienfink, *J. Mater. Res.* **3,** 248 (1988).

rily on the Cu(1) site. It is difficult to locate gallium by x-ray or neutron diffraction because the scattering amplitudes for gallium and copper are similar. We found no reports with direct evidence for the location of gallium in gallium-substituted 123.

The most comprehensive studies of the location of a substituted ion have been made for iron because its local state can be probed using Mossbauer spectroscopy. Unfortunately, considerable disagreement has arisen about the site assignment of the observed quadrupole doublets.[251-260] Two prominent quadrupole doublets are seen at room temperature,[251-254] one with a splitting of 1.05 mm/s and the other with a larger splitting of 1.97 mm/s. Both doublets have a small isomer shift of about -0.1 to -0.2 mm/s. Some researchers suggest that the spectra consist of two doublets with nearly the same splitting but different isomer shifts.[254,257] This assignment seems unlikely because changes in iron or oxygen content then do not affect the intensity of the doublets symmetrically. Rough estimates of the field gradients in the Cu(1) and Cu(2) sites led Tang et al.[251] to assign the larger splitting (1.97 mm/s) to Fe^{3+} ions located in the Cu(1) site and the smaller splitting to the Cu(2) site, whereas Coey et al.[252] reached the opposite conclusion. Assignment on the assumption that the iron is uniformly distributed between sites led Zhou et al.[255] to conclude that the more intense doublet with the smaller splitting came from Fe^{+3} on the more abundant Cu(2) site.

Variations in the Mossbauer spectra with composition have also been studied. For example, Bauminger et al.[256] compared spectra from oxygenated and quenched samples and found changes in the intensities of both doublets. Because it seems unlikely that the distribution of iron in the two sites changes on cooling, and because the oxygen environment

[251]H. Tang, Z. Q. Qiu, Y. Du, G. Xiao, C. L. Chien, and J. C. Walker, *Phys. Rev. B* **36**, 4018 (1987).

[252]J. M. D. Coey and K. Donnelly, *Z. Phys. B* **67**, 513 (1987).

[253]Z. Q. Qiu, Y. W. Du, H. Tang, J. C. Walker, W. A. Bryden, and K. Moorjani, *J. Magnetism and Magnetic Mater.* **69**, L221 (1987).

[254]R. Gómez, S. Aburto, M. L. Marquina, M. Jiménez, V. Marquina, C. Quintanar, T. Akachi, R. Escudero, R. A. Barrio, and D. Rios-Jara, *Phys. Rev. B* **36**, 7226 (1987).

[255]X. Z. Zhou, M. Raudsepp, Q. A. Pankhurst, A. H. Morrish, Y. L. Luo, and I. Maartense, *Phys. Rev. B* **36**, 7230 (1987).

[256]E. R. Bauminger, M. Kowitt, I. Felner, and I. Nowik, *Solid State Commun.* **65**, 123 (1988).

[257]C. W. Kimball, J. L. Matykiewicz, J. Giapintzakis, A. E. Dwight, M. B. Brodsky, M. Slaski, B. D. Dunlap, and F. Y. Fradin, *Physica B* **148**, 309 (1987).

[258]T. Tamaki, T. Komai, A. Ito, Y. Maeno, and T. Fujita, *Solid State Commun.* **65**, 43 (1988).

[259]M. Takano and Y. Takeda, *Jpn. J. Appl. Phys.* **26**, L1862 (1987).

[260]M. W. Dirken, R. C. Thiel, H. H. A. Smit, and H. W. Zandbergen, *Physica C* **156**, 303 (1988).

around the Cu(2) site is unchanged by oxygenation, they proposed that both doublets must come from Fe^{+3} in the Cu(1) site. Furthermore, they suggested that the doublets came from ions in the chain sites coordinated with different numbers of oxygens. A more detailed study of the effect of oxygen content on the spectra, in which the iron content was kept constant and the oxygen content varied,[259] found that the doublet with the largest splitting remained constant for oxygen contents from 6.0 to 6.9, while the second doublet grew with increasing oxygen content. This behavior suggests that the doublet that remains constant comes from iron in the unchanging Cu(2) site and that the change in the second doublet comes from changes in the environment of Fe^{+3} ions on the Cu(1) site. At oxygen contents higher than 6.8, changes in both doublets were observed that were attributed to Fe ions transferring from the Cu(2) site to the Cu(1) site.

In some studies, an additional doublet with a splitting of 0.58 mm/s and an isomer shift of 0.36 mm/s was observed. This doublet has been attributed to Fe^{3+} in Cu(1) sites coordinated with different numbers of oxygens,[252,255-260] Fe^{+3} in octahedrally coordinated sites in the Cu(2) layer,[259] and increased Fe^{+3} coordination at sites in grain boundaries or twins.[252] The assignment of a +3 oxidation state to Fe has also been questioned and the suggestion made that some peaks may be caused by impurity phases.[261]

Diffraction studies of iron-doped materials provide more direct information about the location of iron in the structure. These studies find that the iron substitutes primarily on the Cu(1) site,[250,261,262] thus favoring the assignment of Mossbauer peaks proposed by Bauminger et al.[256] The combination of the Mossbauer and x-ray diffraction data strongly supports the suggestion that iron has several different oxygen coordinations in the chains.[261,262] The observation that Co ions substitute onto the chain sites is also consistent with this idea. The proposal that the higher valence ions change the distribution of oxygen in the basal plane of the structure in order to increase their oxygen coordination therefore seems reasonable.

The coordination of iron in the chains with additional oxygen provides an energetically favorable means for the chains to branch or intersect with each other. Increased iron content should thus cause more branches, loops, and zig-zags (twins) in the chains.[243,263] On average such a

[261]G. Roth, G. Heger, B. Renker, J. Pannetier, V. Caignaert, M. Hervieu, and B. Raveau, *Z. Phys. B* **71**, 43 (1988).

[262]P. Bordet, J. L. Hodeau, P. Strobel, M. Marezio, and A. Santoro, *Solid State Commun.* **66**, 435 (1988).

[263]Y. Oda, H. Fujita, H. Toyoda, T. Kaneko, T. Kohara, I. Nakada, K. Asayama, *Jpn. J. Appl. Phys.* **26**, L1660 (1987).

structure has tetragonal symmetry, but microdomains of the orthorhombic structure are preserved. TEM studies have provided direct evidence for this microdomain formation.[261,262,264] Evidence has also been found for the clustering of iron along $\{110\}$ planes in the crystal, as expected if iron atoms try to maximize their oxygen coordination.[262] The observed microdomain structure varies with iron concentration.[264] Below the critical concentration needed to form the tetragonal phase (as determined by x-ray diffraction), the structure is orthorhombic and the normal twinning behavior is observed, although the twin spacing is finer than that observed in iron-free 123. Above the critical concentration, the microdomain structure forms and becomes finer with increasing iron content. Similar behavior should be observed in cobalt- and gallium-substituted 123 if the explanation for the impurity-induced orthorhombic-to-tetragonal transition in these systems is the same, but so far microstructural studies of these materials have not been reported.

In iron- and cobalt-substituted 123, the a and b lattice parameters converge when only 1–3% of the copper is replaced.[242,244,250,265,266] In gallium-substituted 123, the lattice parameters converge more slowly and the tetragonal phase forms after 10% substitution.[247,267] The a and b lattice parameters also converge with nickel additions, but a multiphase region is reached before the tetragonal phase forms.[244] With zinc additions, the a and b lattice parameters first converge slightly and then the difference between them remains almost constant.[244,246]

The observation of microdomains in iron-doped 123 implies that the material remains orthorhombic beyond the concentration needed to form the tetragonal phase on a macroscopic scale. The question then remains as to how disordered the oxygen is in the orthorhombic microdomains. The macroscopic lattice parameters are deceptive since they reflect the influence of both the oxygen disorder and the strains that arise to accommodate microdomain formation. Indeed it is unclear whether or not there is a true phase transformation from an orthorhombic to a tetragonal structure. If there is an order-disorder transformation, it occurs at a higher iron concentration than that deduced from x-ray measurements of lattice constants. At the higher iron concentrations, where the microdomain size approaches unit cell dimensions, the microdomain model is no longer meaningful and the structure must

[264]Z. Hiroi, M. Takano, Y. Takeda, R. Kanno, and Y. Bando, *Jpn. J. Appl. Phys.* **27,** L580 (1988).

[265]J. Langen, M. Veit, M. Galffy, H. D. Jostarndt, A. Erle, S. Blumenröder, H. Schmidt, and E. Zirngiebl, *Solid State Commun.* **65,** 973 (1988).

[266]H. Obara, H. Oyanagi, K. Murata, H. Yamasaki, H. Ihara, M. Tokumoto, Y. Nishihara, and Y. Kimura, *Jpn. J. Appl. Phys.* **27,** L603 (1988).

[267]M. Hiratani, Y. Ito, K. Miyauchi, and T. Kudo, *Jpn. J. Appl. Phys.* **26,** L1997 (1987).

become truly tetragonal. The concentration at which this occurs is not known, but should depend on how the iron is distributed between the chain and plane sites and the degree to which the iron tends to cluster.

In addition to substitutions affecting oxygen ordering, the oxygen content of the unit cell may increase with increasing gallium, cobalt, or iron substitution to compensate for the higher oxidation state of the ions. Iodometric titration and neutron diffraction measurements of the oxygen content of cobalt- and iron-doped 123 confirm this trend, as seen in the plot of oxygen content versus degree of substitution given in Fig. 23.[244,250] It is possible that oxygen intercalates in the form of peroxide ions as proposed in rare-earth-substituted 123,[250] but at present there is no direct evidence for this.

The reason for the suppression of T_c in transition-element-doped 123 remains obscure. The substitution-induced orthorhombic-to-tetragonal transition suggests strong similarity to the behavior in the rare earth solid solution materials. There is no evidence, however, that the symmetry change itself is linked to the depression of T_c. Indeed, the material remains orthorhombic with zinc substitution, yet zinc shows the strongest effect on T_c. This behavior may reflect the tendency of zinc to substitute into the copper oxygen planes and so directly disrupt the superconductivity. In this respect it is particularly important to reconcile the conflicting data about the location of zinc in the structure.[245,246] Nickel substitution may have the same direct influence on superconductivity, since the existing diffraction data indicate that nickel substitution occurs

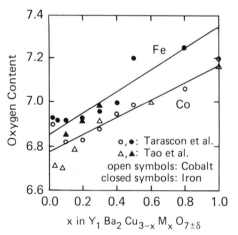

FIG. 23. Total oxygen content of $Y_1Ba_2Cu_{3-x}M_xO_{7\pm\delta}$ as a function of x for $M = $ Co, Fe. Increasing substitution induces intercalation of oxygen into the structure. (Compilation of data from Tarascon et al.[244] and Tao et al.[250])

in the planes.[245] A possible effect of the trivalent iron, cobalt, and gallium ions on T_c is to alter the hole concentration in the material, as has been suggested for $La_{2-x}Sr_xCuO_{4-y}$[238] and the rare earth solid solution materials.[231] Measurements of hole concentrations in the different materials are needed to test this idea.

10. SUBSTITUTIONS OF OTHER CATIONS

Sc, Ca, and Na can partially replace yttrium. For scandium, T_c drops only slightly with up to 75% replacement of yttrium.[268] Larger Sc concentrations then cause a rapid drop in T_c, and the material becomes nonsuperconducting if more than 81% of the yttrium is replaced. Data for Na substitution indicate that replacement of up to 60% of the yttrium is possible, but that T_c continuously drops with increasing substitution.[201,269] To explore further the relationship between hole concentration and T_c, attempts have been made deliberately to dope holes into the $Y_1Ba_2Cu_3O_{7-\delta}$ structure by substituting Ca into the Y site.[223,224,270] Doping with Ca alone depresses T_c, but since the oxygen content decreases with Ca doping, the total hole concentration is also reduced. A more extensive solid solution range for calcium is observed in the $La_{1-x}Ca_xBa_2Cu_3O_{7-\delta}$ and $Nd_{1-x}Ca_xBa_2Cu_3O_{7-\delta}$ systems, and again T_c is depressed on substitution.[223,270]

A wide variety of elements, including Ag,[271–273] Al,[244,274,275] Cr,[240,241,271] Li,[271] Mn,[241,271] Pt,[271] Ti,[240] and V[276] can substitute into the copper sites in $Y_1Ba_2Cu_3O_{7-\delta}$. In each case T_c is suppressed, and, to date, complete replacement of copper without the destruction of super-conductivity has not been achieved. Silver and vanadium appear to have the least deleterious effect on superconductivity, but it is unclear whether

[268]B. R. Zhao, Y. H. Shi, Y. Lu, H. S. Wang, Y. Y. Zhao, and L. Li, *Solid State Commun.* **63,** 409 (1987).

[269]A. Fartash and H. Oesterreicher, *Solid State Commun.* **66,** 39 (1988).

[270]H. Uwe, T. Sakuda, H. Asano, T. S. Han, K. Yagi, R. Harada, M. Iha, and Y. Yokoyama, *Jpn. J. Appl. Phys.* **27,** L577 (1988).

[271]P. Strobel, C. Paulsen, and J. L. Tholence, *Solid State Commun.* **65,** 585 (1988).

[272]C. V. Tomy, A. M. Umarji, D. T. Adroja, S. K. Malik, R. Prasad, N. C. Soni, A. Mohan, and C. K. Gupta, *Solid State Commun.* **64,** 889 (1987).

[273]Y. Saito, T. Noji, A. Endo, N. Higuchi, T. Fujimoto, T. Oikawa, A. Hattori, and K. Furuse, *Jpn. J. Appl. Phys.* **26,** L832 (1987).

[274]J. P. Franck, J. Jung, and M. A. K. Mohamed, *Phys. Rev. B* **36,** 2308 (1987).

[275]T. Siegrist, L. F. Schneemeyer, J. V. Waszczak, N. P. Singh, R. L. Opila, B. Batlogg, L. W. Rupp, and D. W. Murphy, *Phys. Rev. B* **36,** 8365 (1987).

[276]S. X. Dou, A. J. Bourdillon, X. Y. Sun, J. P. Zhou, H. K. Liu, N. Savvides, D. Haneman, C. C. Sorrell, and K. E. Easterling, *J. Phys. C* **21,** L127 (1988).

or not these elements have an appreciable solubility in the structure, as both form multiphase materials.[271–273,276] Substitution of 1–2% aluminum causes the structure to become tetragonal.[244,275] X-ray diffraction studies of Al-doped single crystals found that Al substitutes for Cu in the Cu(1) chain sites.[275] Thus Al behaves in a similar manner to the trivalent and tetravalent transition metal ions.

There are few reports of substitution in the barium site other than the substitution of the larger lanthanide ions that have already been discussed. Most of the work has concentrated on the substitution of strontium for barium.[277–281] Up to 50% of the barium can be replaced by strontium, but the substitution is accompanied by a linear decrease in T_c with concentration.[277,278,280] Attempts to correlate T_c with oxygen content have been unsuccessful.[280] Some argue that the slightly smaller size of strontium perturbs the structure, causing a slight collapse in the copper-oxygen structure around the barium site, which lowers T_c.[278] Potassium had little effect on T_c when up to 10% of the barium was replaced.[282]

11. Oxygen Substitution

There have been several reports of oxygen substitutions causing dramatic increases in T_c. In particular, it was claimed that fluorine substitution produced a material with zero resistance at temperatures as high as 155 K.[283,284] Attempts to reproduce these results, however, have not been successful.[285–287] Solid-state reaction routes resulted in little

[277]T. Wada, S. Adachi, T. Mihara, and R. Inaba, *Jpn. J. Appl. Phys.* **26**, L706 (1987).

[278]B. W. Veal, W. K. Kwok, A. Umezawa, G. W. Crabtree, J. D. Jorgensen, J. W. Downey, L. J. Nowicki, A. W. Mitchell, A. P. Paulikas, and C. H. Sowers, *Appl. Phys. Lett.* **51**, 279 (1987).

[279]T. Wada, S. Adachi, O. Inoue, S. Kawashima, and T. Mihara, *Jpn. J. Appl. Phys.* **26**, L1475 (1987).

[280]A. Ono, T. Tanaka, H. Nozaki, and Y. Ishizawa, *Jpn. J. Appl. Phys.* **26**, L1687 (1987).

[281]Q. R. Zhang, L. Cao, Y. Qian, Z. Chen, Y. Zhao, G. Pang, H. Zhang, J. Xia, M. Zhang, D. Yu, Z. He, S. Sun, M. Fang, and T. Zhang, *Solid State Commun.* **63**, 535 (1987).

[282]I. Felner and B. Barbara, *Solid State Commun.* **66**, 205 (1988).

[283]S. R. Ovshinsky, R. T. Young, D. D. Allred, G. DeMaggio, and G. A. Van der Leeden, *Phys. Rev. Lett.* **58**, 2579 (1987).

[284]X. R. Meng, Y. R. Ren, M. Z. Lin, Q. Y. Tu, Z. J. Lin, L. H. Sang, W. Q. Ding, M. H. Fu, Q. Y. Meng, C. J. Li, X. H. Li, G. L. Qiu, and M. Y. Chen, *Solid State Commun.* **64**, 325 (1987).

[285]R. N. Bhargava, S. P. Herko, and W. N. Osborne, *Phys. Rev. Lett.* **59**, 1468 (1987).

[286]A. K. Tyagi, S. J. Patwe, U. R. K. Rao, and R. M. Iyer, *Solid State Commun.* **65**, 1149 (1988).

[287]P. K. Davies, J. A. Stuart, D. White, C. Lee, P. M. Chaikin, M. J. Naughton, R. C. Yu, and R. L. Ehrenkaufer, *Solid State Commun.* **64**, 1441 (1987).

or no fluorine substitution. When partial substitution was achieved, T_c remained the same or was slightly depressed. Sulphur[288] and nitrogen[289-291] substitution have also been reported to raise T_c, but again the results have not been confirmed and the effects are short lived. Another report indicates that some sulphur substitution can take place but that the substitution has no effect on T_c.[292]

In addition to studies of oxygen substitution by different atomic species, several studies were made of ^{18}O substitution to determine if the isotope shift in T_c expected for phonon-mediated superconductivity occurs in these materials. By a process of reduction followed by reannealing in ^{18}O gas, as much as 90% of the ^{16}O can be exchanged in the $Y_1Ba_2Cu_3O_{7-\delta}$ structure.[293-299] In studies where 75 to 90% of the ^{16}O was exchanged, decreases in T_c too small to be significant were observed.[293,294] In a third study, a significant decrease in T_c of 0.3—0.4 K on 90% exchange was detected, indicating that phonons may play an important role in the pairing mechanism.[295] It was suggested that the absence of an effect in earlier studies was caused by preferential substitution of the isotope into sites that did not influence the superconducting behavior.[296] Replies to this comment noted that this was unlikely in view of the large fraction of ^{16}O that was exchanged[297] and the evidence from Raman spectroscopy that a substantial fraction of the substituted ^{18}O occupies sites in the copper-oxygen planes.[298] Later results confirmed the presence of a small shift but again disagreed as to

[288]K. N. R. Taylor, D. N. Matthews, and G. J. Russell, *J. Cryst. Growth* **85,** 628 (1987).

[289]D. N. Matthews, A. Bailey, R. A. Vaile, G. J. Russell, and K. N. R. Taylor, *Nature* **328,** 786 (1987).

[290]D. D. Sarma, C. T. Simmons, and G. Kaindl, *Nature* **330,** 213 (1987).

[291]K. N. R. Taylor, G. J. Russell, D. N. Matthews, A. Bailey, and R. A. Vaile, *Nature,* **330,** 214 (1987).

[292]I. Felner, I. Nowik, and Y. Yeshurun, *Phys. Rev. B* **36,** 3923 (1987).

[293]B. Batlogg, R. J. Cava, A. Jayaraman, R. B. van Dover, G. A. Kourouklis, S. Sunshine, D. W. Murphy, L. W. Rupp, H. S. Chen, A. White, K. T. Short, A. M. Mujsce, and E. A. Rietman, *Phys. Rev. Lett.* **58,** 2333 (1987).

[294]L. C. Bourne, M. F. Crommie, A. Zettl, H. C. Loye, S. W. Keller, K. L. Leary, A. M. Stacy, K. J. Chang, M. L. Cohen, and D. E. Morris, *Phys. Rev. Lett.* **58,** 2337 (1987).

[295]K. J. Leary, H. C. Loye, S. W. Keller, T. A. Faltens, W. K. Ham, J. N. Michaels, and A. M. Stacy, *Phys. Rev. Lett.* **59,** 1236 (1987).

[296]M. Grimsditch, T. O. Brun, R. Bhadra, B. Dabrowski, D. G. Hinks, J. D. Jorgensen, M. A. Beno, J. Z. Liu, H. B. Schüttler, C. U. Segre, L. Soderholm, B. W. Veal, and I. K. Schuller, *Phys. Rev. Lett.* **60,** 752 (1988).

[297]A. Zettl and J. Kinney, *Phys. Rev. Lett.* **60,** 753 (1988).

[298]B. Batlogg, R. J. Cava, and M. Stavola, *Phys. Rev. Lett.* **60,** 754 (1988).

[299]H. Katayama-Yoshida, T. Hirooka, A. J. Mascarenhas, Y. Okabe, T. Takahashi, T. Sasaki, A. Ochiai, T. Suzuki, J. I. Pankove, T. Ciszek, and S. K. Deb, *Jpn. J. Appl. Phys.* **26,** L2085 (1987).

the magnitude of the shift.[299–301] Preparation of materials enriched in ^{135}Ba, ^{136}Ba, ^{63}Cu, or ^{65}Cu failed to detect any isotope shift associated with the cation lattice.[299,302–304] Although the detection of an isotope shift on oxygen substitution implies that phonons are involved in the pairing mechanism, the small magnitude of the shift suggests that additional interactions are also involved.

VI. Related Structures

The main reasons for studying related superconducting structures are largely the same as those for studying structural variants of 123, namely to determine which structural features play an important role in superconductivity and to look for materials with better superconducting properties. In this section, we describe the basic structures of three related families of copper-oxide-based superconductors: $La_{2-x}M_xCuO_{4-y}$ (M = Ba, Sr, or Ca), $Bi_2Ca_{n-1}Sr_2Cu_nO_{2n+4}$ (n = 1, 2, or 3), and $Tl_mCa_{n-1}Ba_2Cu_nO_{2(n+1)+m}$ (m = 1 or 2, n = 1, 2, or 3).

12. LANTHANUM COPPER OXIDES

Bednorz and Müller[1,2] first discovered high-temperature superconductivity in barium-doped lanthanum copper oxide, $La_{2-x}Ba_xCuO_{4-y}$, a material that was studied previously by Michel and Raveau[305] for electrocatalytic applications. The room temperature structure of this phase is isomorphic to K_2NiF_4, body-centered tetragonal with space group $I4/mmm$. The structure is made by stacking alternate layers of perovskite, $LaCuO_3$, and rock salt, LaO, along the c axis such that the copper sites in one perovskite layer are aligned with the lanthanum sites in the next perovskite layer (Fig. 24).

A tetragonal-to-orthorhombic phase transition occurs in $La_{2-x}M_xCuO_{4-y}$ at low temperatures and/or low dopant

[300]H. C. Loye, K. J. Leary, S. W. Keller, W. K. Ham, T. A. Faltens, J. N. Michaels, and A. M. Stacy, *Science* **238**, 1558 (1987).

[301]D. E. Morris, R. M. Kuroda, A. G. Markelz, J. H. Nickel, and J. Y. T. Wei, *Phys. Rev. B* **37**, 5936 (1988).

[302]L. C. Bourne, A. Zettl, T. W. Barbee III, and M. L. Cohen, *Phys. Rev. B.* **36**, 3990 (1987).

[303]T. Hidaka, T. Matsui, and Y. Nakagawa, *Jpn. J. Appl. Phys.* **27**, L553 (1988).

[304]Q. Lin, Y. N. Wei, Q. W. Yan, G. H. Chen, P. L. Zhang, Z. G. Shen, Y. M. Ni, Q. S. Yang, C. X. Liu, T. S. Ning, J. K. Zhao, Y. Y. Shao, S. H. Han, and J. Y. Li, *Solid State Commun.* **65**, 869 (1988).

[305]C. Michel and B. Raveau, *Rev. Chim. Miner.* **21**, 407 (1984).

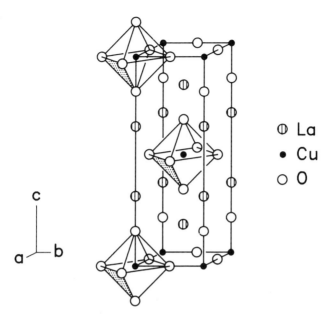

⊕ La
• Cu
○ O

FIG. 24. The tetragonal La_2CuO_4 structure.

concentrations.[306-309] At the transition, adjacent CuO_6 octahedra tilt about the [110] direction in opposite directions, thereby buckling the CuO_2 planes and doubling the unit cell size. The orthorhombic unit cell, $\sim\sqrt{2}\,a_p \times \sim\sqrt{2}\,a_p \times c$, is related to the tetragonal cell, $a_p \times a_p \times c$, by a 45° rotation about the tetragonal c axis. By convention, for the orthorhombic space group *Cmca*, the long axis of the unit cell is the b axis (whereas it is the c axis in the tetragonal unit cell). The tetragonal-to-orthorhombic transition temperature changes dramatically with doping. For example, it varies from 533 K in undoped La_2CuO_{4-y}[306] to 12 K in $La_{1.8}Sr_{0.2}CuO_{4-y}$.[309] Fleming *et al.*[309] found that the structure is orthorhombic in the superconducting state and that the largest volume fraction of ideal diamagnetism occurs for compositions in which the tetragonal-to-orthorhombic transition temperature coincides with the

[306]J. M. Longo and P. M. Raccah, *J. Solid State Chem.* **6**, 526 (1973).
[307]V. B. Grande, Hk. Müller-Buschbaum, and M. Schweizer, *Z. Anorg. Allg. Chem.* **428**, 120 (1977).
[308]J. D. Jorgensen, H. B. Schüttler, D. G. Hinks, D. W. Capone II, K. Zhang, M. B. Brodsky, and D. J. Scalapino, *Phys. Rev. Lett.* **58**, 1024 (1987).
[309]R. M. Fleming, B. Batlogg, R. J. Cava, and E. A. Rietman, *Phys. Rev. B* **35**, 7191 (1987).

superconducting transition temperature. Fleming *et al.*[309] and others[310,311] concluded that the latter result is fortuitous; there does not appear to be a direct connection between this lattice instability and superconductivity. For clarity, it should be noted that the tetragonal-to-orthorhombic transition in $La_{2-x}M_xCuO_{4-y}$ caused by tilting of the CuO_6 octahedra is not related in any way to the orthorhombic-to-tetragonal transition in 123 caused by oxygen disordering on the basal copper planes.

At least one additional structural transition occurs in $La_{2-x}M_xCuO_{4-y}$ at very low temperatures. Early experimental and theoretical studies[312–314] found evidence for structural instability, possibly due to a monoclinic distortion. A more recent study[315] proposed that there is a transformation to a new tetragonal structure with space group $P4_2/ncm$. The role of these proposed transitions in superconductivity has yet to be established.

Like 123, the precise structure and resulting properties of $La_{2-x}M_xCuO_{4-y}$ depend sensitively on processing.[1,2,316–320] Many of the early studies examined the effects of changing the alkaline earth dopant and its concentration. Superconducting transition temperatures of 37, 32, and 17 K were found for Sr, Ba, and Ca, respectively, for $x \sim 0.15$.[320] It rapidly became evident, however, that oxygen and lanthanum deficiencies also affect superconducting properties. This point was brought out in studies of undoped $La_{2-\Delta}CuO_{4-y}$, which can be either a 40 K superconductor[321–323] or an antiferromagnetic insulator,[324–329] depending

[310]W. Weber, *Phys. Rev. Lett.* **58**, 1371 (1987).

[311]P. Böni, J. D. Axe, G. Shirane, R. J. Birgeneau, D. R. Gabbe, H. P. Jenssen, M. A. Kastner, C. J. Peters, P. J. Picone, and T. R. Thurston, *Phys. Rev. B* **38**, 185 (1988).

[312]S. C. Moss, K. Forster, J. D. Axe, H. You, D. Hohlwein, D. E. Cox, P. H. Hor, R. L. Meng, and C. W. Chu, *Phys. Rev. B* **35**, 7195 (1987).

[313]R. V. Kasowski, W. Y. Hsu, and F. Herman, *Solid State Commun.* **63**, 1077 (1987).

[314]D. McK. Paul, G. Balakrishnan, N. R. Bernhoeft, W. I. F. David, and W. T. A. Harrison, *Phys. Rev. Lett.* **58**, 1976 (1987).

[315]J. D. Axe, D. E. Cox, K. Mohanty, H. Moudden, A. R. Moodenbaugh, Y. Xu, and T. R. Thurston, *IBM J. Res. Develop.* (1989) (in press).

[316]S. Uchida, H. Takagi, K. Kitazawa, and S. Tanaka, *Jpn. J. Appl. Phys.* **26**, L1 (1987).

[317]H. Takagi, S. Uchida, K. Kitazawa, and S. Tanaka, *Jpn. J. Appl. Phys.* **26**, L123 (1987).

[318]C. W. Chu, P. H. Hor, R. L. Meng, L. Gao, Z. J. Huang, and Y. Q. Wang, *Phys. Rev. Lett.* **58**, 405 (1987).

[319]R. J. Cava, R. B. van Dover, B. Batlogg, and E. A. Rietman, *Phys. Rev. Lett.* **58**, 408 (1987).

[320]J. G. Bednorz, K. A. Müller, and M. Takashige, *Science* **236**, 73 (1987).

[321]J. Beille, R. Cabanel, C. Chaillout, B. Chevallier, G. Demazeau, F. Deslandes, J. Etourneau, P. Lejay, C. Michel, J. Provost, B. Raveau, A. Sulpice, J. L. Tholence, and R. Tournier, *C.R. Acad. Sci. Ser. 2* **304**, 1097 (1987).

[322]K. Sekizawa, Y. Takano, H. Takigami, S. Tasaki, and T. Inaba, *Jpn. J. Appl. Phys.* **26**, L840 (1987).

on the lanthanum and oxygen concentrations. Thus, the preparation of thermodynamically well-defined samples of these doped ternary oxides is just as difficult as the preparation of well-defined quaternary 123 samples, if not more so. The discovery of the 123 materials preempted many more quantitative studies of processing-structure-property relationships in the lanthanum copper oxides.

13. BISMUTH-CONTAINING SUPERCONDUCTORS

In May 1987, Michel et al.[330] reported the discovery of superconductivity between 7 and 22 K in the Bi–Sr–Cu–O system. Because of the intense interest in the 90 K materials at that time, their report did not attract widespread interest. However, attention quickly focussed on the bismuth-containing superconductors in January 1988 when Maeda et al.[4] and Chu et al.[331] reported that adding Ca to the Bi–Sr–Cu–O system produced material that was superconducting above liquid nitrogen temperature. Three superconducting oxides were subsequently identified in the Bi–Ca–Sr–Cu–O system:[332–347] $Bi_2Sr_2Cu_1O_{6+x}$ ($T_c = 7 - 22$ K), $Bi_2Ca_1Sr_2Cu_2O_{8+x}$ ($T_c \sim 85$ K), and $Bi_2Ca_2Sr_2Cu_3O_{10+x}$ ($T_c \sim 110$ K). For brevity, these phases will be referred to as Bi2021, Bi2122, and Bi2223, respectively. The structures consist of perovskite-like units containing one, two, or three CuO_2 planes sandwiched between Bi–O bilayers (Fig. 25(d), (e), and (f)). Bi2021, the phase responsible for the results reported by Michel et al.[330] remains the least studied of the three because of its low transition temperature.[344,345]

[323]P. M. Grant, S. S. P. Parkin, V. Y. Lee, E. M. Engler, M. L. Ramirez, J. E. Vazquez, G. Lim, R. D. Jacowitz, and R. L. Greene, *Phys. Rev. Lett.* **58**, 2482 (1987).

[324]Y. Yamaguchi, H. Yamauchi, M. Ohashi, H. Yamamoto, N. Shimoda, M. Kikuchi, and Y. Syono, *Jpn. J. Appl. Phys.* **26**, L447 (1987).

[325]R. L. Greene, H. Maletta, T. S. Plaskett, J. G. Bednorz, and K. A. Müller, *Solid State Commun.* **63**, 379 (1987).

[326]D. Vaknin, S. K. Sinha, D. E. Moncton, D. C. Johnston, J. Newsam, C. R. Safinya, and H. E. King, Jr., *Phys. Rev. Lett.* **58**, 2802 (1987).

[327]S. Mitsuda, G. Shirane, S. K. Sinha, D. C. Johnston, M. S. Alvarez, D. Vaknin, and D. E. Moncton, *Phys. Rev. B* **36**, 822 (1987).

[328]T. Freltoft, J. P. Remeika, D. E. Moncton, A. S. Cooper, J. E. Fischer, D. Harshman, G. Shirane, S. K. Sinha, and D. Vaknin, *Phys. Rev. B* **36**, 826 (1987).

[329]D. C. Johnston, J. P. Stokes, D. P. Goshorn, and J. T. Lewandowski, *Phys. Rev. B* **36**, 4007 (1987).

[330]C. Michel, M. Hervieu, M. M. Borel, A. Grandin, F. Deslandes, J. Provost, and B. Raveau, *Z. Phys. B* **68**, 421 (1987).

[331]C. W. Chu, J. Bechtold, L. Gao, P. H. Hor, Z. J. Huang, R. L. Meng, Y. Y. Sun, Y. Q. Wang, and Y. Y. Xue, *Phys. Rev. Lett.* **60**, 941 (1988).

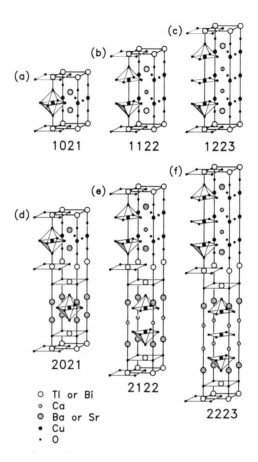

FIG. 25. Nominal unit cells for the bismuth- and thallium-containing superconductors: (a) $Tl_1Ba_2Cu_1O_5$, (b) $Tl_1Ca_1Ba_2Cu_2O_7$, (c) $Tl_1Ca_2Ba_2Cu_3O_9$, (d) $Tl_2Ba_2Cu_1O_6$, (e) $Tl_2Ca_1Ba_2Cu_2O_8$, and (f) $Tl_2Ca_2Ba_2Cu_3O_{10}$. Only the 2021, 2122, and 2223 structures form in the Bi–Ca–Sr–Cu–O system, whereas all six structures form in the Tl–Ca–Ba–Cu–O system. (Beyers et al.[366])

[332]R. M. Hazen, C. T. Prewitt, R. J. Angel, N. L. Ross, L. W. Finger, C. G. Hadidiacos, D. R. Veblen, P. J. Heaney, P. H. Hor, R. L. Meng, Y. Y. Sun, T. Q. Wang, Y. Y. Xue, Z. J. Huang, L. Gao, J. Bechtold, and C. W. Chu, *Phys. Rev. Lett.* **60,** 1174 (1988).

[333]Y. Bando, T. Kijima, T. Kitami, J. Tanaka, F. Izumi, and M. Yokoyama, *Jpn. J. Appl. Phys.* **27,** L358 (1988).

[334]Y. Matsui, H. Maeda, Y. Tanaka, and S. Horiuchi, *Jpn. J. Appl. Phys.* **27,** L361 (1988).

[335]E. Takayama-Muromachi, Y. Uchida, A. Ono, F. Izumi, M. Onoda, Y. Matsui, K. Kosuda, S. Takekawa, and K. Kato, *Jpn. J. Appl. Phys.* **27,** L365 (1988).

[336]M. A. Subramanian, C. C. Torardi, J. C. Calabrese, J. Gopalakrishnan, K. J. Morrissey, T. R. Askew, R. B. Flippen, U. Chowdhry, and A. W. Sleight, *Science* **239,** 1015 (1988).

Bi2122 has been the most extensively studied bismuth-containing superconductor.[332-344,347-358] The perovskite-like unit in Bi2122 is remarkably similar to that in 123; the Ca and Sr cations in Bi2122 play the same roles as the Y and Ba cations in 123. The linear chains in the 123 structure are replaced by the Bi–O bilayers, proving that linear chains are not essential for high-temperature superconductivity. Early on there was speculation that the atomic arrangements in Bi2122, especially in the

[337]D. R. Veblen, P. J. Heaney, R. J. Angel, L. W. Finger, R. M. Hazen, C. T. Prewitt, N. L. Ross, C. W. Chu, P. H. Hor, and R. L. Meng, *Nature* **332**, 334 (1988).

[338]J. M. Tarascon, Y. Le Page, P. Barboux, B. G. Bagley, L. H. Greene, W. R. McKinnon, G. W. Hull, M. Giroud, and D. M. Hwang, *Phys. Rev. B* **37**, 9382 (1988).

[339]T. M. Shaw, S. A. Shivashankar, S. J. La Placa, J. J. Cuomo, T. R. McGuire, R. A. Roy, K. H. Kelleher, and D. S. Yee, *Phys. Rev. B* **37**, 9856 (1988).

[340]S. A. Sunshine, T. Siegrist, L. F. Schneemeyer, D. W. Murphy, R. J. Cava, B. Batlogg, R. B. van Dover, R. M. Fleming, S. H. Glarum, S. Nakahara, R. Farrow, J. J. Krajewski, S. M. Zahurak, J. V. Waszczak, J. H. Marshall, P. Marsh, L. W. Rupp, Jr., and W. F. Peck, *Phys. Rev. B* **38**, 893 (1988).

[341]T. Kijima, J. Tanaka, Y. Bando, M. Onoda, and F. Izumi, *Jpn. J. Appl. Phys.* **27**, L369 (1988).

[342]P. Bordet, J. J. Capponi, C. Chaillout, J. Chenavas, A. W. Hewat, E. A. Hewat, J. L. Hodeau, M. Marezio, J. L. Tholence, and D. Tranqui, *Physica C* **153–155**, 623 (1988).

[343]P. Bordet, J. J. Capponi, C. Chaillout, J. Chenavas, A. W. Hewat, E. A. Hewat, J. L. Hodeau, M. Marezio, J. L. Tholence, and D. Tranqui, *Physica C* **156**, 189 (1988).

[344]J. B. Torrance, Y. Tokura, S. J. La Placa, T. C. Huang, R. J. Savoy, and A. I. Nazzal, *Solid State Commun.* **66**, 703 (1988).

[345]G. Van Tendeloo, H. W. Zandbergen, and S. Amelinckx, *Solid State Commun.* **66**, 927 (1988).

[346]H. W. Zandbergen, Y. K. Huang, M. J. V. Menken, J. N. Li, K. Kadowaki, A. A. Menovsky, G. Van Tendeloo, and S. Amelinckx, *Nature* **332**, 620 (1988).

[347]H. W. Zandbergen, P. Groen, G. Van Tendeloo, J. Van Landuyt, and S. Amelinckx, *Solid State Commun.* **66**, 397 (1988).

[348]Y. Matsui, H. Maeda, Y. Tanaka, and S. Horiuchi, *Jpn. J. Appl. Phys.* **27**, L372 (1988).

[349]E. A. Hewat, M. Dupuy, P. Bordet, J. J. Capponi, C. Chaillout, J. L. Hodeau, and M. Marezio, *Nature* **333**, 53 (1988).

[350]R. L. Withers, J. S. Anderson, B. G. Hyde, J. G. Thompson, L. R. Wallenberg, J. D. Fitzgerald, and A. M. Stewart, *J. Phys. C* **21**, L417 (1988).

[351]M. Onoda, A. Yamamoto, E. Takayama-Muromachi, and S. Takekawa, *Jpn. J. Appl. Phys.* **27**, L833 (1988).

[352]P. L. Gai and P. Day, *Physica C* **152**, 335 (1988).

[353]H. W. Zandbergen, W. A. Groen, F. C. Mijlhoff, G. Van Tendeloo, and S. Amelinckx, *Physica C* **156**, 325 (1988).

[354]S. Ikeda, H. Ichinose, T. Kimura, T. Matsumoto, H. Maeda, Y. Ishida, and K. Ogawa, *Jpn. J. Appl. Phys.* **27**, L999 (1988).

[355]G. Van Tendeloo, H. W. Zandbergen, J. Van Landuyt, and S. Amelinckx, *Appl. Phys. A* **46**, 153 (1988).

[356]G. S. Grader, E. M. Gyorgy, P. K. Gallagher, H. M. O'Bryan, D. W. Johnson, Jr., S. Sunshine, S. M. Zahurak, S. Jin, and R. C. Sherwood, *Phys. Rev. B* **38**, 757 (1988).

[357]H. Niu, N. Fukushima, and K. Ando, *Jpn. J. Appl. Phys.* **27**, L1442 (1988).

[358]A. K. Cheetham, A. M. Chippindale, and S. J. Hibble, *Nature* **333**, 21 (1988).

FIG. 26. High-resolution TEM image along the [100] directions in Bi2122 showing the incommensurate modulation. Lattice distortions and so-called "Bi-concentrated bands" (marked B) are easily seen. (Matsui *et al.*[348])

Bi–O bilayers, were the same as those in Aurivillius[359] phases, but this was shown later to be incorrect. Weak bonding between the Bi–O bilayers results in easy cleavage between these planes. The precise atomic arrangements in Bi2122 are more complicated than those indicated by the tetragonal subcell ($a_p \times a_p \times c$) in Fig. 25(d). TEM studies[332–334,336–339] found that the structure has an orthorhombic subcell ($\sim\sqrt{2}a_n \times \sim\sqrt{2}a_p \times c$) with a 25.8 Å ($\sim 5\sqrt{2}a_p$) incommensurate superlattice modulation present along the b axis (Fig. 26). Numerous studies[348–353] have tried to identify the source of the incommensurate structure, but the precise atomic arrangements giving rise to the modulation have yet to be unambiguously determined. Proposed sources include displacement of the Bi cations from their ideal sites to accommodate the mismatch between the perovskite-like units and the Bi–O bilayers,[347] incorporation of extra oxygen in the Bi–O bilayers,[349,353] periodic substitution of Bi by Sr, Ca, or vacancies,[339,348] and puckering of the CuO_2 planes that results in rotation of the individual CuO_5 polyhedra.[352] Two additional factors

[359]B. Aurivillius, *Arkiv. Kemi.* **1**, 463 (1949); **1**, 499 (1949); **2**, 519 (1950); and **5**, 39 (1952).

will complicate a complete determination of this crystal structure. First, high densities of intergrowths of perovskite-like units containing either one, three, or four CuO_2 planes are commonly found in Bi2122.[332,333,337,354,355] Second, there is a solid solubility range between Ca and Sr that may not extend to the ideal $Bi_2Ca_1Sr_2Cu_2O_8$ composition.[338,356,357] For the idealized $Bi_2Ca_1Sr_2Cu_2O_8$ structure, the average copper valence is +2 and the material is expected to be a Mott-Hubbard insulator. Several mechanisms to increase the copper valence have been proposed, including excess oxygen in the Bi–O bilayers[349,353] and Sr deficiency,[358] but this question cannot be resolved without better structure data.

The processing conditions required to form single phase or single crystal Bi2223 have yet to be determined.[360–362] Consequently, the structure and properties of this phase have not been studied extensively. The first Bi–Ca–Sr–Cu–O samples displayed sharp drops in resistance at ~110 K, but did not reach zero resistance above 100 K.[4,331] Susceptibility measurements indicated that two superconducting transitions were present in most samples, one at 85 K and the other at 110 K. Subsequent studies identified Bi2122 as the 85 K phase and linked the 110 K transition to the presence of the Bi2223 phase.[346,347] The connectivity between Bi2223 grains appears to be quite poor, because it is difficult to reach zero resistance at 110 K even in samples in which Bi2223 is the predominant phase. Unless the difficulties in preparing Bi2223 can be overcome, there is little chance for scientific or technical progress in studies of this phase.

14. THALLIUM-CONTAINING SUPERCONDUCTORS

Sheng and Hermann's[5] discovery of superconductivity above liquid nitrogen temperature in the Tl–Ba–Cu–O system in January 1988 was initially overshadowed by the breakthroughs in the Bi–Ca–Sr–Cu–O system. While numerous researchers struggled to achieve zero resistivity above 100 K in the bismuth system, Sheng and Hermann added Ca to their system and produced a Tl–Ca–Ba–Cu–O mixture that reached zero resistance at 107 K.[6] This result sparked numerous investigations of the

---

[360]J. M. Tarascon, Y. Le Page, L. H. Greene, B. G. Bagley, P. Barboux, D. M. Hwang, G. W. Hull, W. R. McKinnon, and M. Giroud, *Phys. Rev. B* **38**, 2504 (1988).

[361]H. Nobumasa, K. Shimizu, Y. Kitano, and T. Kawai, *Jpn. J. Appl. Phys.* **27**, L846 (1988).

[362]J. L. Tallon, R. G. Buckley, P. W. Gilberd, M. R. Presland, I. W. M. Brown, M. E. Bowden, L. A. Christian, and R. Goguel, *Nature* **333**, 153 (1988).

phases present in the Tl–Ca–Ba–Cu–O system.[363–384] Two superconduct-ing phases were identified in Sheng and Hermann's[6] samples by Hazen *et al.*,[363] $Tl_2Ca_1Ba_2Cu_2O_{8+x}$ and $Tl_2Ca_2Ba_2Cu_3O_{10+x}$. Then Parkin *et al.*[364] changed the processing conditions to greatly increase the amount of the $Tl_2Ca_2Ba_2Cu_3O_{10+x}$ phase and produced a material with bulk supercon-

[363]R. M. Hazen, L. W. Finger, R. J. Angel, C. T. Prewitt, N. L. Ross, C. G. Hadidiacos, P. J. Heaney, D. R. Veblen, Z. Z. Sheng, A. El Ali, and A. M. Hermann, *Phys. Rev. Lett.* **60,** 1657 (1988).

[364]S. S. P. Parkin, V. Y. Lee, E. M. Engler, A. I. Nazzal, T. C. Huang, G. Gorman, R. Savoy, and R. Beyers, *Phys. Rev. Lett.* **60,** 2539 (1988).

[365]S. S. P. Parkin, V. Y. Lee, A. I. Nazzal, R. Savoy, R. Beyers, and S. J. La Placa, *Phys. Rev. Lett.* **61,** 750 (1988).

[366]R. Beyers, S. S. P. Parkin, V. Y. Lee, A. I. Nazzal, R. Savoy, G. Gorman, T. C. Huang, and S. La Placa, *Appl. Phys. Lett.* **53,** 432 (1988).

[367]S. S. P. Parkin, V. Y. Lee, A. I. Nazzal, R. Savoy, T. C. Huang, G. Gorman, and R. Beyers, *Phys. Rev. B* **38,** 6531 (1988).

[368]M. A. Subramanian, J. C. Calabrese, C. C. Torardi, J. Gopalakrishnan, T. R. Askew, R. B. Flippen, K. J. Morrissey, U. Chowdhry, and A. W. Sleight, *Nature* **332,** 420 (1988).

[369]C. C. Torardi, M. A. Subramanian, J. C. Calabrese, J. Gopalakrishnan, E. M. McCarron, K. J. Morrissey, T. R. Askew, R. B. Flippen, U. Chowdhry, and A. W. Sleight, *Phys. Rev. B* **38,** 225 (1988).

[370]C. C. Torardi, M. A. Subramanian, J. C. Calabrese, J. Gopalakrishnan, K. J. Morrissey, T. R. Askew, R. B. Flippen, U. Chowdhry, and A. W. Sleight, *Science* **240,** 631 (1988).

[371]S. Iijima, T. Ichihashi, and Y. Kubo, *Jpn. J. Appl. Phys.* **27,** L817 (1988).

[372]S. Iijima, T. Ichihashi, Y. Shimakawa, T. Manako, and Y. Kubo, *Jpn. J. Appl. Phys.* **27,** L837 (1988).

[373]S. Iijima, T. Ichihashi, Y. Shimakawa, T. Manako, and Y. Kubo, *Jpn. J. Appl. Phys.* **27,** L1054 (1988).

[374]S. Iijima, T. Ichihashi, Y. Shimakawa, T. Manako, and Y. Kubo, *Jpn. J. Appl. Phys.* **27,** L1061 (1988).

[375]D. S. Ginley, E. L. Venturini, J. F. Kwak, R. J. Baughman, M. J. Carr, P. F. Hlava, J. E. Schirber, and B. Morosin, *Physica C* **152,** 217 (1988).

[376]B. Morosin, D. S. Ginley, E. L. Venturini, P. F. Hlava, R. J. Baughman, J. F. Kwak, and J. E. Schirber, *Physica C* **152,** 223 (1988).

[377]B. Morosin, D. S. Ginley, P. F. Hlava, M. J. Carr, R. J. Baughman, J. E. Schirber, E. L. Venturini, and J. F. Kwak, *Physica C* **152,** 413 (1988).

[378]J. D. Fitz Gerald, R. L. Whithers, J. G. Thompson, L. R. Wallenberg, J. S. Anderson, and B. G. Hyde, *Phys. Rev. Lett.* **60,** 2797 (1988).

[379]P. Haldar, A. Roig-Janicki, S. Sridhar, and B. C. Giessen, *Mater. Lett.* (1988) (in press).

[380]H. W. Zandbergen, G. Van Tendeloo, J. Van Landuyt, and S. Amelinckx, *Appl. Phys. A* **46,** 233 (1988).

[381]Y. Shimakawa, Y. Kubo, T. Manako, Y. Nakabayashi, and H. Igarashi, *Physica C* **156,** 97 (1988).

[382]A. W. Hewat, E. A. Hewat, J. Brynestad, H. A. Mook, and E. D. Specht, *Physica C* **152,** 438 (1988).

[383]A. W. Hewat, P. Bordet, J. J. Capponi, C. Chaillout, J. Chenavas, M. Godinho, E. A. Hewat, J. L. Hodeau, and M. Marezio, *Physica C* **156,** 369 (1988).

[384]E. A. Hewat, P. Bordet, J. J. Capponi, C. Chaillout, J. Chenavas, M. Godinho, A. W. Hewat, J. L. Hodeau, and M. Marezio, *Physica C* **156,** 375 (1988).

ductivity at 125 K, the highest superconducting transition temperature yet found.

Six perovskite-related oxides have been identified in the Tl–Ca–Ba–Cu–O system thus far: $Tl_1Ba_2Cu_1O_5$, $Tl_1Ca_1Ba_2Cu_2O_7$, $Tl_1Ca_2Ba_2Cu_3O_9$, $Tl_2Ba_2Cu_1O_6$, $Tl_2Ca_1Ba_2Cu_2O_8$, and $Tl_2Ca_2Ba_2Cu_3O_{10}$ (Fig. 25). For brevity, these phases will be referred to as Tl1021, Tl1122, Tl1223, Tl2021, Tl2122, and Tl2223. Both the size and the separation of the Cu perovskite-like units can be independently varied in the thallium-containing superconductors. Tl2021, Tl2122, and Tl2223 are made up of Cu perovskite-like units containing one, two, and three CuO_2 planes separated by Tl–O *bilayers*, respectively, and are thallium analogs to the nominal Bi2021, Bi2122, and Bi2223 structures. Conversely, Tl1021, Tl1122, and Tl1223 are made up of Cu perovskite-like units containing one, two, and three CuO_2 planes separated by Tl–O *monolayers*, respectively, and have no bismuth analogs. All of the oxides have a tetragonal structure at room temperature. The oxides with Tl–O monolayers have primitive tetragonal cells, whereas the oxides with Tl–O bilayers have body-centered tetragonal cells. Tl2021 has two polymorphs, one face-centered orthorhombic and the other body-centered tetragonal. The nominal compositions and structures of these oxides are summarized in Table VI.[366]

Superlattice modulations have been observed in all of the structures except Tl1021. For Tl1122 and Tl1223 the wave vectors are $\langle 0.29, 0, 0.5 \rangle$ type,[365–367] while for Tl2122 and Tl2223 the wave vectors are $\langle 0.17, 0, 1 \rangle$ type.[366,367,378,380,382] The symmetry of these crystals remains tetragonal if the modulations are two-dimensional, but is lowered to orthorhombic if the modulations are one-dimensional. Both Tl2021 polymorphs contain superlattice modulations with an approximate wave vector $\langle \overline{0.16}, 0.08, 1 \rangle$ in the tetragonal $a_p \times a_p$ setting [and $\langle 0.08, 0.24, 1 \rangle$ in the orthorhombic $\sqrt{2}a_p \times \sqrt{2}a_p$ setting].[366,367,384] Taking the superlattice into account lowers the symmetry of these two structures to monoclinic, with the c axis being the unique axis. The superlattice modulations in the thallium-containing superconductors are distinct from those found in the related bismuth-containing superconductors and are much weaker in intensity. As for the bismuth-containing superconductors, the source of these modulations is still being investigated. Local distortions in the Tl–O layers are believed to be the most probable cause of the modulations in the materials with bilayer and trilayer Cu perovskite-like units. Recent neutron diffraction and high-resolution TEM studies[383,384] suggest that the modulations in both Tl2021 polymorphs arise from ordered vacancies on one-eighth of the Tl and O sites in the Tl–O layers.

The predominant defects in crystals with bilayer or trilayer Cu perovskite-like units are intergrowths.[364,366,367,371–373,380] High-resolution

TABLE VI. Measured Properties of Tl–Ca–Ba–Cu Oxides. (Beyers et al.[366])

Name	Composition	Crystal Structure	Lattice Parameters (Å)	Superlattice Wave Vector	T_c(K)
1021	$Tl_{1.2}Ba_2Cu_{0.7}O_{4.8}$	prim. tetragonal	$a = 3.869(2)$ $c = 9.694(9)$	[1]	2[2]
1122	$Tl_{1.1}Ca_{0.9}Ba_2Cu_{2.1}O_{7.1}$	prim. tetragonal	$a = 3.8505(7)$ $c = 12.728(2)$	$\langle 0.29, 0, 0.5 \rangle$	80
1223	$Tl_{1.1}Ca_{1.8}Ba_2Cu_{3.0}O_{9.7}$	prim. tetragonal	$a = 3.8429(6)$ $c = 15.871(3)$	$\langle 0.29, 0, 0.5 \rangle$	110
2021	$Tl_{1.9}Ba_2Cu_{1.1}O_{6.4}$	f.c. orthorhombic	$a = 5.445(2)$ $b = 5.492(1)$ $c = 23.172(6)$	$\langle \overline{0.08}, 0.24, 1 \rangle$	2[2]
	$Tl_{1.8}Ca_{0.02}Ba_2Cu_{1.1}O_{6.3}$	b.c. tetragonal[3]	$a = 3.8587(4)$ $c = 23.152(2)$	$\langle \overline{0.16}, 0.08, 1 \rangle$	20[4]
2122	$Tl_{1.7}Ca_{0.9}Ba_2Cu_{2.3}O_{8.1}$	b.c. tetragonal	$a = 3.857(1)$ $c = 29.39(1)$	$\langle 0.17, 0, 1 \rangle$	108
2223	$Tl_{1.6}Ca_{1.8}Ba_2Cu_{3.1}O_{10.1}$	b.c. tetragonal	$a = 3.822(4)$ $c = 36.26(3)$	$\langle 0.17, 0, 1 \rangle$	125

[1] No superlattice modulations have been observed in these crystals thus far.
[2] Non-metallic or weakly metallic samples with no superconducting transition observed down to 4.2 K.
[3] Taking the superlattice into account lowers the symmetry of this structure to monoclinic with the c axis being the unique axis.
[4] Refs. 363 and 369 report T_c = 80 K for tetragonal 2021.

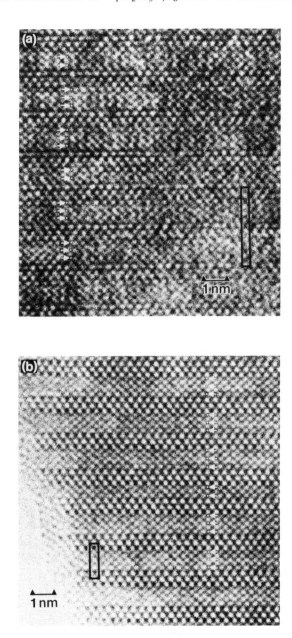

FIG. 27. High-resolution TEM images along the [100] direction in (a) Tl2223 and (b) Tl1223, showing bilayer Cu perovskite intergrowths. The black rectangles outline the unit cells and the white arrows point out the copper columns. (Beyers et al.[366])

TEM studies found a wide variety of intergrowths, each corresponding to the addition or deletion of a perovskite-like unit or a Tl–O layer from the ideal structure. The superconducting transition temperatures in these crystals vary with the density of stacking defects. For example, Parkin et al.[364] found that defect-free Tl2223 crystals have a T_c of 125 K, whereas Tl2223 crystals containing a high density of randomly distributed bilayer Cu perovskite intergrowths have a T_c of only 118 K [Fig. 27(a)]. Similarly, Beyers et al.[366] reported that randomly distributed Tl1122 intergrowths reduce the T_c of Tl1223 crystals from 110 K to 100 K [Fig. 27(b)] and the T_c of Tl2122 crystals from 108 K to 95 K. It is difficult to say whether it is the local change in structure or the change in composition produced by the intergrowths that affects the superconducting transition temperature because these are concurrent changes.

From a scientific viewpoint, the most interesting thallium-containing superconductors may turn out to be those containing single CuO_2 layers, Tl2021 and Tl1021. Hazen et al.[363] identified Tl2021 as the 80 K superconductor in Sheng and Hermann's[5] Tl–Ba–Cu–O samples. Torardi et al.[369] verified this result and found Tl2021 to be tetragonal with no superlattice modulations. On the other hand, Beyers et al.[366] found that tetragonal Tl2021 contained superlattice modulations and was only a 20 K superconductor, whereas orthorhombic Tl2021 was metallic, but not superconducting down to 4 K. No intergrowths were observed in either of the Tl2021 polymorphs examined by Beyers et al.[366] The structural differences between the 0, 20, and 80 K Tl2021 superconductors may provide important clues to the origin of superconductivity in these oxides. Wide variations in the transition temperatures in materials with the Tl1021 structure may also occur: Beyers et al.[366] found Tl1021 was not superconducting down to 4 K, whereas Haldar et al.[379] reported a transition temperature of 50 K in $(Tl, Bi)_1(Sr, Ca)_2Cu_1O_{4.5+x}$, which has the Tl1021 structure.

It is an idealization to treat the thallium-containing superconductors as line compounds (see Table VI).[366,367,368,370,382,383] The stacking defects are one way to accommodate off-stoichiometry in the metal cations. Microprobe analysis consistently shows thallium deficiency in the Tl–O bilayer phases, possibly indicating Tl–O monolayer intergrowths, whereas the reverse is found in the Tl–O monolayer phases.[366,367] Additionally, there may be deficiencies or substitutions on certain cation sites.[368,370,382,383] Diffraction studies of large single crystals are required before any definitive conclusions regarding site occupancy can be drawn. Furthermore, neutron diffraction and thermogravimetric studies of single phase or single crystal samples are needed to determine the absolute oxygen content in these oxides and its variation with processing. These

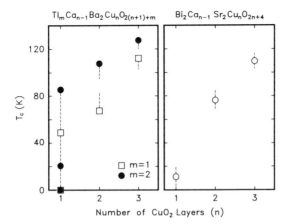

FIG. 28. Superconducting transition temperature versus the number of CuO_2 layers in the thallium- and bismuth-containing superconductors.

studies are required before more can be said about the average Cu oxidation state. Note, however, that the average Cu oxidation state in the ideal Tl–O monolayer structures varies from +3 in Tl1021 (not superconducting) to +2.33 in 1223 ($T_c = 110$ K), whereas the average Cu oxidation state is +2 for all of the ideal Tl–O bilayer structures.

In both the bismuth- and thallium-containing superconductors, the most obvious correlation between structure and properties is that the superconducting transition temperature increases with the number of CuO_2 layers in the perovskite-like unit (Fig. 28). Earlier comparisons of the 123 and $La_{2-x}M_xCuO_{4-y}$ structures led to speculation that this might be the case. Theorists are currently trying to determine if this trend can be explained by a simple density of states argument—i.e., increasing the number of CuO_2 layers increases the electronic density of states, which in turn increases T_c in a BCS formalism[385,386]—or if more elaborate explanations are required. The wide variation in superconducting properties in the Tl2021 phase alone implies that more than just the number of CuO_2 layers must be included in a rigorous theory. Experimentalists, on the other hand, are trying to fabricate materials with greater than three CuO_2 layers, such as $Tl_1Ca_3Ba_2Cu_4O_{11}$ and $Tl_2Ca_3Ba_2Cu_4O_{12}$. The observation of isolated intergrowths with greater than three CuO_2 layers in high-resolution images offers some hope that these structures may yet

[385]F. Herman, R. V. Kasowski, and W. Y. Hsu, *Phys. Rev. B* **38,** 204 (1988).
[386]R. V. Kasowski, W. Y. Hsu, and F. Herman, *Phys. Rev. B* **38,** 6470 (1988).

be formed in bulk quantities. Indeed, Ihara *et al.*[387,388] recently reported forming $Tl_1Ca_3Ba_2Cu_4O_{11}$ as a bulk, 122 K superconductor.

VII. Summary and Future Work

It is evident from the variety of behaviors exhibited by 123 after different processing conditions that the structure has considerable flexibility. The oxygen stoichiometry can be varied over a wide range, and different cations can be substituted at least partially into every site. Certain cation substitutions have extensive solid solution ranges. Structural and physical changes occur on heating and cooling, and numerous structural defects form. Because of this variability in the 123 structure, it is not surprising that a considerable amount of the experimental data from different researchers is quantitatively and in some cases qualitatively in conflict. The variability of the structure highlights the need for carefully controlled experiments to resolve these conflicts. The complexity of the structure renders difficult, if not impossible, the preparation of a series of samples in which a single aspect of the structure is changed. For example, substitution of even small amounts of iron can change the oxygen ordering, oxygen content, twin structure, and magnetic ordering in the structure. Thus characterization of only one aspect of the material can lead to misleading correlations. There is a clear need for further studies in which several complementary characterization techniques are applied to each sample in the study.

In spite of these difficulties, many aspects of the 123 structure have been resolved. The basic structure is well understood. The mechanism of the high-temperature orthorhombic-to-tetragonal transformation has been determined. The major defects in the structure have been identified and characterized.

A major challenge has been to link the basic features of the structure to the superconducting behavior. Most work in this area has focussed on determining the role of the chains versus the planes in superconductivity. Early speculation, based on the observation that only the orthorhombic form of $Y_1Ba_2Cu_3O_{7-\delta}$ is superconducting, suggested that continuous copper oxygen chains in the basal plane of the 123 structure were essential for superconductivity. It is now known that tetragonal forms of

---

[387]H. Ihara, R. Sugise, M. Hirabayashi, N. Terada, M. Jo, K. Hayashi, A. Negishi, M. Tokumoto, Y. Kimura, and T. Shimomura, *Nature* **334**, 51° (1988).

[388]H. Ihara, R. Sugise, K. Hayashi, N. Terada, M. Jo, M. Hirabayashi, A. Negishi, N. Atoda, H. Oyanagi, T. Shimomura, and S. Ohashi, *Phys. Rev. B* (1988) (in press).

the 123 structure can superconduct. Although short copper oxygen chains may still exist in the basal copper planes of the tetragonal 123 superconductors, they are probably not essential for superconductivity. The discovery of tetragonal 123 superconductors implies that it is the two-dimensional CuO_2 planes that are the essential structural feature needed for superconductivity. This conclusion is strongly supported by the observation of high-temperature superconductivity in the Bi–Ca-Sr-Cu–O and Tl–Ca–Ba–Cu–O structures, which contain copper-oxygen planes but no chains. A particularly compelling observation in this respect is the finding of 80 K superconductivity in the $Tl_1Ca_1Ba_2Cu_2O_7$ structure, which is essentially identical to the 123 structure except for the substitution of Tl for Cu in the basal plane sites and the replacement of Y by Ca.

The question remains as to what role the chains play in superconductivity. The experiments on reduced or cation-substituted materials show that the major structural effect of reduction or substitution is to change the oxygen arrangement in the basal copper plane. The structure of the CuO_2 planes remains unchanged unless direct substitution for copper in the planes occurs. These observations imply that the chains' role in superconductivity is primarily electronic. The problem, then, is to determine the mechanisms by which changes in oxygen content or substitution of aliovalent cations are electronically compensated for. Hole formation, oxygen intercalation, and peroxide formation have all been implicated as possible compensation mechanisms. Of these only hole concentration has been linked directly to T_c. Analysis of the total hole concentration requires independent measurements of each of these quantities. The problem is further complicated by the fact that there are two copper sites in the structure whose oxidation states can vary independently. Compensation for a substituted cation might change the concentration of holes in one location in the structure but not the other. Tokura et al.[223] used this possibility to rationalize the transition temperatures observed in samples with a wide range of oxygen contents and total charge on the non-copper cations. They suggested that hole formation occurs more easily in the basal copper plane so that the basal plane acts as a reservoir for charge at low hole concentrations. Only after a threshold concentration of holes forms is superconductivity induced in the CuO_2 planes.

The possibility that certain cation substitutions have a direct structural effect on superconductivity can not be completely ruled out. For example, the rapid suppression of T_c by the substitution of isovalent cations such as Zn or Ni suggests that they directly influence T_c by replacing the Cu in the CuO_2 planes. However, the conflicting reports about the location of Zn in the structure need to be resolved. The decrease in T_c caused by the substitution of Sr for Ba also appears to be structural in origin, although an alternate explanation in which the difference in polarizability of the Ba and Sr ions changes the interaction between holes in the structure has been suggested.[389] Strong correlations between changes in certain bond lengths and T_c that occur with oxygen intercalation or cobalt substitution have also been noted. It is unclear, however, whether these changes are coincidental or are directly linked to T_c, since the charge compensation in the structure also changes in each case. The nature of any such direct structural effect remains unknown.

Defects in the structure, especially the twins, have been speculated to be directly linked to the superconducting behavior. This speculation seems unfounded, given the observation of superconductivity in untwinned tetragonal crystals. Defects, however, clearly influence secondary superconducting properties. In particular, the grain boundaries are responsible for the low critical current densities observed in polycrystalline 123. The sensitivity of critical current densities to defects probably results from the short coherence length in these materials. The disruption of superconductivity by grain boundaries is probably further aggravated by the large unit cell that necessitates the formation of defects with large cores of distorted structure. The challenge now is to develop fabrication techniques that minimize these detrimental effects. Twins and faults distort the structure less severely, and it seems likely that their influence on critical current densities will be small. Experiments in which the twin or fault densities are varied in materials with constant composition and oxygen content are needed to determine what effect these defects have on superconducting properties. It is also important to recognize that variations in defect structure and microstructure can influence the behavior during heat or oxygen treatments and possibly alter the distribution of substituted elements. Characterization of defect structures is therefore also essential in preparing materials for controlled experiments.

The same types of experiments used to determine the relationships among processing, structure, and superconductivity in 123 will surely be performed on the bismuth- and thallium-containing superconductors. For these quinary oxides, the preparation of samples under reproducible, thermodynamically well-defined conditions will be especially difficult. Consequently, the need to use complementary characterization techniques to derive quantitative processing-structure-property relationships will become all the more essential.

[389]M. Ronay and D. M. Newns, *Phys. Rev. B* (1988) (submitted).

ACKNOWLEDGMENTS

We thank our many colleagues within IBM Research for their stimulating discussions and collaborations throughout the past two years. We thank David Clarke, Frank Herman, Sam LaPlaca, Merril Shafer, and Jerry Torrance for their comments on the manuscript. One of the authors (R.B.) thanks Byung Tae Ahn, Turgut Gür, and Robert Huggins at Stanford University for their invaluable collaboration on the solid-state ionic studies and Robert Sinclair, also at Stanford, for his advice and for the use of his electron microscope.

Notes Added in Proof

Three new oxides have been discovered that may provide important clues to the role of various structural features in high-temperature superconductivity. First, Mattheiss et al.[1] and Cava et al.[2] found superconductivity near 30 K in a copperless oxide, $Ba_{1-x}K_xBiO_{3-\delta}$ ($x \sim 0.4$). This oxide has a cubic perovskite structure, with none of the two-dimensional features observed in the copper-based superconducting oxides. In $Ba_{1-x}K_xBiO_{3-\delta}$, potassium substitution on the barium site is used to dope the parent compound, $BaBiO_3$, whereas lead substitution on the bismuth site had been used previously to make the first known superconducting oxide, $BaPb_xBi_{1-x}O_3$ ($T_c = 13$ K).[3] It remains an open question whether the pairing mechanism in these oxides is the same as that in the copper-containing oxides.[4-7] The lack of magnetic ordering in $Ba_{1-x}K_xBiO_{3-\delta}$ appears to rule out a magnetically mediated pairing mechanism in the material.[7]

Second, Siegrist et al.[8] synthesized $Ca_{0.88}Sr_{0.14}CuO_2$, a structure that consists of infinite CuO_2 planes separated by Ca and Sr atoms. Its copper valence is +2 and the material is an insulator. It remains to be seen whether this material can be made metallic or superconducting by

[1]L. F. Mattheiss, E. M. Gyorgy, and D. W. Johnson Jr., *Phys. Rev. B* **37**, 3745 (1988).

[2]R. J. Cava, B. Batlogg, J. J. Krajewski, R. C. Farrow, L. W. Rupp, Jr., A. E. White, K. T. Short, W. F. Peck, Jr., and T. Y. Kometani, *Nature* **332**, 814 (1988).

[3]A. W. Sleight, J. L. Gillson, and P. E. Bierstedt, *Solid State Commun.* **17**, 27 (1975).

[4]T. M. Rice, *Nature* **332**, 780 (1988).

[5]D. G. Hinks, B. Dabrowski, J. D. Jorgensen, A. W. Mitchell, D. R. Richards, Shiyou Pei, and Donglu Shi, *Nature* **333**, 836 (1988).

[6]Shiyou Pei, N. J. Zaluzec, J. D. Jorgensen, B. Dabrowski, D. G. Hinks, A. W. Mitchell, and D. R. Richards, *Phys. Rev. B* (1988) (in press).

[7]Y. J. Uemura, B. J. Sternlieb, D. E. Cox, J. H. Brewer, R. Kadono, J. R. Kempton, R. F. Kiefl, S. R. Kreitzman, G. M. Luke, T. Riseman, D. L. Williams, W. J. Kossler, X. H. Yu, C. E. Stronach, M. A. Subramanian, J. Gopalakrishnan, and A. W. Sleight, *Nature* **335**, 151 (1988).

[8]T. Siegrist, S. M. Zahurak, D. W. Murphy, and R. S. Roth, *Nature* **334**, 231 (1988).

doping. If the material can be doped but does not become superconducting, then this result would imply that five- or six-coordinated, Jahn-Teller distorted copper ions, not just CuO_2 planes per se, are an essential structural feature for high-temperature superconductivity.[9]

Third, Cava et al.[10] discovered superconductivity near 70 K in a series of copper oxides with the general formula $Pb_2Sr_2ACu_3O_{8+\delta}$ where A = Y, La, Pr, Nd, Eu, Dy, Ho, Tm, or Lu plus Sr or Ca. This structure contains the same square pyramidal copper-oxygen planes that are found in 123 and in the bismuth- and thallium- containing superconductors.

[9]C. Greaves, *Nature* **334,** 193 (1988).

[10]R. J. Cava, B. Batlogg, J. J. Krajewski, L. W. Rupp, Jr., L. F. Schneemeyer, T. Siegrist, R. B. van Dover, P. Marsh, W. F. Peck, Jr., P. K. Gallagher, S. H. Glarum, J. H. Marshall, R. C. Farrow, J. V. Waszczak, R. Hull, and P. Trevor, *Nature* **336,** 211 (1988).

SOLID STATE PHYSICS, VOLUME 42

Electronic Structure of Copper-Oxide Superconductors

K. C. HASS

Research Staff
Ford Motor Company
Dearborn, Michigan

I. Introduction

Despite a year and a half of unprecedented activity, the origin of high-temperature superconductivity in $La_{2-x}Ba_xCuO_4$ ($T_c = 35$ K),[1] $YBa_2Cu_3O_7$ ($T_c = 90$ K),[2] and other copper-oxide ceramics[3] remains unclear. Many theories have been proposed, including both conventional[4]

[1] J. G. Bednorz and K. A. Müller, *Z. Phys.* B**64**, 189 (1986).
[2] M. K. Wu, J. R. Ashburn, C. J. Torng, P. H. Hor, R. L. Meng, L. Gao, Z. J. Huang, Y. Q. Wang, and C. W. Chu, *Phys. Rev. Lett.* **58**, 908 (1987).
[3] M. A. Subramanian, C. C. Toradi, J. C. Calabrese, J. Gopalakrishnan, K. J. Morissey, T. R. Askew, R. B. Flippen, U. Chowdhry, and A. W. Sleight, *Science* **239**, 1015 (1988).
[4] J. Bardeen, L. N. Cooper, and J. R. Schrieffer, *Phys. Rev.* **108**, 1175 (1957).

(phonon-mediated electron pairing) and unconventional[5] (excitonic, magnetic, etc.) mechanisms, and many remain viable at the time of this writing (6/88).

An important prerequisite for understanding the nature of the superconductivity in these materials is a detailed knowledge of their electronic structure. Considerable evidence now exists that even the normal state properties of the new copper oxides differ substantially from those of most previously known superconductors and are difficult to reconcile with conventional one-electron band theory. Among the more surprising properties of these materials are (1) highly anisotropic, loosely packed crystal structures; (2) carrier concentrations well below those of typical metals; and (3) rich phase diagrams including closely related insulating and magnetic phases. The last of these, in particular, has prompted many researchers, beginning with Anderson,[6] to suggest that the valence electrons in these systems are strongly correlated. The importance of correlation effects and the appropriate model for describing them remain subjects of intense controversy. Some insight has been obtained from previous experience with other strongly correlated systems such as Mott–Hubbard insulators[7] (e.g., NiO), valence-fluctuation compounds[8] (e.g., SmB_6) and heavy fermion metals[9] (e.g., UPt_3). A fully satisfactory description of the new superconductors, however, may require new concepts in many-body theory.[6]

The present chapter will attempt to consolidate much of the extensive literature which has already appeared on the electronic structure of the $La_{2-x}M_xCuO_{4-y}$ (M = Ba, Sr, Ca) and $RBa_2Cu_3O_{7-\delta}$ (R = Y, Eu, Gd, . . .) systems (hereafter referred to as 2-1-4 and 1-2-3, respectively). Emphasis will be placed on the various theoretical models that have been proposed, beginning with state-of-the-art band structure calculations (Part III) and proceeding to more idealized model Hamiltonian treatments of correlation effects (Part IV). The relationships between different models and the implications of existing spectroscopic data (Part V) will also be examined. In view of the rapid progress still being made in this field, it would be

[5]For a brief introduction to the various classes of theories, see T. M. Rice, *Z. Phys.* **B67**, 141 (1987).

[6]P. W. Anderson, *Science* **235**, 1196 (1987); P. W. Anderson, *Proc. of the Enrico Fermi Summer School, Frontiers and Borderlines in Many-Particle Physics*—Varenna, July, 1987, North-Holland, Amsterdam, 1988.

[7]D. Adler, *in* "Solid State Physics" (F. Seitz, D. Turnbull, and H. Ehrenreich, eds.), Vol. 21, p. 1. Academic Press, New York, 1968; B. H. Brandow, *Adv. in Phys.* **26**, 651 (1977).

[8]C. M. Varma, *Rev. Mod. Phys.* **48**, 219 (1976).

[9]P. A. Lee, T. M. Rice, J. W. Serene, L. J. Sham, and J. W. Wilkins, *Comments Cond. Mat. Phys.* **12**, 99 (1986); P. Fulde, J. Keller, and G. Zwicknagl, *in* "Solid State Physics" (H. Ehrenreich and D. Turnbull, eds.), Vol. 41, p. 1. Academic Press, New York, 1988.

premature to draw any definite conclusions at this time. Part VI of this review will nevertheless offer a few perspective comments on what now seems to be well understood about the electronic structure of the 2-1-4 and 1-2-3 copper oxides and what key questions remain.

II. Overview of Basic Physical Properties

To set the stage, it is useful to review briefly some of the relevant structural, electrical, and magnetic properties of the 2-1-4 and 1-2-3 systems. More detailed information on these and other properties can be found elsewhere in this volume and in recent conference proceedings.[10]

1. $La_{2-x}M_xCuO_{4-y}$

At high temperatures, all of the 2-1-4 copper oxides exhibit the body-centered tetragonal K_2NiF_4 structure[11] (space group I4/mmm or D_{4h}^{17}) shown in Fig. 1. Each Cu atom in this structure sits at the center of an elongated $Cu-O_6$ octahedron constructed from four strongly bonded O1 neighbors at a distance of 1.9 Å and two more weakly bonded O2 neighbors at a distance of 2.4 Å. The structure alternates along the c axis between a square planar $Cu-O_2$ layer and two buckled (La, M)–O layers. The layered nature of the structure gives rise to highly anisotropic electrical and magnetic properties.[10] The superconductivity is usually assumed to originate in the $Cu-O_2$ planes since this is the only structural element common to all high T_c materials.[3]

Figure 2 shows a schematic phase diagram for $La_{2-x}Sr_xCuO_{4-y}$. Qualitatively similar results have been obtained for $M =$ Ba or Ca, but $M =$ Sr yields the highest T_c. The construction of this figure is designed to illustrate the rich behavior that must be accounted for in any comprehensive theory of these materials.[12-16] Quantitative details will probably

[10]For example, *Proc. of the XVIIIth Int. Conf. on Low Temp. Phys.*, in *Jpn. J. Appl. Phys.* **26**, Sup. 26-3 (1987) and *Proc. of the Adriatico Res. Conf. on High Temp. Superconductivity*, in *Int. J. Mod. Phys. B*1 (1987).

[11]A brief review of the structural properties of 2-1-4 and 1-2-3 compounds can be found in J. D. Jorgensen, *Jpn. J. Appl. Phys.* **26**, Sup. 26-3, 2017 (1987).

[12]R. M. Fleming, B. Batlogg, R. J. Cava, and E. A. Rietman, *Phys. Rev. B35*, 7191 (1987).

[13]D. Vaknin, S. K. Sinha, D. E. Moncton, D. C. Johnston, J. M. Newsam, C. R. Safinya, and H. E. King, *Phys. Rev. Lett.* **58**, 2802 (1987).

[14]D. C. Johnston, J. P. Stokes, D. P. Goshorn, and J. T. Lewandowski, *Phys. Rev. B36*, 4007 (1987).

[15]T. Freltoft, J. E. Fischer, G. Shirane, D. E. Moncton, S. K. Sinha, D. Vaknin, J. P. Remeika, A. S. Cooper, and D. Harshman, *Phys. Rev. B36*, 826 (1987).

[16]M. W. Shafer, T. Penney, and B. L. Olson, *Phys. Rev. B36*, 4047 (1987).

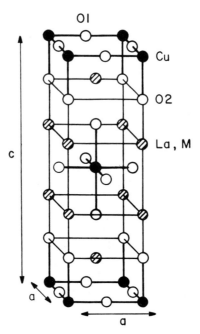

FIG. 1. Body-centered tetragonal structure of $La_{2-x}M_xCuO_{4-y}$ ($M = $ Ba, Sr, Ca). Lattice vectors given by $(a, 0, 0)$, $(0, a, 0)$, and $(a/2, a/2, c/2)$. La and M atoms assumed to be randomly distributed over sites denoted by cross-hatched circles. Cu and O1 atoms form horizontal Cu-O$_2$ "planes."

continue to evolve as better characterized samples with fewer uncertainties in the absolute concentrations of O vacancies and other defects become available. Doping beyond $x = 0.15$, for example, is now known to cause significant O depletion,[16] which makes the measured properties in this region particularly sensitive to preparation conditions.

Below the dashed line in Fig. 2, each Cu–O$_6$ "octahedron" tilts slightly about one of the original [110] axes. An orthorhombic distortion results (space group Cmca or D_{2h}^{18}) which is characterized by an approximately $\sqrt{2}\,a \times \sqrt{2}\,a$ doubling of the unit cell in the basal plane.[17] The new a and b axes, which lie along the original [110] and [1$\bar{1}$0] directions, differ by less than 1%. The transition appars to be electrically neutral,[18] although

[17]J. D. Jorgensen, H.-B. Schüttler, D. G. Hinks, D. W. Capone, K. Zhang, M. B. Brodsky, and D. J. Scalapino, *Phys. Rev. Lett.* **58**, 1024 (1987).

[18]N. P. Ong, Z. Z. Wang, J. Clayhold, J. M. Tarascon, L. H. Greene, and W. R. McKinnon, *Phys. Rev.* **B35**, 8807 (1987).

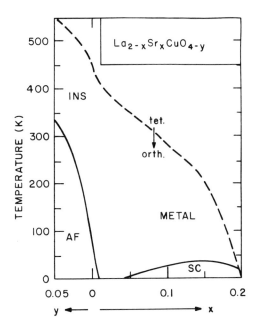

FIG. 2. Schematic phase diagram for $La_{2-x}Sr_xCuO_{4-y}$. Horizontal axis assumes $y = 0$ when $x \neq 0$ and $x = 0$ when $y \neq 0$. AF = antiferromagnet; SC = superconductor. Dashed curve separates tetragonal and orthorhombic structures. Insulator (INS)–metal boundary not shown because of experimental uncertainties. Diagram based on experimental data from Fleming et al., Phys. Rev. B35, 7191 (1987); Vaknin et al., Phys. Rev. Lett. 58, 2802 (1987); Johnston et al., Phys. Rev. B36, 4007 (1987); Freltoft et al., Phys. Rev. B36, 826 (1987); and Shafer et al., Phys. Rev. B36, 4047 (1987).

superconductivity has only been observed in the orthorhombic phase. Additional structural transitions have also been reported for some alloy compositions,[11] but these results have yet to be fully confirmed and characterized.

The parent compound La_2CuO_{4-y} does not exhibit superconductivity, except perhaps in trace amounts.[19,20] At low temperatures, this material is an antiferromagnetic (AF) insulator[13–15] with a Neel temperature T_N for three-dimensional AF ordering that increases with increasing y; the maximum staggered magnetization in the Neel state is approximately

[19]R. L. Greene, H. Maletta, T.-S. Plaskett, J. G. Bednorz, and K. A. Müller, Solid State Commun. 63, 379 (1987).
[20]P. M. Grant, S. S. P. Parkin, V. Y. Lee, E. M. Engler, M. L. Ramirez, J. E. Vazquez, G. Lim, R. D. Jacowitz, and R. L. Greene, Phys. Rev. Lett. 58, 2482 (1987).

$0.5 \mu_B/Cu$. Above T_N, no true long-range order exists, although the spins within the two-dimensional $Cu-O_2$ planes exhibit instantaneous AF correlations[21] over distances as long as 200 Å. Similar spin correlations have also been observed[22] for $x > 0$ with a correlation length that decreases as $x^{-1/2}$. The substitution of Sr for La causes a rapid suppression of the three-dimensional AF state[22] and may result in the creation of a low-temperature spin-glass phase.[23] Hall measurements[16,18] indicate that this substitution also introduces p-type carriers with a concentration roughly proportional to x. For relatively pure La_2CuO_4, the resulting holes are localized and display hopping conductivity that is consistent with a short localization length.[24] For larger x, the system becomes metallic and is paramagnetic at high temperatures.[19] Bulk superconductivity occurs for $x > 0.04$ with T_c exhibiting a broad maximum near 40 K for $x \simeq 0.15$.

Formal valence considerations suggest that the insulating ground state of La_2CuO_4 results from the purely ionic configurations La^{3+}, Cu^{2+}, and O^{2-}. The magnetic properties for $x \simeq 0$ are indeed qualitatively consistent with anisotropic Heisenberg interactions between spin-1/2 Cu^{2+} (d^9) ions. The substitution of M^{2+} ions and/or introduction of O vacancies in this picture implies a hole concentration in $La_{2-x}M_xCuO_{4-y}$ of $(x - 2y)/cell$. The character of such holes is often assumed to be that of Cu^{3+} (d^8) ions with no magnetic moments. Although this picture does have some heuristic value, it will be seen that such an oversimplified model should not be interpreted too literally.

2. $RBa_2Cu_3O_{7-\delta}$

The more complex crystal structure[11] of the 1-2-3 compounds $RBa_2Cu_3O_7$ is shown in Fig. 3. This structure has orthorhombic symmetry (with space group Pmmm or D_{2h}^1) due to the presence of Cu1–O1 chains along the b axis and is usually characterized as that of an oxygen-deficient

[21]G. Shirane, Y. Endoh, R. J. Birgeneau, M. A. Kastner, Y. Hidaka, M. Oda, M. Suzuki, and T. Murakami, *Phys. Rev. Lett.* **59**, 1613 (1987); Y. Endoh, K. Yamada, R. J. Birgeneau, D. R. Gabbe, H. P. Jenssen, M. A. Kastner, C. J. Peters, P. J. Picone, T. R. Thurston, J. M. Tranquada, G. Shirane, Y. Hidaka, M. Oda, Y. Enomoto, M. Suzuki, and T. Murakami, *Phys. Rev.* **B37**, 7443 (1988).

[22]R. J. Birgeneau, D. R. Gabbe, H. P. Jenssen, M. A. Kastner, P. J. Picone, T. R. Thurston, G. Shirane, Y. Endoh, M. Sato, K. Yamada, Y. Hidaka, M. Oda, Y. Enomoto, M. Suzuki, and T. Murakami, *Phys. Rev.* **B38**, 6614 (1988).

[23]A. Aharony, R. J. Birgeneau, A. Coniglio, M. A. Kastner, and H. E. Stanley, *Phys. Rev. Lett.* **60**, 1330 (1988).

[24]M. A. Kastner, R. J. Birgeneau, C. Y. Chen, Y. M. Chiang, D. R. Gabbe, H. P. Jenssen, T. Junk, C. J. Peters, P. J. Picone, T. Thio, T. R. Thurston, and H. L. Tuller, *Phys. Rev.* **B37**, 111 (1988).

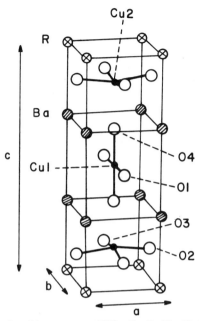

FIG. 3. Unit cell of orthorhombic structure of $RBa_2Cu_3O_7$ ($R = Y, Eu, \ldots$). Lattice vectors given by $(a, 0, 0)$, $(0, b, 0)$, and $(0, 0, c)$. Each set of Cu2, O2, and O3 atoms forms part of a slightly buckled Cu–O_2 "plane." Cu1 and O1 atoms form Cu–O "chains" along the b axis.

perovskite. Each Cu1 atom is fourfold coordinated, having two O1 neighbors at a distance of 1.94 Å and two O4 neighbors at 1.85 Å. The unit cell also contains two Cu2 atoms, which again form Cu–O_2 planes (which are now slightly buckled) with their two O2 and two O3 neighbors at roughly 1.95 Å. Each Cu2 atom is also loosely bonded to one O4 atom along the c axis at a distance of 2.30 Å. The rocksalt-like Ba–O4 layer is similar to the (La, M)–O layers in Fig. 1. The complete absence of O from the R layers separates the material into Cu–O_2/Ba–O/Cu–O/Ba–O/Cu–O_2 slabs, which again produce highly anisotropic electrical and magnetic properties.[10]

The phase diagram for $YBa_2Cu_3O_{7-\delta}$, sketched in Fig. 4, is remarkably similar to that of Fig. 2. With increasing hole concentration, associated here with a decrease in δ, the system again transforms from an antiferromagnetic insulator[25] to a metal that exhibits superconductivity at

[25]J. M. Tranquada, D. E. Cox, W. Kunnmann, H. Moudden, G. Shirane, M. Suenaga, P. Zolliker, D. Vaknin, S. K. Sinha, M. S. Alvarez, A. J. Jacobson, and D. C. Johnston, *Phys. Rev. Lett.* **60**, 156 (1988); J. M. Tranquada, A. H. Moudden, A. I. Goldman, P. Zolliker, D. E. Cox, G. Shirane, S. K. Sinha, D. Vaknin, D. C. Johnston, M. S. Alvarez, A. J. Jacobson, J. T. Lewandowski, and M. Newsam, *Phys. Rev.* **B38**, 2477 (1988).

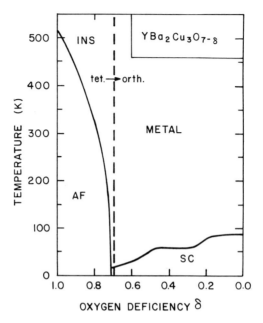

FIG. 4. Schematic phase diagram for $YBa_2CuO_{7-\delta}$ based on experimental data referenced in text. Notation same as in Fig. 2.

temperatures as high as 93 K (for $\delta = 0$). An important complication not addressed in Fig. 4 is the fact that the structure and properties of $YBa_2Cu_3O_{7-\delta}$ depend not only on δ, but also on the distribution of O among a large number of possible sites. The variation from $\delta = 0$ to $\delta = 1$ affects only the O content in the Cu1–O1 layers.[11] The results shown here were obtained under conditions in which, for $\delta < 0.7$, the Cu1–O1 chains remain largely intact with none of the O sites along the a axis perpendicular to the chains (not labeled in Fig. 3) becoming occupied.[26,27] When this latter condition is not met,[11] the observed T_c decreases more rapidly with δ and does not exhibit the 60 K plateau[26] shown in Fig. 4 for $0.3 \leq \delta \leq 0.5$. An increase beyond $\delta = 0.7$ in the present case causes the remaining O in the Cu1–O1 layers to become equally distributed along the a and b axes.[16] The symmetry thus changes at this point from orthorhombic to tetragonal. While it was initially

[26]R. J. Cava, B. Batlogg, C. H. Chen, E. A. Rietman, S. M. Zahurak, and D. Werder, Nature 329, 423 (1987).
[27]Even under these conditions there are still many possible arrangements of the O, with varying degrees of order.

thought that such a loss of orthorhombic symmetry was by itself sufficient to destroy superconductivity, more recent experiments[28] in the vicinity of $\delta = 0.7$ and on Fe- and Co-substituted systems contradict this hypothesis.

A formal valence picture for $RBa_2Cu_3O_{7-\delta}$ suggests a hole concentration of $(1 - 2\delta)$/cell relative to a purely ionic R^{3+}, Ba^{2+}, Cu^{2+}, and O^{2-} configuration. The increase in hole concentration from 0.15/Cu in $La_{1.85}M_{0.15}CuO_4$ to 0.33/Cu in $RBa_2Cu_3O_7$ may be an important factor in the higher T_c of the 1-2-3 compound.[16] The distribution of holes between $Cu–O_2$ planes and $Cu–O$ chains is also expected to play a major role.[28a] The fact that the metallic and superconducting regimes in Fig. 4 extend beyond $\delta = 0.5$, for example, is believed to be due to an increase in the hole concentration in the planes due to the conversion of some Cu^{2+} to Cu^+ in the chains (cf. Part V).

Most theorists now believe that the superconductivity does occur preferentially in the $Cu–O_2$ planes, with the remaining layers providing a source of carriers and influencing the three-dimensional phase coherence.[5,29] Strong support for this hypothesis is provided by the observation that T_c is much more sensitive to substitutions for the Cu2 atoms than for Cu1.[30] That the superconducting electrons are almost completely decoupled from the R layers is suggested by the remarkable similarity between T_c values in almost all $RBaCuO$ compounds, even those in which the R^{3+} ions are magnetic.[31]

III. One-Electron Band Structures

3. Spin-Restricted Calculations

The earliest attempts[32–43] to describe the electronic structures of the 2-1-4 and 1-2-3 systems were based on ab initio, one-electron band

[28]R. J. Cava, B. Batlogg, C. H. Chen, E. A. Rietman, S. M. Zahurak, and D. Werder, *Phys. Rev.* B**36**, 5719 (1987); I. Feldner, I. Nowik, and Y. Yeshurun, *Phys. Rev.* B**36**, 3923 (1987); P. F. Miceli, J. M. Tarascon, L. H. Greene, P. Barboux, F. J. Rotella, and J. D. Jorgensen, *Phys. Rev.* B**37**, 5932 (1988).

[28a]J. Zaanen, A. T. Paxton, O. Jepsen, and O. K. Anderson, *Phys. Rev. Lett.* **60**, 2685 (1988).

[29]Z. Tešanović, *Phys. Rev.* B**36**, 2364 (1987), and *Phys. Rev.* B**38**, 2489 (1988).

[30]G. Xiao, M. Z. Cieplak, A. Gavrin, F. H. Streitz, A. Bakshai, and C. L. Chien, *Phys. Rev. Lett.* **60**, 1446 (1988).

[31]P. H. Hor, R. L. Meng, Y. Q. Wang, L. Gao, Z. J. Huang, H. Bechtold, K. Forster, and C. W. Chu, *Phys. Rev. Lett.* **58**, 1891 (1987); J. M. Tarascon, W. R. McKinnon, L. H. Greene, G. W. Hull, and E. M. Vogel, *Phys. Rev.* B**36**, 226 (1987).

[32]L. F. Mattheiss, *Phys. Rev. Lett.* **58**, 1028 (1987).

structure calculations performed within the local density approximation[44] (LDA). This self-consistent field, density-functional theory approach has recently been developed to a high degree of precision. The spin-independent results discussed in the first part of this section are only weakly dependent on the particular choice of computational scheme [e.g., linear augmented plane waves[32] (LAPW) and linear muffin tin orbital[41] (LMTO)], and exchange-correlation potential.

a. $La_{2-x}M_xCuO_{4-y}$

Figure 5 shows the LAPW bands calculated by Mattheiss[32] for the parent compound La_2CuO_4 in the tetragonal K_2NiF_4 structure. The inset to the figure shows a portion of the $x-y$ plane in reciprocal space containing the high symmetry points $\Gamma\,(0, 0, 0)$, $X\,(\pi/a, \pi/a, 0)$ and $Z\,(0, 0, \pi/c) = (2\pi/a, 0, 0)$. Most of the discussion which follows will be concerned with the 17-band complex between -7 and $+2\,eV$, which is composed primarily of strongly hybridized Cu 3d and O 2p states. The nearly two-dimensional behavior of this complex, which can be seen from the relatively weak dispersion along the Λ direction, indicates an essential decoupling of the $Cu-O_2$ planes in Fig. 2 by the La layers. Each La donates three electrons to the Cu–O complex but otherwise makes little contribution to the occupied states. The La 5d and 4f states are primarily responsible for the unoccupied bands beginning at $2\,eV$ in Fig.

[33]J. Yu, A. J. Freeman, and J.-H. Xu, *Phys. Rev. Lett.* **58,** 1035 (1987); J.-H. Xu, T. J. Watson-Yang, J. Yu, and A. J. Freeman, *Phys. Lett.* A**120,** 489 (1987); C. L. Fu and A. J. Freeman, *Phys. Rev.* B**35,** 8861 (1987); A. J. Freeman, J. Yu, and C. L. Fu, *Phys. B***36,** 7111 (1987).

[34]W. E. Pickett, H. Krakauer, D. A. Papaconstantopoulos, and L. L. Boyer, *Phys. Rev.* B**35,** 7252 (1987).

[35]K. Takegahara, H. Harima, and A. Yanase, *Jpn. J. Appl. Phys.* **26,** L352 (1987).

[36]R. V. Kasowski, W. Y. Hsu, and F. Herman, *Solid State Commun.* **63,** 1077 (1987).

[37]L. F. Mattheiss and D. R. Hamann, *Solid State Commun.* **63,** 395 (1987).

[38]S. Massida, J. Yu, A. J. Freeman, and D. D. Koelling, *Phys. Lett.* A**122,** 198 (1987); J. Yu, S. Massida, A. J. Freeman, and D. D. Koelling, *Phys. Lett.* **122,** 203 (1987).

[39]W. Y. Ching, Y. Xu, G.-L. Zhao, K. W. Wong, and F. Zandiehnadem, *Phys. Rev. Lett.* **59,** 1333 (1987).

[40]F. Herman, R. V. Kasowski, and W. Y. Hsu, *Phys. Rev.* B**36,** 6904 (1987).

[41]T. Fujiwara and Y. Hatsugai, *Jpn. J. Appl. Phys.* **26,** L716 (1987).

[42]W. M. Temmerman, G. M. Stocks, P. J. Durham, and P. A. Sterne, *J. Phys.* F**17,** L135 (1987).

[43]D. W. Bullett and W. G. Dawson, *J. Phys.* C**20,** L853 (1987).

[44]For recent reviews, see J. Callaway and N. H. March, *in* "Solid State Physics" (H. Ehrenreich, D. Turnbull, and F. Seitz, eds.), Vol. 38, p. 136. Academic Press, New York, 1984, and papers in "Theory of the Inhomogeneous Electron Gas" (S. Lundquist and N. H. March, eds.), Plenum Press, New York, 1983.

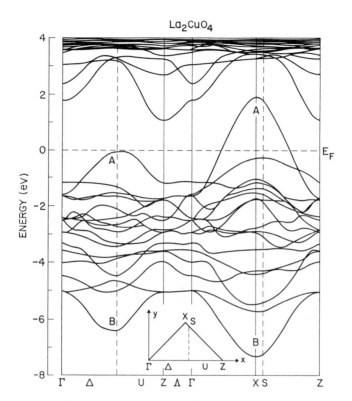

FIG. 5. Calculated LDA energy bands for La_2CuO_4 along high symmetry directions in body-centered tetragonal Brillouin zone. Portion of x–y plane in extended zone scheme shown in inset. [After L. F. Mattheiss, *Phys. Rev. Lett.* **58**, 1028 (1987)].

5 and the relatively flat bands near 4 eV, respectively. The Cu 4s states also contribute to these higher energy states, leaving a total of approximately 33 electrons to fill the 17-band Cu d–O p complex.

The character of the hybridized Cu d–O p bands is illustrated in Fig. 6. The Cu d states are split into four components by the presence of a tetragonal crystal field.[45] If each Cu were at the center of an undistorted $Cu-O_6$ octahedron, the overall symmetry would be cubic and the two e_g states (x^2–y^2 and $3z^2$–r^2) would be degenerate, as would the three t_{2g} states (xy, yz, and zx). The tetragonal nature of La_2CuO_4 is driven by a cooperative Jahn–Teller mechanism[46] that splits the degeneracy of the

[45]J. B. Goodenough, in *Proc. of the 1987 Fall Meeting of the Mater. Res. Soc.,* to be published.

[46]K. Terakura, H. Ishida, K. T. Park, A. Yanase, and N. Hamada, *Jpn. J. Appl. Phys.* **26,** L512 (1987).

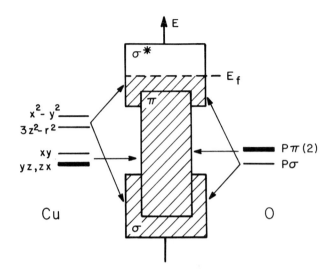

FIG. 6. Schematic energy level diagram for crystal field split Cu d and O p orbitals and resulting molecular orbital interpretation of LDA bands in Fig. 5. Shaded regions denote occuped states. σ and σ^* correspond, respectively, to bonding and antibonding bands resulting from nearest neighbor pdσ interactions. The π bands are constructed from nonbonding states and bonding and antibonding states resulting from nearest-neighbor pdπ interactions.

single unoccupied d state in the nominal Cu^{2+} configuration. In a purely ionic picture, each Cu^{2+} would then have a single d hole of x^2-y^2 symmetry. The d states of e_g character, however, hybridize strongly with the p_σ orbitals on the neighboring O. (The subscript σ denotes the p orbital which points along the Cu–O bond.) This hybridization gives rise to broad bonding (σ) and antibonding (σ^*) bands that account for most of the total 9 eV bandwidth in Fig. 5. The relatively flat bands near the center of the Cu d–O p complex are composed of nonbonding states as well as additional π-bonding and antibonding states that result from the much weaker hydridization between d orbitals of t_{2g} character and O p_π orbitals (the p orbitals perpendicular to the Cu–O bond).

Contrary to experiment,[17,24] this spin-restricted picture predicts metallic behavior in La_2CuO_4. The Fermi level E_f lies near the top of the Cu d–O p complex in Fig. 5 and is crossed by a single half-filled antibonding band labeled "A". Mattheiss[32] has demonstrated that this band can be reasonably described within a simple two-dimensional tight-binding model which includes only the Cu $d(x^2-y^2)$ orbital and the p_σ orbitals on the O1 atoms (cf. Fig. 1) in the Cu–O plane. For nearest-neighbor hybridization only, described by $V = -\sqrt{3}\,V_{pd\sigma}/2$, where $V_{pd\sigma}$ is the usual Slater–Koster two-center integral,[47,48] this model gives rise to a single nonbonding band at energy ε_p and bonding $(-)$ and antibonding $(+)$ bands at

$$E_{\pm}(\mathbf{k}) = \frac{\varepsilon_p + \varepsilon_d}{2} \pm \left\{ \left(\frac{\varepsilon_p - \varepsilon_d}{2} \right)^2 + 4V^2 \left(\sin^2 \frac{k_x a}{2} + \sin^2 \frac{k_y a}{2} \right) \right\}^{1/2}; \quad (3.1)$$

ε_p and ε_d are the p and d on-site energies, respectively. In the special case $\varepsilon_p = \varepsilon_d = \varepsilon$, which seems well satisfied[32] in La$_2$CuO$_4$, the antibonding band reduces to

$$E_A(\mathbf{k}) = \varepsilon + 2V\{1 - (\cos k_x a + \cos k_y a)/2\}^{1/2}, \quad (3.2)$$

which provides an excellent representation of band A for the parameter values $\varepsilon = -3.2\,\text{eV}$ and $V = 1.6\,\text{eV}$. An even simpler description of this band can be obtained[6,49] by considering a single Wannier orbital (composed of hybridized Cu d and O p states) with x^2-y^2 symmetry about each Cu site with on-site energy $\bar{\varepsilon}$ and nearest neighbor hopping t. For $\bar{\varepsilon} = \varepsilon + 2V$ and $t = V/4$, the resulting dispersion relation

$$E_A(\mathbf{k}) = \bar{\varepsilon} - 2t(\cos k_x a + \cos k_y a) \quad (3.3)$$

is essentially identical to Eq. (3.2) in the region near the Fermi surface, which in either model is defined by $\cos k_x a + \cos k_y a = 0$.

A schematic drawing of this idealized Fermi surface is shown in Fig. 7 in an extended zone scheme. The shaded electron pockets centered at Γ are equal in area to the hole pockets centred at X and are perfectly nested along the (110) and (1$\bar{1}$0) directions.[17,32,33] The existence of such nesting suggests the possibility of a gap opening at E_f due to a Peierls-like, commensurate charge density wave (CDW) distortion with wave vector $\mathbf{Q_1}$ or $\mathbf{Q_2}$ (with magnitude $2k_f$). The observed orthorhombic distortion in La$_2$CuO$_4$ was initially believed to be a manifestation of this effect and the origin of nonmetallic behavior.[17,32,33] More careful analysis revealed, however, that the resulting structure still has too high a symmetry to produce a gap because the two Cu sites per unit cell remain equivalent.[35,36,43,50,51] Calculated LDA bands for the orthorhombic

[47] J. C. Slater and G. F. Koster, *Phys. Rev.* **94**, 1498 (1954).

[48] W. A. Harrison, "Electronic Structure and the Properties of Solids." W. H. Freeman and Co., San Francisco, 1980.

[49] Y. Hasegawa and H. Fukuyama, *Jpn. J. Appl. Phys.* **26**, L322 (1987).

[50] W. Weber, *Phys. Rev. Lett.* **58**, 1371 (1987).

[51] M.-H. Whangbo, M. Evain, M. A. Beno, and J. M. Williams, *Inorg. Chem.* **26**, 1829 (1987); S. Barišić, I. Batistić, and J. Friedel, *Europhys. Lett.* **3**, 1231 (1987).

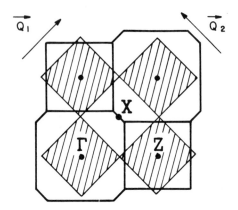

FIG. 7. Idealized Fermi surface associated with half-filled pdσ antibonding band (A in Fig. 5; Eq. (3.2) or (3.3)) in x–y plane of extended zone scheme. Electron pockets shaded. Nesting vectors given by $\mathbf{Q_1} = (\pi/a, \pi/a, 0)$ and $\mathbf{Q_2} = (-\pi/a, \pi/a, 0)$.

phase[36,43] confirm this conclusion and yield results that are virtually indistinguishable from those which would be obtained by folding the bands in Fig. 5 into the correspondingly smaller Brillouin zone.[52] While it is still possible that an additional, as yet undiscovered, crystallographic distortion could occur with the appropriate symmetry to produce a gap, the observed nonmetallic nature of La_2CuO_4 is more likely associated with its low temperature AF ordering (cf. Sections 4–7), which some have suggested represents the formation of a commensurate spin density wave[49,53,54] (SDW).

The full LDA Fermi surface differs from the idealized result in Fig. 7 in that it contains slightly smaller hole pockets and is weakly convex relative to the X point.[32,33,35] This curvature has been attributed[55,56] to a direct hopping between neighboring O p orbitals, which is neglected in the simplified tight-binding bands of Eqs. (3.2) and (3.3). The magnitude of this hopping will be seen in Section 6 to be an important factor in distinguishing between different strong correlation models. Harrison's universal tight-binding scheme[48] suggests that the nearest neighbor O p hopping parameters could be as large as a few eV. A detailed

[52]The small effects of the orthorhombic distortion will be neglected in the remaining discussion. Some interesting speculations on their possible significance can be found in D. C. Mattis and M. P. Mattis, *Phys. Rev. Lett.* **59**, 2780 (1987).

[53]P. A. Lee and N. Read, *Phys. Rev. Lett.* **58**, 2691 (1987).

[54]J. R. Schrieffer, X.-G. Wen, and S.-C. Zhang, *Phys. Rev. Lett.* **60**, 944 (1988).

[55]H. Eschrig and G. Seifert, *Solid State Commun.* **64**, 521 (1987).

[56]R. S. Markiewicz, submitted for publication.

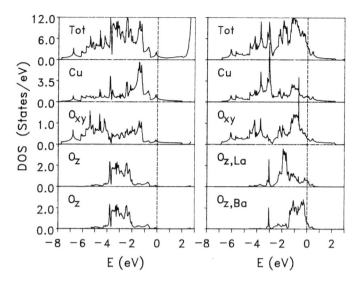

FIG. 8. Total and projected LDA densities of states (DOS) per atom for La_2CuO_4 (left panel) and an ordered $LaBaCuO_4$ alloy (right panel). [After W. E. Pickett, H. Krakauer, D. A. Papaconstantopoulos, and L. L. Boyer, *Phys. Rev. B35*, 7252 (1987).]

tight-binding analysis of constrained LMTO results by McMahan et al.[57] gives the slightly smaller values $V_{pp\sigma} = 1.0$ eV and $V_{pp\pi} = -0.3$ eV. Direct O–O hopping of this magnitude has also been extracted from other realistic tight-binding parametrizations[58,59] and is consistent with previous experience with other transition metal oxides.[7] In addition to its effect on the Fermi surface, such hopping can also account[57] for the ~4 eV width of the Cu d–O p complex at Γ in Fig. 5 and for the fact[35,57] that the lowest band "B" in the LDA results from a combination of O p orbitals with s-like symmetry about the Cu sites and not from the pdσ bonding band [− sign in Eq. (3.2)], which was incorrectly identified by Mattheiss.[32] (This latter band is actually the second lowest at X.)

The density of states of La_2CuO_4, shown in the left panel of Fig. 8,[34] contains a van Hove saddle-point singularity at an energy E_s just below the Fermi level. This slightly broadened logarithmic peak, $N(E) \propto \ln|E - E_s|$, results from the nearly two-dimensional character of the antibonding band A.[32–35] Even at the top of this peak, the total density of states is

[57]A. K. McMahan, R. M. Martin, and S. Satpathy, *Phys. Rev. B38* (1988).

[58]W. Weber, *Z. Phys. B70*, 323 (1988) and Ref. 50.

[59]D. A. Papaconstantopoulos, M. J. Deweet, and W. E. Pickett, in *Materials Research Society Symposium Proc.* (M. B. Brodsky, R. C. Dynes, K. Kitazawa, and H. L. Tuller, eds.), Vol. 99, p. 447, MRS, Pittsburgh, 1988 and private communication.

quite low compared to that in most previous "high T_c" materials (e.g., the A-15 compounds).[60] Calculated $N(E_f)$ values in La_2CuO_4 range from 1.1 to 1.65 states/eV-cell.[32–35,42,43] The projected densities in Fig. 8 show that almost all of this comes from the in-plane Cu and O (O_{xy} here or O1 in Fig. 1) which mix strongly throughout the 9 eV complex. The out-of-plane O (O_z or O2), by contrast, has most of its spectral weight in a much narrower region below E_f due to the longer Cu–O2 bond length. The average ground state configuration,[43] which is expected to be reasonably accurate in the LDA, corresponds to a Cu valence between 1+ and 2+ and an average O valence between −1 and −2.

Despite the failure of the above results to explain the insulating character of the stoichiometric parent compound, the spin-restricted LDA bands may still be relevant to the "normal" metallic state of $La_{2-x}M_xCuO_4$ alloys. The simplest method for extrapolating to the alloy is to assume that each M^{2+} ion contributes one less electron than La^{3+} to the otherwise unperturbed Cu–O bands.[32–34] The Fermi level in such a "rigid band model" moves to lower energy with increasing x. At a critical x value of about 0.14–0.17, E_f coincides with the van Hove singularity,[33,34,56] at which point $N(E_f)$ is a maximum and is almost a factor of two larger than for $x = 0$. The fact that this is also the composition range which exhibits the highest T_c's suggests that $N(E_f)$ does play an important role, as in the conventional Bardeen–Cooper–Schrieffer (BCS) theory of superconductivity.[4] Calculations[61] of anisotropic transport coefficients based on rigid LDA bands are in reasonable agreement with single crystal Hall data.[62]

More realistic alloy calculations have been performed within the virtual crystal[63] and coherent potential approximations[64] (CPA) and by considering ordered $La_{2-x}M_xCuO_4$ supercells.[34,65] In all cases, simple rigid band behavior is well obeyed in the vicinity of E_f for all relevant x values. The right side of Fig. 8, for example, shows the LDA density of states for an ordered $LaBaCuO_4$ alloy constructed by replacing one of the La atoms in the unit cell by Ba.[34] Even in this case for which $x = 1$, the height and

[60]B. M. Klein, L. L. Boyer, D. A. Papaconstantopoulos, and L. F. Mattheiss, *Phys. Rev.* **B18,** 6411 (1978).

[61]P. B. Allen, W. E. Pickett, and H. Krakauer, *Phys. Rev.* **B36,** 3926 (1987); and *Phys. Rev.* **B37,** 7482 (1988).

[62]M. Suzuki and T. Murakami, *Jpn. J. Appl. Phys.* **26,** L524 (1987).

[63]B. A. Richert and R. E. Allen, *Phys. Rev.* **B37,** 7869 (1988).

[64]D. A. Papaconstantopoulos, W. E. Pickett, and M. J. Deweert, *Phys. Rev. Lett.* **61,** 211 (1988).

[65]R. A. DeGroot, H. Gutfreund, and M. Weger, *Solid State Commun.* **63,** 451 (1987); K. Schwarz, *Solid State Commun.* **64,** 421 (1987).

shape of the van Hove singularity associated with band A are essentially unchanged from the $x = 0$ result, although the Fermi level now lies well below this peak. Noticeable changes do occur, however, elsewhere in the bands. In particular, the spectral weight on Cu is shifted to lower energy and the spectral weights on the two O_z atoms, which are now inequivalent, are shifted to higher energy and become partially metallic. The lower binding energies of the O_z states in layers containing Ba result from the weaker attractive potential induced by Ba^{2+} compared to La^{3+}.

The effects of O vacancies have been examined in a similar manner.[64,66,67] Here a rigid band description, in which the Fermi level rises in energy with decreasing O content, fails completely. Kasowski *et al.*[67] have shown that for an ordered array of vacancies in La_2CuO_{4-y}, band A breaks up into much narrower mini-bands, which the authors propose as an explanation for the enhanced tendency towards antiferromagnetism for $y > 0$. Recent CPA calculations by Papaconstantopoulos *et al.*[64] indicate that in the disordered case, a small concentration of vacancies has little effect, but a larger concentration introduces a slight broadening throughout the Cu–O complex and shifts the Fermi level to lower energies, in contrast to the rigid band prediction.

b. $YBa_2Cu_3O_{7-\delta}$

State-of-the-art LDA band structures for the protypical 1-2-3 system $YBa_2Cu_3O_7$ are generally more complicated than for La_2CuO_4 because of the more complex unit cell.[37–43,68,69] The two systems nevertheless exhibit a number of remarkable similarities. The conduction bands in $YBa_2Cu_3O_7$ are again ~9 eV wide and arise predominantly from strongly hybridized Cu d–O p states. The 36 bands in this complex are filled by a total of 68 electrons, including the valence electrons of the Y and Ba. The resulting Y^{3+} and Ba^{2+} states contribute primarily to unoccupied bands separated by a gap from the Cu–O complex. The absence of significant valence charge in the Y layers produces a negligible dispersion along k_z (<0.1 eV) in the Cu–O bands. This decoupling of the conduction electrons from the Y site is also responsible for the remarkable insensitivity of the superconducting properties to the particular choice of rare earth ion.[31]

Figure 9 shows the LDA bands calculated by Freeman and coworkers[38] for $YBa_2Cu_3O_7$ near the top of the Cu–O complex. Since k_z has little

[66]B. A. Richert and R. E. Allen, *Jpn. J. Appl. Phys.* **26,** Sup. 26-3, 989 (1987).
[67]R. V. Kasowski, W. Y. Hsu, and F. Herman, *Phys. Rev. B***36,** 7248 (1987).
[68]H. Krakauer, W. E. Pickett, and R. E. Cohen, to be published (*J. Supercond.*).
[69]B. Szpunar and V. H. Smith, *Phys. Rev. B***37,** 7525 (1988).

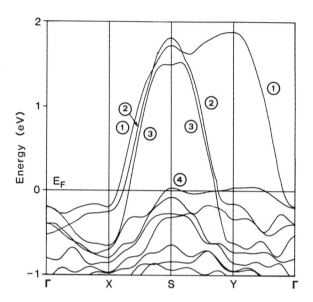

FIG. 9. Calculated LDA energy bands for $YBa_2Cu_3O_7$ at the top of the Cu–O complex in central plane of Brillouin zone. [After J. Yu, S. Massida, A. J. Freeman, and D. D. Koelling, *Phys. Lett.* **A122**, 203 (1987).

effect, the bands are plotted only in the basal plane of the Brillouin zone defined by the high symmetry points $\Gamma\,(0, 0, 0)$, $X\,(\pi/a, 0, 0)$, $Y\,(0, \pi/b, 0)$, and $S\,(\pi/a, \pi/b, 0)$. Of primary interest here are the four bands that intersect the Fermi level whose origins are worth exploring in some detail. Two of these bands—② and ③—exhibit an appreciable (and nearly identical) dispersion along X–S and Y–S indicative of two-dimensional behavior. The other two—① and ④—are also strongly dispersed along X–S but their relatively flat variation along Y–S suggests a more one-dimensional character.

Bands ② and ③ are remarkably similar to the single band A that intersects the Fermi level in La_2CuO_4 (Fig. 5). As in the 2-1-4 case, these bands are associated primarily with the Cu–O_2 planes (of which there are now two per cell) and are derived from $pd\sigma$ antibonding combinations of Cu2 $d(x^2-y^2)$, O2 p_x and O3 p_y orbitals. A reasonable description[37] of these bands is again provided by the simple tight-binding expressions (3.2) or (3.3) (with $k_y a$ replaced by $k_y b$), with parameter values also similar to those in La_2CuO_4.

Bands ① and ④, on the other hand, originate in the Cu–O chains. Like the plane bands, band ① represents a $pd\sigma$ antibonding combination of Cu1 d and neighboring O p_σ states. A minimal basis for describing this band consists of a Cu1 d (y^2-z^2) orbital, an O1 p_y orbital, and O4 p_z orbitals above and below the chain.[37] For equal p and d on-site energies ε, a nearest-neighbor hybridization V in this case leads to the one-dimensional (1D) antibonding band

$$E_{1D}(\mathbf{k}) = \varepsilon + 2V\{1 - \cos k_y b)/2\}^{1/2}. \tag{3.4}$$

In analogy with Eq. (3.3), a single-band Wannier description, with on-site energy $\bar{\varepsilon}$ and hopping t, produces the slightly different result

$$E_{1D}(\mathbf{k}) = \bar{\varepsilon} - 2t \cos k_y b. \tag{3.5}$$

Either of these expressions provides a reasonable description of band ①, although the parameters must be modified slightly[70] compared to those used to describe bands ② and ③. (Using the same parameters places band ① about 0.6 eV too low compared to bands ② and ③.) The need for slightly different chain parameters is not surprising in view of the different electrostatic potentials, bond lengths, and interactions with neglected orbitals in the chains and planes.

Band ④ in Fig. 9 is derived from Cu1 d (zy), O1 p_z and O4 p_y orbitals. This band can also be modeled by using Eq. (3.4) or (3.5) with a smaller hybridization parameter, assumed to result from nearest-neighbor pdπ interactions, which are much weaker than pdσ interactions.[48] Also contributing, according to Krakauer et al.,[68] are ppσ interactions between O1 and O4 orbitals, similar to the direct O–O hopping discussed for 2-1-4 compounds.

Although the plane and chain bands are largely independent, the detailed behavior in Fig. 9 indicates a weak coupling between the three pdσ bands. This interaction can be described by the 3×3 matrix

$$\begin{pmatrix} E_A(\mathbf{k}) & 0 & t_\perp \\ 0 & E_A(\mathbf{k}) & t_\perp \\ t_\perp & t_\perp & E_{1D}(\mathbf{k}) \end{pmatrix},$$

where $t_\perp$ is the coupling between neighboring plane and chain Wannier functions. Band ② corresponds in this model to the noninteracting band at energy $E_A(\mathbf{k})$, which results from a combination of plane states with odd parity under z reflection.[38] Band ③ is derived primarily from the analogous combination with even parity.[38] The hybridization between

---

[70]For example, the on-site energy $\bar{\varepsilon}$ in Eq. (3.5) can be shifted up by an amount $2t$ compared to the on-site energy in Eq. (3.3).

even parity states and states in the chain is responsible for the small splitting between bands ② and ③ in Fig. 9 and for the anticrossing behavior of bands ① and ③ along S–Y.

The larger number of bands crossing the Fermi level in $YBa_2Cu_3O_7$ gives rise to a more complicated Fermi surface than in 2-1-4 systems.[37,38,68] The states above E_f in Fig. 9 can accommodate four holes that are distributed unevenly among bands ①–④. (Recall that the 36 Cu–O bands are filled with 68 electrons.) Band ② (③) is slightly less (more) than half full while band ① (④) is almost completely empty (full). The portions of the Fermi surface resulting from the two-dimensional bands ② and ③ are not strongly nested along the (110) and (1$\bar{1}$0) directions as in Fig. 4. These portions are partially nested along the (100) and (010) directions and the one-dimensional band ① is also nested along (010). Commensurate CDW or SDW formation is not expected, however, since the spanning vectors do not correspond to reciprocal lattice vectors.

The flatness of band ④ makes both the Fermi surface and the value of $N(E_f)$ difficult to calculate accurately and strongly dependent on δ. Theoretical $N(E_f)$ values[37–43,68,69] for $\delta = 0$ range from 3.4 to 6.6 states/eV-cell. On a per Cu atom basis, these values are quite comparable to those in La_2CuO_4 and are again small compared to most conventional superconductors.

Calculated LDA densities of states for $\delta = 0$ contain a sharp peak due to band ④ at less than 0.1 eV below the Fermi level.[37–43,68,69] A simple rigid band model predicts an increase in E_f with increasing δ and thus a shift away from this peak to a lower value of $N(E_f)$. The effects of O vacancies are extremely important in this material since even the best 1-2-3 superconductors are believed to have $\delta \simeq 0.1$. As in the 2-1-4 case, the validity of a rigid band treatment of these effects is highly questionable. The removal of an O atom represents a large, localized perturbation that may be particularly significant in 1-2-3 materials, since it causes the disruption of a one-dimensional Cu–O chain.[28a,71]

Full LDA calculations for $YBa_2Cu_3O_6$, which contains no Cu–O chains, have been reported by several groups.[40,41,43,69,72] No one-dimensional bands occur in this system but the two pdσ antibonding bands associated with the Cu–O$_2$ planes still cross the Fermi level. The situation is thus similar to that of La_2CuO_4, where the LDA incorrectly predicts metallic behavior while the material is actually an AF insulator.

[71]B. A. Richert and R. E. Allen, in *Materials Research Society Symposium Proc.* (M. B. Brodsky, R. C. Dynes, K. Kitazawa, and H. L. Tuller, eds.), Vol. 99, p. 463, MRS, Pittsburgh, 1988.

[72]W. M. Temmerman, Z. Szotek, P. J. Durham, G. M. Stocks, and P. A. Sterne, *J. Phys. F***17**, L319 (1987).

One final LDA result that should be mentioned is that the ground-state charge densities in $YBa_2Cu_3O_{7-\delta}$ are inconsistent with an integral valence picture in which the $\delta = 0.5$ compound consists entirely of Cu^{2+} and O^{2-}, with the additional holes for $\delta = 0$ residing on the Cu1 sites. The actual LDA ground state[40,42,68] for $\delta = 0$ corresponds to a Cu valence of less than +2 on both the Cu1 and Cu2 sites and an average O valence again between -1 and -2.

4. Spin-Polarized Calculations

The insulating character of the nonsuperconducting compounds La_2CuO_4 and $YBa_2Cu_3O_6$ is clearly related in some way to the low-temperature AF ordering observed in these systems. Slater[73] first showed in 1951 how to modify a band structure calculation for an AF material by considering a different self-consistent one electron potential for each direction of spin. An electron with spin parallel to the net spin at a given site feels a more attractive potential in this approach than an electron with spin antiparallel to the net spin. Since an AF state does not have the full translational symmetry of the underlying lattice, the new periodicity folds the bands into a smaller Brillouin zone and introduces new splittings at the zone boundaries. An insulating gap may thus occur that would not be present in a spin-restricted calculation.

Modern spin-polarized band calculations are generally performed within the local spin density approximation[44] (LSDA). Several such calculations have been reported recently for La_2CuO_{4-y}, with most results in agreement that the LSDA *does not* produce a stable AF ground state with an insulating gap,[74-76] even for $y > 0$. (The one exception to this is composed of the pseudopotential calculations of Shiraishi *et al.*)[77] The unit cell of AF–La_2CuO_4 is the same as that of the orthorhombically-distorted K_2NiF_4 structure. The introduction of magnetic moments of opposite sign on the two Cu sites in this cell does produce a splitting of the folded version of band A in Fig. 5 at every point in k space. The

[73]J. C. Slater, *Phys. Rev.* **82**, 538 (1951).

[74]T. C. Leung, X. W. Wang, and B. N. Harmon, *Phys. Rev. B***37**, 384 (1988); Y. Hatsugai and T. Fujiwara, *Solid State Commun.* **65**, 1271 (1988); P. A. Sterne and C. S. Wang, *Phys. Rev. B***37**, 7472 (1988); J. Kubler, V. Eyert, and J. Sticht, *Physica C***153–155**, 1237 (1988); V. Eyert, J. Sticht, and J. Kubler, *Helv. Phys. Acta* **61**, 496 (1988).

[75]G. Y. Guo, W. M. Temmerman, and G. M. Stocks, *J. Phys. C***21**, L103 (1988).

[76]J. Zaanen, O. Jepsen, O. Gunnarsson, A. T. Paxton, O. K. Anderson, and A. Svane, *Physica C***153–155**, 1636 (1988).

[77]K. Shiraishi, A. Ohiyama, N. Shima, T. Nakayama, and H. Kamimura, *Solid State Commun.* **66**, 629 (1988).

magnitude of the calculated splitting[74-76] ($\sim$0.1 eV) is too small relative to the bandwidth, however, to open up a gap at the Fermi level; the calculated moments of $0.1 \mu_B$/Cu are also too small compared to experiment.[13-15] Guo et al.[75] have shown that a finite gap of 0.15 eV can be obtained in the LSDA by artificially imposing the much larger moments of $0.38 \mu_B$/Cu. Weinert and Fernando[78] have suggested that the true insulating ground state may correspond to a more complicated AF ordering with magnetic moments on the O sites instead of the Cu.

Only one realistic LSDA calculation has been reported to date for $YBa_2Cu_3O_6$. Szpunar et al.,[79] prior to the experimental determination[25] of the AF structure of this material, did find a stable insulating ground state in this system with a gap of <0.1 eV, alternating moments of $0.38 \mu_B$ on the Cu2 sites and no moments on the Cu1 sites. Nonmagnetic Cu1 sites were also predicted by LSDA cluster calculations for Cu1–O1 chains.[80] The observed AF structure[25] is qualitatively consistent with these results. The experimental Cu2 moments ($\sim$0.5 μ_B) are slightly larger in magnitude, however, and the true insulating gap (cf. Section 11) is believed to be much larger than in the LSDA.

The limited success of the above results is similar to that encountered previously[81] in LSDA calculations for the prototypical AF insulators MnO, FeO, CoO, and NiO. In these systems, which are generally assumed to be strongly correlated,[7] Terakura et al.[81] found that the LSDA could only account for the insulating behavior of MnO and NiO and, even then, the calculated band gaps were much too small.[82]

The principal limitation of any one-electron theory, including an exact density-functional treatment, is the failure to account properly for the different potentials of occupied and unoccupied states.[7,44] A conduction electron moves in an N-electron potential associated with a neutral solid while a hole moves in an (N-1)-electron, ionic potential. The assumption in the LDA (or LSDA) that both types of excitation can be described by the same local potential is responsible for the well known "band gap problem" in ordinary wide-band semiconductors.[44,83] As states become more localized, as in systems with narrow d bands, the magnitudes of the

[78]M. Weinert and G. W. Fernando, submitted for publication.

[79]B. Szpunar, V. H. Smith, and R. W. Smith, *Physica C* **152**, 91 (1988).

[80]H. Chen, J. Callaway, and P. K. Misra, *Phys. Rev. B***36**, 8863 (1987); and *Phys. Rev. B***38**, 195 (1988).

[81]K. Terakura, A. R. Williams, T. Oguchi, and J. Kubler, *Phys. Rev. Lett.* **52**, 1839 (1984); K. Terakura, T. Oguchi, A. R. Williams, and J. Kubler, *Phys. Rev. B***30**, 4734 (1984).

[82]G. A. Sawatzky and J. W. Allen, *Phys. Rev. Lett.* **53**, 2339 (1984).

[83]M. S. Hybertson and S. G. Louie, *Phys. Rev. B***34**, 5390 (1986) and references therein.

discrepancies between density-functional theory eigenvalues and true quasiparticle excitation energies become even larger.

Recently, Zaanen *et al.*[76] have shown that some of the difficulty in achieving a stable insulating ground state in AF–La_2CuO_4 can be eliminated by considering self-interaction corrections[84] to the LSDA. Techniques also exist in some cases for calculating reasonable excitation energies using the LSDA (e.g., from total energy differences[85] or Slater's transition state construct)[44] but only by sacrificing the simplicity of a pure one-electron description.

IV. Correlated Electron Models

The analogy to other transition-metal oxides suggests that the physics of the new copper-oxide superconductors may be dominated, not by one-electron effects, but by correlated electronic behavior resulting from a large intraatomic Coulomb repulsion U_d between Cu d electrons.[6] In this picture, La_2CuO_4 and $YBa_2Cu_3O_6$ are assumed to be insulators, in part, because of a Mott–Hubbard correlation gap[7,86] between occupied and unoccupied Cu d states. Strong correlation also alters the character of low-energy excitations in the metallic phases of these materials. This section will examine some of the more popular model Hamiltonians that have been proposed to treat these effects. There is a growing suspicion that one or more of these models may hold the key to a purely electronic mechanism[5] for superconductivity that may be operative in both high T_c and heavy fermion[9] materials.

5. SINGLE-BAND HAMILTONIAN

The simplest model for studying correlation effects, and the one first proposed by Anderson[6] for the 2-1-4 and 1-2-3 copper oxides, is the single-band Hubbard Hamiltonian[86]

$$\mathcal{H} = -t \sum_{\langle ij \rangle \sigma} (C_{i\sigma}^+ C_{j\sigma} + \text{h.c.}) + U \sum_i n_{i\uparrow} n_{i\downarrow}. \qquad (5.1)$$

Here $c_{i\sigma}^+$ is a creation operator for an electron with spin σ in a Wannier orbital at site i, $n_{i\sigma} = c_{i\sigma}^+ c_{i\sigma}$, t is a nearest neighbor hopping integral ($\langle ij \rangle$

[84]J. P. Perdew and A. Zunger, *Phys. Rev.* B**23**, 5048 (1981).
[85]A. E. Carlsson, *Phys. Rev.* B**31**, 5178 (1985).
[86]J. Hubbard, *Proc. Roy. Soc.* (London) A**276**, 238 (1963).

denotes a given nearest neighbor pair), and U is a Coulomb repulsion between two electrons on the same site. Equation (5.1) is generally assumed to model the states closest to the Fermi level that are associated with a single Cu–O$_2$ plane; the Wannier functions thus correspond to the same antibonding combinations of Cu d (x^2-y^2) and O p$_\sigma$ orbitals that give rise to band A in Fig. 5. In the limit $U = 0$, Eq. (5.1) reduces to Eq. (3.3) with $\bar{\varepsilon} = 0$. Band A is then reasonably modeled by choosing t in the range 0.4–0.5 eV.

For a nominally (Cu^{2+})–$(O^{2-})_2$ plane (corresponding to $x = 0$ in La$_{2-x}M_x$CuO$_4$ or $\delta \approx 0.3$–0.5 in RBa$_2$Cu$_3$O$_{7-\delta}$), each Wannier orbital in Eq. (5.1) contains an average of one electron. Although conventional band theory always predicts metallic behavior in this case, the two-dimensional Hubbard model correctly accounts for the insulating character of these materials for any $U > 0$. In the itinerant limit, $U \ll t$, the nested Fermi surface in Fig. 7 is unstable to the formation of a commensurate SDW;[49,53,54] this opens up a gap at the Fermi level and provides a possible explanation for the small moments (<1 μ_B) observed in the AF phase.[13–15,25] In the localized limit, $U \gg t$, a Mott–Hubbard correlation gap exists independently of any spin ordering due to the large energy penalty for double occupancy on any site.[6] Equation (5.1) then maps directly onto a two-dimensional $S = 1/2$ Heisenberg Hamiltonian

$$\mathcal{H} = J \sum_{\langle ij \rangle} \mathbf{S}_i \cdot \mathbf{S}_j, \tag{5.2}$$

with an AF superexchange[87] interaction $J = 4t^2/U$. In this picture, the small experimental moments are attributed to quantum fluctuation effects resulting from the low spin and low dimensionality.[88]

Which of these limits applies to the 2-1-4 and 1-2-3 copper oxides is unclear at the moment.[89] From studies of other Cu compounds,[90] one would expect the magnitude of U_d for the Cu 3d electrons to be in the range of 5 to 15 eV, which would be roughly consistent with experimental values[88,91] of J (of the order of 1000 K (100 meV) in both systems).[92] The value of U does not correspond directly to U_d, however, since the

[87]P. W. Anderson, in "Solid State Physics" (F. Seitz and D. Turnbull, eds.), Vol. 14, p. 99. Academic Press, New York, 1963.

[88]S. Chakravarty, B. I. Halperin, and D. R. Nelson, *Phys. Rev. Lett.* **60**, 1057 (1988).

[89]There are also models based on "negative U" values, e.g., J. A. Wilson, *J. Phys.* **C21**, 2067 (1988).

[90]M. R. Thuler, R. L. Benbow, and Z. Hurych, *Phys. Rev.* **B26**, 669 (1982).

[91]K. B. Lyons, P. A. Fleury, L. F. Schneemeyer, and J. V. Waszczak, *Phys. Rev. Lett.* **60**, 732 (1988); K. B. Lyons, P. A. Fleury, J. P. Remeika, and T. J. Negran, *Phys. Rev.* **B37**, 2353 (1988).

Wannier functions are not pure Cu d states. As will be seen in Section 6, it is also possible that Eq. (5.1) is more properly viewed as an "effective" single-band model;[93] in this case, the appropriate parameter values are even less certain and may not be simply related to J.

The limit $U \ll t$ forms the basis of the so-called "spin-bag" mechanism for superconductivity proposed by Schrieffer *et al.*[54] This model assumes that even the presence of short-range SDW order produces a "pseudogap" at the Fermi level that may vary both spatially and temporally. The pseudogap is suppressed locally by the introduction of a hole (e.g., by substituting Sr for La in La_2CuO_4), which causes the hole to become self-trapped in a region known as a "bag." Schrieffer *et al.*[54] argue that such a bag provides an attractive potential to an additional hole (or holes),[68] which leads to a superconducting ground state.

Much more attention[6,94,95] has been paid to the localized limit $U \gg t$. Anderson first proposed this for La_2CuO_4 very early on,[6] and suggested that the ground state of Eq. (5.2) for a two-dimensional square lattice is not the classical Neel state, but is rather a quantum spin-liquid or "resonating valence bond" (RVB) state.[96] A single valence bond in this picture is defined as a singlet state constructed from any pair of spins, which need not be nearest neighbors. The RVB ground state then corresponds to a coherent superposition of all possible configurations of valence bonds.[96] Such a state is itself a precise singlet and has no long-range order and no fixed spin direction at each site. Since it is now known that both the 2-1-4 and 1-2-3 systems do exhibit true AF long-range order,[13–15,25] it must be that the RVB state is not the lowest energy state in the insulating regime. The RVB concept may still be relevant to the metallic phase,[97] however, since the AF order is destroyed by a small concentration of holes.

A finite hole concentration in the large U/t limit is conveniently examined by performing a canonical transformation to eliminate the

[92]The slightly smaller (~10%) values of J obtained in 1-2-3 compounds compared to 2-1-4 compounds has been attributed by S. R. Ovshinsky, S. J. Hudgens, R. L. Lintvedt, and D. B. Rorabacher, *Mod. Phys. Lett.* **B1**, 275 (1987), to the slight buckling of the Cu-O$_2$ planes in Fig. 3.

[93]F. C. Zhang and T. M. Rice, *Phys. Rev.* **B37**, 3759 (1988).

[94]G. Baskaran, Z. Zou, and P. W. Anderson, *Solid State Commun.* **63**, 973 (1987); H. Fukuyama and K. Yosida, *Jpn. J. Appl. Phys.* **26**, L371 (1987); A. E. Ruckenstein, P. J. Hirschfeld, and A. J. Appel, *Phys. Rev.* **B36**, 857 (1987); M. Cyrot, *Solid State Commun.* **62**, 821 (1987); M. Cyrot, *Solid State Commun.* **63**, 1015 (1987).

[95]H. Fukuyama and Y. Hasegawa, *Physica B&C* **148**, 2104 (1987).

[96]P. W. Anderson, *Mat. Res. Bull.* **8**, 153 (1973).

[97]P. W. Anderson, G. Baskaran, Z. Zou, and T. Hsu, *Phys. Rev. Lett.* **58**, 2790 (1987).

possibility of two of the remaining electrons occupying the same site.[98] To lowest order in t/U, this yields the effective Hamiltonian

$$\mathcal{H} = -t \sum_{\langle ij \rangle \sigma} \{(1 - n_{i-\sigma})C_{i\sigma}^+ C_{j\sigma}(1 - n_{j-\sigma}) + \text{h.c.}\} + J \sum_{\langle ij \rangle} (\mathbf{S}_i \cdot \mathbf{S}_j - n_i n_j/4),$$

(5.3)

where $n_i = n_{i\uparrow} + n_{i\downarrow}$.

Equation (5.3) describes metallic behavior since there are both spin and charge degrees of freedom. The restricted hopping in the first term destroys the AF long-range order and increases the effective mass (m^*) of the small number of free holes.[99] In a Fermi liquid picture, this mass enhancement is associated with a narrowing of the entire quasiparticle band[9] since, according to Luttinger's theorem,[100] the volume enclosed by the Fermi surface must be the same as in the noninteracting case $(U = 0)$. An extreme case of such a band narrowing and mass enhancement $(m^*/m > 1000)$ is believed to occur in heavy fermion systems. A more moderate mass enhancement $(m^*/m \simeq 3\text{--}10)$ has been proposed for the copper oxides[19,101] to explain several discrepancies between band theory predictions and the results of magnetic susceptibility, specific heat, and optical measurements.

The Fermi liquid picture assumes that the second term in Eq. (5.3), which represents the potential energy, is of secondary importance compared to the kinetic energy.[95] The potential energy leads to interactions between quasiparticles, which, in mean field theory, result in a superconducting ground state with extended singlet pairing.[94] In RVB theory, by contrast, the potential energy term is assumed to dominate.[95] The elementary excitations relative to an RVB ground state are of two principal types, which, for topological reasons, have anomalous statistics.[6,97,102] An unpaired spin acts as an uncharged, spin-1/2 fermion known as a spin soliton, or "spinon." The hole produced by removing such a spin acts as a charge e, spin-zero, boson, or "holon." A physical hole is a combination of a spinon and a holon. While this theory has led to some interesting formal developments[103] and proposals for novel superconducting mechanism (e.g., Bose condensation of holons)[6,97,102,104]

[98]J. E. Hirsch, *Phys. Rev. Lett.* **54**, 1317 (1985).

[99]W. F. Brinkman and T. M. Rice, *Phys. Rev. B* **2**, 4302 (1970).

[100]J. M. Luttinger, *Phys. Rev.* **119**, 1153 (1960).

[101]G. A. Thomas, J. Orenstein, D. H. Rapkine, M. Capizzi, A. J. Millis, R. N. Bhatt, L. F. Schneemeyer, and J. V. Waszczak, *Phys. Rev. Lett.* **61**, 1313 (1988).

[102]S. A. Kivelson, D. S. Rokhsar, and J. P. Sethna, *Phys. Rev. B* **35**, 8865 (1987).

[103]V. Kalmeyer and R. B. Laughlin, *Phys. Rev. B* **59**, 2095 (1987); P. B. Wiegmann, *Phys. Rev. Lett.* **60**, 821 (1988).

[104]J. M. Wheatley, T. C. Hsu, and P. W. Anderson, *Phys. Rev. B* **37**, 5897 (1988).

it has not yet made much direct contact with experiment.[105] Since the relative importance of the kinetic energy term increases with hole concentration, an RVB description is most likely to be relevant in the low doping regime.[95]

Whether or not the less than half-filled, single-band Hubbard model, without further approximations, leads directly to superconductivity remains an open question.[106] As will be discussed below, it is likely that this question has only limited relevance to the copper oxides, since there is now considerable evidence that the "internal electronic structure" of the $Cu-O_2$ Wannier functions, which is neglected in Eqs. (5.1) and (5.3), must also be considered. Nevertheless, the lack of definitive results in even this simplest of models clearly illustrates a disturbing deficiency in our present understanding of correlation effects in general.

6. Extended-Hubbard and Anderson Hamiltonians

In contrast to the prediction of the single-band Hubbard model, most existing spectroscopic data for the 2-1-4 and 1-2-3 copper oxides (cf. Part V) indicate that the holes responsible for metallic behavior in these systems have almost pure O p character and very little Cu d component. The need to include ligand p orbitals explicitly has also been recognized recently in the context of other transition metal compounds, including the MnO-NiO series.[82,107] Generalized Hubbard models for the 2-1-4 and 1-2-3 copper oxides based on this idea generally fall into two categories. The more popular approach,[108] which is common to all of the models described in this section, assumes that the most important O p orbitals are those of σ character. Models based on O p_{π} orbitals[109,110] will be discussed in Section 7.

a. Emery Model

One of the first and most general models in the first category is the extended Hubbard Hamiltonian proposed by Emery.[108] To define this

[105]P. W. Anderson and Z. Zou, *Phys. Rev. Lett.* **60**, 132 (1988); see, however, C. Kallin and A. J. Berlinsky, *Phys. Rev. Lett.* **60**, 2556 (1988) and the critical comments in Ref. 111.

[106]S. A. Trugman, *Phys. Rev.* **B37**, 1597 (1988); J. E. Hirsch and H. Q. Lin, *Phys. Rev.* **B37**, 5070 (1988).

[107]J. Zaanen, G. A. Sawatzky, and J. W. Allen, *Phys. Rev. Lett.* **55**, 418 (1985) and *J. Mag. Magn. Mater.* **54–57**, 607 (1986); J. Zaanen and G. A. Sawatzky, *Can. J. Phys.* **65**, 1262 (1987).

[108]V. J. Emery, *Phys. Rev. Lett.* **58**, 2794 (1987).

[109]Y. Guo, J. M. Langlois, and W. A. Goddard, *Science* **239**, 896 (1988); G. Chen, and W. A. Goddard, *Science* **239**, 899 (1988).

[110]R. J. Birgeneau, M. A. Kastner, and A. Aharony, *Z. Physik* **B71**, 57 (1988).

model, it is convenient to switch to a hole representation in which the vacuum state consists of completely filled Cu d and O p orbitals. The Hamiltonian for a single Cu–O_2 plane may then be written as

$$\mathcal{H} = \varepsilon_d^0 \sum_{i\sigma} d_{i\sigma}^+ d_{i\sigma} + \varepsilon_p^0 \sum_{l\sigma} p_{l\sigma}^+ p_{l\sigma} + \sum_{\langle il \rangle \sigma} t_{il}(d_{i\sigma}^+ p_{l\sigma} + \text{h.c.})$$

$$+ U_d \sum_i n_{i\uparrow}n_{i\downarrow} + U_p \sum_l n_{l\uparrow}n_{l\downarrow} + U_{pd} \sum_{\langle il \rangle} n_i n_l, \tag{6.1}$$

where i and l refer to Cu and O sites, respectively, $d_{i\sigma}^+$ creates a Cu d (x^2-y^2) hole of spin σ, and $p_{l\sigma}^+$ creates an O p_x or p_y hole (whichever points toward the neighboring Cu) of spin σ. In addition to the intraatomic Coulomb repulsion U_d between Cu holes, Eq. (6.1) also contains an intraatomic repulsion U_p between O p holes and an inter-atomic repulsion U_{pd} between holes on neighboring Cu and O sites. Nearest-neighbor hopping is described by $t_{il} = \pm V^0$ with the sign deter-mined by symmetry to be positive (negative) if site l lies along the $+x$ or $-y$ $(-x$ or $+y)$ direction relative to site i.

Equation (6.1) reduces trivially to the single-band Hubbard model of Section 5 in the limits (1) $U_{pd} = 0$, $U_d = U_p = U$ and (2) U_{pd}, U_d, $V^0 \ll \Delta \equiv \varepsilon_p^0 - \varepsilon_d^0$. In case (1), the relevant Wannier functions are again the same antibonding combinations of Cu d and O p states that describe band A in the LDA. In case (2), the O p hole states are so energetically unfavorable that only the Cu d orbitals need to be considered.[111]

In general, the insulating state analogous to the half-filled condition in Section 5 is obtained from Eq. (6.1) by introducing one hole per Cu–O_2 complex. For $\Delta > 0$, such holes are localized primarily on the Cu sites (+2 valence) in the ground state. The stability of this ionic configuration is enhanced by the U_{pd} term in Eq. (6.1), which represents the Madelung energy in this model.[111] The lowest energy excitations of such a system are again described by the spin-1/2 Heisenberg Hamiltonian, Eq. (5.2), although the superexchange interaction is now modified (cf. Section 6b). For $U_d > \Delta + 2U_{pd}$, the conductivity gap is determined primarily by the charge transfer energy Δ associated with the process $Cu^{2+}O^{2-} \rightarrow Cu^+O^-$. In this parameter range, which is believed to apply to the copper oxides,[108] the insulating materials are characterized as "charge transfer insulators," as opposed to pure Mott–Hubbard insulators, in the general classification scheme for transition metal compounds developed by Zaanen et al.[107]

The allowance for O p states within the correlation gap of the Cu d

[111]C. M. Varma, to be published (Proc. Schloss Mautendorf).

states is the key ingredient of the Emery model. Under these circumstances, any additional holes introduced by doping the insulating material will have primarily O p character, which will be seen in Part V to be consistent with experiments. The new holes are free to move through virtual hopping processes involving the Cu sites.[108] The hybridization between Cu d and O p states gives rise to a strong AF exchange interaction[108,112] between Cu d and O p holes that is sufficient to destroy the AF long-range order produced by the weaker Cu–Cu superexchange. Many authors, beginning with Emery,[108] have suggested that this new interaction leads to a magnetically induced attraction between O p holes, which, in turn, leads to superconductivity.[112]

An alternative superconducting mechanism based on Eq. (6.1) has been proposed by Varma *et al.*[111,113] In this theory, the attractive interaction arises from an exchange of "charge-transfer excitons" (e.g., $Cu^{2+}O^{2-} \to Cu^{+}O^{-}$ excitations whose energy is lowered by an amount U_{pd} in the presence of neighboring O p holes). A unique aspect of this theory is that it does not require that the metallic ion be magnetic, but only that there be a low density of carriers and a near degeneracy between the O p orbitals and the valence orbitals of the metal; a similar mechanism could thus also contribute to the superconductivity in $BaPb_xBi_{1-x}O_3$ and in other non-Cu-containing oxides.[114]

The appropriate values of the parameters in Eq. (6.1) are known only approximately at present. One useful constraint is obtained by viewing the simple one-electron, tight-binding bands of Eq. (3.1) as the result of a Hartree–Fock, mean-field treatment of the two-particle Coulomb interactions.[108] This establishes the explicit relationships[115]

$$\varepsilon_d = \varepsilon_d^0 + U_d \langle n_i \rangle /2 + 4U_{pd} \langle n_l \rangle, \qquad (6.2a)$$

$$\varepsilon_p = \varepsilon_p^0 + U_p \langle n_l \rangle /2 + 2U_{pd} \langle n_i \rangle, \qquad (6.2b)$$

and

$$V = V^0 + U_{pd} \langle d_{i\sigma}^{+} p_{l\sigma} \rangle, \qquad (6.2c)$$

where $\langle \ \rangle$ denotes an expectation value in the Hartree–Fock ground state. The equality between ε_p and ε_d discussed in Part III in the case of a

[112] J. E. Hirsch, *Phys. Rev. Lett.* **59**, 228 (1987).

[113] C. M. Varma, S. Schmitt-Rink, and E. Abrahams, *Solid State Commun.* **62**, 681 (1987); C. M. Varma, S. Schmitt-Rink, and E. Abrahams, *in* "Theories of High Temperature Superconductors" (J. W. Halley, ed.), p. 211. Addison-Wesley, Redwood City, 1988.

[114] R. J. Cava, B. Batlogg, J. J. Krajewski, R. Farrow, L. W. Rupp, A. E. White, K. Short, W. F. Peck, and T. Kometani, *Nature* **332**, 814 (1988).

[115] A. M. Oleś, J. Zaanen, and P. Fulde, *Physica B&C* **148**, 260 (1987).

half-filled pdσ antibonding band (1/2 hole on Cu, 1/4 on each O) yields the result $\Delta = (2U_d - U_p)/4$. Emery's assumption that $0 < \Delta < U_d$ is thus consistent with the expectation that U_d represents the largest energy in the problem.

In view of the nearly closed shell nature of both the Cu and O ions, the Coulomb parameters U_d, U_p, and U_{pd} are expected to be relatively poorly screened[116] from their atomic values of >10 eV. Several attempts have been made recently to calculate these parameters by using the LDA. McMahan et al.,[57] Schlüter et al.,[117] and (C. F.) Chen et al.[118] all used a total energy approach in which the occupations of the Cu d and O p shells were allowed to vary; the one-electron results were again assumed to be equivalent to a mean-field treatment of Eq. (6.1). Zaanen et al.[76] and (H.) Chen et al.[80] employed an alternative approach based on LDA transition state calculations. In all cases, the resulting parameters were in the ranges $U_d \simeq 5.5-12$ eV, $U_p \simeq 4.8-14$ eV, and $U_{pd} \simeq 0.6-4$ eV. Reasonable Δ values based on these parameters and the constraint in the preceding paragraph are of the order of 0 to 3 eV. An upper bound for V^0 is given by the tight-binding hopping parameter $V = 1.6$ eV.

b. Simplified Emery Model

To make further progress, many authors[112,119,120] have considered a simplified version of Eq. (6.1) with $U_p = U_{pd} = 0$. The neglect of U_p is not believed to be a serious limitation since, for the small hole concentrations of interest, the probability of two holes occupying the same O site is small.[119] Whether U_{pd} can also be safely neglected is less clear since recent numerical simulations by Hirsch et al.[121] suggest that this parameter plays a crucial role whenever it exceeds V^0 by a factor of about 2 to 4.

The Emery model for $U_p = U_{pd} = 0$ corresponds to an Anderson lattice Hamiltonian[9] in which the uncorrelated O p states have zero intrinsic bandwidth. A further simplification, to a single Cu–O$_n$ cluster,[122] has

[116]E. B. Stechel and D. R. Jennison, *Phys. Rev. B***38**, 4632 (1988).

[117]M. Schlüter, M. S. Hybertsen and N. E. Christensen, *Physica C***153–155**, 1217 (1988).

[118]C. F. Chen, X. W. Wang, T. C. Leung, and B. N. Harmon, submitted for publication.

[119]P. A. Lee, G. Kotliar and N. Read, *in* "Theories of High Temperature Superconductivity" (J. W. Halley, ed.), p. 235. Addison-Wesley, Redwood City, 1988; G. Kotliar, P. A. Lee, and N. Read, *PhysicaC* **153–155**, 538 (1988).

[120]K. Miyake, T. Matsuura, K. Sano, and Y. Nagaoka, *J. Phys. Soc. Jpn.* **57**, 722 (1988).

[121]J. E. Hirsch, S. Tang, E. Loh, and D. J. Scalapino, *Phys. Rev. Lett.* **60**, 1668 (1988).

[122]The appropriate value of n would be 4 for a Cu-O$_2$ plane, or 6 for a distorted Cu-O$_6$ octahedron.

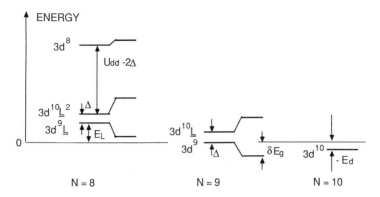

FIG. 10. Schematic energy level diagram for the configuration interaction, CuO_n cluster model description of the one-hole ground state ($N = 9$) and corresponding two-hole ($N = 8$) and zero-hole ($N = 10$) states. Shifts due to hybridization shown explicitly. Energy parameters and configuration notation defined in text. [After Z. Shen, J. W. Allen, J. J. Yeh, J.-S. Kang, W. Ellis, W. E. Spicer, I. Linadu, M. B. Maple, Y. D. Dalichaouch, M. S. Torikachvili, J. Z. Sun, and T. H. Geballe, *Phys. Rev.* **B36**, 8414 (1987).]

been widely used in the analysis of spectroscopic data for the 2-1-4 and 1-2-3 systems.[123,124] The electronic structure of such a cluster is easily calculated by using a configuration interaction (CI) approach.[125,126] Figure 10 shows an example of a schematic energy level diagram obtained in this manner by Shen *et al.*[124] for a cluster containing two ($N = 8$), one ($N = 9$) or zero ($N = 10$) holes. The value of N refers to the number of d electrons in a nominal Cu d^N, O p^6 configuration.

The insulating ground state for $N = 9$ is a linear combination of d^9 and $d^{10}L$ configurations, where L denotes an O p (ligand) hole in a state with d ($x^2 - y^2$) symmetry about the Cu site. In terms of the parameters introduced in Eq. (6.1), the energy of a pure d^9 configuration is ε_d^0 and the energy of a $d^{10}L$ configuration is ε_p^0. The interaction between these two configurations is described by the matrix element $T = \langle d^9 | \mathcal{H} | d^{10}L \rangle$, which is simply a numerical factor of order one[127] times the hopping

[123] A. Fujimori, E. Takayama-Muromachi, and Y. Uchida, *Solid State Commun.* **63**, 857 (1987); A. Fujimori, E. Takayama-Muromachi, Y. Uchida, and B. Okai, *Phys. Rev.* **B35**, 8814 (1987).

[124] Z. Shen, J. W. Allen, J. J. Yeh, J.-S. Kang, W. Ellis, W. E. Spicer, I. Lindau, M. B. Maple, Y. D. Dalichaouch, M. S. Torikachvili, J. Z. Sun, and T. H. Geballe, *Phys. Rev.* **B36**, 8414 (1987).

[125] G. van der Laan, C. Westra, C. Haas, and G. A. Sawatzky, *Phys. Rev.* **B23**, 4369 (1981); A. Fujimori and F. Minami, *Phys. Rev.* **B30**, 957 (1984).

[126] L. C. Davis, *J. Appl. Phys.* **59**, R25 (1986).

[127] The value of this factor depends on the precise definition of L in a particular case and on possible corrections for d level degeneracy (see, for example, Ref. 124).

integral V^0. The hybridization lowers the energy of the $N = 9$ ground state from that of a pure d^9 configuration by an amount

$$\delta E_g = (\Delta^2 + T^2)/2 - \Delta/2. \tag{6.3}$$

The number of d electrons in the hybridized ground state is given by $n_d = 9 \cos^2 \theta + 10 \sin^2 \theta$, where $\tan(2\theta) = 2T/\Delta$.

The energies associated with $N = 8$ and $N = 10$ states are obtained in a similar manner.[124] The former are linear combinations of d^8, $d^9\underline{L}$ and $d^{10}\underline{L}^2$ configurations and are accessible either by doping or by the removal of an electron in a photoemission process. The $N = 10$ state corresponds to the d^{10} "vacuum" configuration in the Emery model that may be reached from the $N = 9$ ground state by injecting an extra electron in an inverse photoemission process. The symbols E_L, U_{dd}, and E_d in Fig. 10 correspond to the energies ε_p^0, U_d, and ε_d^0, respectively. The usefulness of this figure for extracting microscopic parameters from experimental photoemission and inverse photoemission spectra will be discussed further in Section 8.

In the limit $V^0 \ll U_d, \Delta$, the superexchange interaction resulting from the simplified Emery model is given by[124]

$$J = \frac{4(V^0)^4}{\Delta^2} \left(\frac{1}{\Delta} + \frac{1}{U_d} \right). \tag{6.4}$$

The second term here corresponds to the usual Anderson superexchange[87] $(4b^2/U_d$, where $b = (V^0)^2/\Delta)$, but the first term represents an additional contribution[107,128] from virtual processes that do not involve the double occupancy of a Cu site. The first term dominates in the limit $\Delta \ll U_d$. In this regime, several authors have suggested that the parameters Δ and b in the simplified Emery Hamiltonian should be viewed as an "effective U and t," respectively, in a single-band Hubbard model.[123,124] A formal derivation of such a mapping has recently been presented by Zhang and Rice;[93] additional comments can be found in Maekawa et al.[129]

The appropriate parameters for the copper oxides are more likely in the intermediate coupling regime $V^0 \simeq \Delta$. Equation (6.4) is then no longer valid and the superexchange interaction must be calculated numerically.[124] The $U_d = \infty$ limit of the Anderson lattice Hamiltonian for

[128]B. E. Larson, K. C. Hass, H. Ehrenreich, and A. E. Carlsson, *Phys. Rev.* B37, 4137 (1988).
[129]S. Maekawa, T. Matsuura, Y. Isawa, and H. Ebisawa, *Physica* C152, 133 (1988).

arbitrary V^0 and Δ was recently examined by Lee et al.[119] using a "slave boson" approach[9] and by Miyake et al.[120] using a Gutzwiller method.[9] Both groups obtained a Fermi liquid solution in the metallic regime with the lowest energy quasiparticle excitations having the same form as Eq. (3.1). A "renormalization" of the parameters ε_d and V in each case describes a narrowing of the partially occupied band and a mass enhancement factor (m^*/m) which varies inversely with the concentration of "free" holes (e.g., x in $La_{2-x}Sr_xCuO_4$). The quasiparticles themselves are mostly spin excitations on the Cu sites. As in the single-band Hubbard model, the Fermi liquid picture is expected break down at small doping levels due to the onset of AF order.

c. Inclusion of O–O Hopping

A potentially important addition to the Emery model is a term describing direct O–O hopping.[57,58,116,130] Without this term, hole motion can only occur through the Cu sites[108] and may be severely impeded by the presence of strong correlation. Direct hopping between neighboring O p_σ orbitals may be described by the single parameter $V_p = (V_{pp\sigma} - V_{pp\pi})/2$, or $E_{xy}(\frac{1}{2}\frac{1}{2}0)$ in the usual Slater–Koster notation.[47,48] For $U_p = U_{pd} = V^0 = 0$, the resulting O p bands, in a hole representation, have the intrinsic dispersion

$$E(\mathbf{k}) = \varepsilon_p^0 \pm 4V_p \sin(k_x a/2) \sin(k_y a/2), \qquad (6.5)$$

with an overall bandwidth $W = 8V_p$. The band extrema occur at the corners $(\pi/a, \pi/a)$ of the two-dimensional Brillouin zone. Only the lower energy hole states, associated with the minus sign in Eq. (6.5), have the appropriate symmetry to hybridize with the Cu $d(x^2-y^2)$ orbitals. Estimates of V_p based on the tight-binding parametrizations[57,58,116] discussed in Section 3 range from 0.5 to 0.65 eV.

Figure 11 shows a schematic energy level diagram for $U_p = U_{pd} = V^0 = 0$ in the presence of a nonzero V_p, a large U_d and a finite concentration of O p holes. The addition of a nonzero hybridization V^0 between the Cu d and the O p states then yields the general case of the Anderson lattice Hamiltonian. The complexity of this model is well known from previous studies of heavy fermion[9] and mixed-valence[8] systems. The proximity of ε_d^0 to the Fermi level in Fig. 11 suggests that the copper oxides are closer to having mixed-valence character (mixed between d^9 and d^{10}), although the nearly filled O p band represents an additional complication compared to most f-electron metals.[8]

[130]J. Zaanen and A. M. Oleś, Phys. Rev. B37, 9423 (1988).

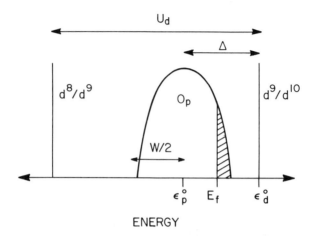

ENERGY

FIG. 11. Pictorial representation of Anderson–Hamiltonian description of copper oxides before hybridization. Hole (electron) energies increase to the left (right). O p band, of intrinsic width W, lies within correlation gap (U_d) of Cu d states. Charge transfer gap Δ defined by separation between centroid of O p band and Cu d^{10} ionization energy (ε_d^0). Shaded region represents occupied O p hole states in metallic regime.

Zaanen and Oleś[130] have performed a systematic study of this model for small V^0/Δ by using canonical perturbation theory.[131] They find that the interactions between O p holes and localized Cu spins are favorable for spin polaron formation but not for superconductivity. Several groups[132–134] have reported mean-field calculations for a more general range of parameters in a somewhat simpler model based on square O p bands and a k-independent Cu d-O p hybridization. An approximate treatment by Newns and Rasolt[132] for $U_d = \infty$ was found to lead to a normal state Fermi liquid with $m^*/m \simeq 5$ and a superconducting ground state with T_c proportional to the concentration of free holes; the basic superconducting mechanism is similar to that which is believed to occur in heavy fermion systems.[9] Houghton and Sudbo[133] reached qualitatively similar conclusions by obtaining an exact mean-field solution for finite U_d. Similar results have also been obtained by Newns et al.[134] starting from a more realistic Anderson lattice Hamiltonian for La$_{2-x}$Sr$_x$CuO$_4$ constructed from a detailed tight-binding parametrization.[59]

[131]K. A. Chao, J. Spalek, and A. M. Oleś, *J. Phys.* **C10**, L271 (1977); *Phys. Rev.* **B18**, 3453 (1987).

[132]D. M. Newns, *Phys. Rev.* **B36**, 5595 (1987); D. M. Newns, P. C. Pattnaik, and M. Rasolt, *Phys. Rev.* **B38**, 6513 (1988).

[133]A. Houghton and A. Sudbo, *Phys. Rev.* **B38**, 7037 (1988).

[134]D. M. Newns, P. Pattnaik, M. Rasolt, and D. A. Papaconstantopoulos, *Physica C* **153–155**, 1287 (1988).

The impurity limit of a similar realistic Anderson Hamiltonian, based on slightly different parameters, has recently been examined by McMahan et al.[57] These authors consider only the pure La_2CuO_4 compound and obtain a solution by using the approach of Gunnarson and Schönhammer.[135] The results represent an improvement over the simple cluster model of Fig. 10 for the interpretation of high-energy spectroscopic data. The most important differences in the full impurity model are the presence of a continuum of $N = 8$ states throughout the $d^9\underline{L}$ and $d^{10}\underline{L}^2$ region in Fig. 10 and the splitting off of one or more discrete states at lower energies. These latter states have essentially singlet $d^9\underline{L}$ character and would also be expected to exhibit dispersion in a full lattice model.[57] The presence of such split-off states in an impurity calculation is a well-known consequence of strong hybridizationn.[107]

By far the most ambitious model yet considered for $Cu-O_2$ planes is that of Stechel and Jennison.[116] This model has all of the ingredients of the full Emery model of Eq. (6.1) but also includes a direct O–O hopping term and a direct exchange interaction between neighboring Cu and O holes. Stechel and Jennison demonstrate the consistency of their particular choice of parameters by performing detailed superexchange calculations. They then construct an effective Hamiltonian to examine the behavior of normal state quasiparticles in the metallic phase. These quasiparticles are referred to as "spin-hybrids" to reflect the strong coupling between O p-like holes and local spin configurations. The authors[116] conclude that this model also leads to correlation-induced superconductivity since the effective interaction between such quasiparticles turns out to be attractive.

d. Weber d–d Excitation Model

A basic assumption of the Emery model is that only the Cu d orbitals of (x^2-y^2) symmetry are relevant to the low energy excitations. Recently, Weber and coworkers[58,136] have argued that one should also include the possibility of a single $d(x^2-y^2)$ hole being excited to a higher energy orbital with different symmetry. Such a "d–d" excitation conserves the Cu valence and thus does not cost an energy U_d. Instead, the actual excitation energy is determined by the crystal field splitting between Cu d states (cf. Fig. 6, left side), which is of the order of one eV. Transitions of this kind are dipole-forbidden but have still been observed optically in a variety of transition metal compounds.[9,107] Some evidence for their

[135]O. Gunnarsson and K. Schönhammer, Phys. Rev. B28, 4315 (1987).
[136]W. Weber, to be published (Proc. Schloss Mauterndorf); A. L. Shelankov, Y. Zotos, and W. Weber, Physica C153–155, 1307 (1988).

existence in 1-2-3 compounds has recently been provided by the optical transmission measurements of Geserich *et al.*[137] (cf. Section 11); the lowest energy structure observed in this study, at 0.5 eV, is attributed by the authors to an x^2-y^2 to $3z^2-r^2$ excitation. The impurity model calculations of McMahan *et al.*[57] also predict that such a d–d excitation should occur within the charge transfer gap of La_2CuO_4.

The specific model proposed by Weber[58] includes both the Cu d(x^2-y^2) and ($3z^2-r^2$) orbitals in an Emery-like model that neglects U_p but allows direct O–O hopping. The on-site energies of the Cu d orbitals differ by the Jahn–Teller splitting δE_{JT}. Both the hopping integrals and the Coulomb interactions between nearest-neighbor Cu d and O p states are allowed to differ for the two different symmetry d orbitals. Since U_{pd} is expected to be smaller for d orbitals of ($3z^2-r^2$) symmetry, it may be energetically favorable for a Cu d(x^2-y^2) hole to make a transition to the ($3z^2-r^2$) state in the presence of a neighboring O p hole. Weber[58] has suggested that this "polarization" effect can lead to the effective attraction between O p holes that may be responsible for high-temperature superconductivity. The detailed behavior of the underlying Hamiltonian is extremely complicated, however, and has only been worked out in certain limiting cases.[58,136] An ironic feature of this mechanism is that the scale factor for T_c is determined by the Jahn–Teller splitting δE_{JT}, which is the aspect of the copper oxides that first drew the attention of Bednorz and Müller.[1]

7. ROLE OF O p_π ORBITALS

Once one accepts that the fact that, because of a large value of U_d, the correct starting point for the 2-1-4 and the 1-2-3 copper oxides is closer to Fig. 11 than to Figs. 5 and 9, it is no longer clear that only the O p_σ orbitals are relevant. In a purely ionic picture of $(Cu^{2+})–(O^{2-})_2$ planes, in fact, the crystal field splitting of the O p states (cf. Fig. 6) makes it more favorable by about one eV to create a hole on an in-plane O p_π orbital ($p_\pi^\parallel$) than on an O p_σ orbital; an out-of-plane O p_π hole ($p_\pi^\perp$) would have intermediate energy.[45,110,138]

The importance of O $p_\pi^\parallel$ orbitals is also suggested by the quantum chemical calculations of Goddard and coworkers.[109] These calculations

[137]H. P. Geserich, G. Schreiber, J. Geerk, H. C. Li, G. Linker, W. Assmus, and W. Weber, *Europhys. Lett.* **6**, 227 (1988); H. P. Geserich, G. Sheiber, J. Geerk, H. C. Li, W. Weber, H. Romberg, N. Nücker, and J. Fink, to be published (*Proc. Schloss Matuerndorf*).

[138]F. J. Adrian, *Phys. Rev.* **B37**, 2326 (1988).

were performed for small clusters of Cu and O atoms in geometries characteristic of Cu–O_2 planes and Cu–O chains. Neighboring ions out to 8 Å were also included in a point charge approximation. The electronic structures of the clusters were calculated using the "generalized valence bond" (GVB) method,[139] which can be viewed as either a self-consistent Heitler–London scheme or as an approximate treatment of configuration interaction corrections to Hartree–Fock theory. The principal conclusion of this study is that the removal of an electron from an insulating AF configuration creates an O $p_\pi^\parallel$ hole that is *ferromagnetically* coupled to the preexisting Cu^{2+} holes. The interaction is larger in magnitude than Cu–Cu superexchange and is ferromagnetic because of the orthogonality between the O $p_\pi^\parallel$ and Cu $d(x^2-y^2)$ orbitals.[87] The O $p_\pi^\parallel$ oribtals on different sites overlap sufficiently to give rise to a band of hole states,[109,110] which for a Cu–O_2 plane, can again be described by Eq. (6.5). Superconducting mechanisms based on the interactions between O $p_\pi^\parallel$ carriers and Cu spins have been proposed by Chen and Goddard[109] and by Birgeneau *et al.*[110]

Proponents of O p_σ models would argue that neither the crystal field argument nor the GVB calculations make adequate allowance for the strong Cu $d(x^2-y^2)$–O p_σ hybridization.[116] That such hybridization can lower the energy of holes with predominatly O p_σ character can be seen from the fact that the true $N = 8$ ground state in Fig. 10 lies well below the energy of a pure $3d^9L$ configuration; a much smaller shift would occur for O p_π states.

The problem of determining the relative energies of the O p_σ and O $p_\pi^\parallel$ holes is clearly a subtle one which has not yet been definitively addressed either theoretically or experimentally. Both of the realistic Anderson Hamiltonian studies[57,134] discussed in Section 6c suggest that the O p_σ hole states are the first to become occupied, but, beyond a certain doping level, both O p_σ and O p_π holes will be present. In a k-space picture in which one allows for nearest neighbor O–O hopping, the distinction between the two types of states is only meaningful for low hole concentrations anyway; at higher energies in the hole band, the two types of states become strongly mixed.[57,140] In addition, other O orbitals also start to play a role,[57,109] particularly the p_z orbitals on the out-of-plane O. (Note, for example, the shift in the O_z projected densities of states in Fig.

---

[139]W. A. Goddard and T. C. McGill, *J. Vac. Sci. Technol.* **16**, 1308 (1979).

[140]The hopping between a p_σ orbital (p_x) at $(-a/2, 0, 0)$ and a p_π orbital (p_x) at $(0, a/2, 0)$ is described by the Slater–Koster parameter $E_{xx}(\frac{1}{2}\frac{1}{2}0) = (V_{pp\sigma} + V_{pp\pi})/2 \equiv V_p'$. This produces a coupling of the form $4V_p' \cos(k_x a/2) \cos(k_y a/2)$ between the pure p_σ and p_π bands of opposite sign in Eq. (6.5) which vanishes at the band minimum ($k_x, k_y = \pi/a$) but becomes increasingly important at higher hole energies.

8 upon Ba substitution.) Possible implications of an overlap of O p_σ and O p_π states at the Fermi level for superconductivity have been discussed by Goodenough.[45]

V. Implications of Spectroscopic Data

The preceding theoretical developments have to a large extent gone hand in hand with the accumulation of extensive spectroscopic data on the new superconductors. After much confusion in the early literature, due in part to difficulties associated with the preparation and characterization of sintered samples, a reasonably consistent picture has now emerged from experiment, particularly from high-energy probes such as photoemission and x-ray absorption. This section will review some of the principal results obtained to date by using these and other techniques and will discuss the significance of these results in light of the proposed theories. Earlier reviews of the experimental literature on the 2-1-4 and 1-2-3 copper oxides have been given by Kurtz[141] and Wendin.[142] The basic theory underlying the analysis of spectroscopic data for transition-metal compounds was reviewed by Davis[126] in 1986.

8. PHOTOEMISSION AND INVERSE PHOTOEMISSION

In principle, photoemission and inverse photoemission spectroscopy provide the most direct probes of the energies of occupied and unoccupied states, respectively. An important caveat to keep in mind when analyzing photoemission data for the new superconductors is the extreme surface sensitivity of this technique (electron mean free paths are of the order of 5 to 10 Å for photon energies of <100 eV and 20 Å for x-ray energies) and the difficulty of preparing clean, stoichiometric surfaces in the new materials.[141] Most of the data discussed below have been obtained on sintered polycrystalline samples for which clean surfaces have been exposed either by fracturing or scraping in ultra-high vacuum. The good agreement obtained by different groups and the few single crystal studies[143–145] that have been reported support the belief that

[141]R. L. Kurtz, in "Thin Film Processing and Characterization of High Temperature Superconductors" (J. M. E. Harper, R. J. Colton, and L. C. Feldman, eds.), AIP Conf. Proc. No. 165, p. 222. AIP, New York, 1988.

[142]G. Wendin, J. de Physique, Colloque C9, C9-1157 (1987).

[143]N. G. Stoffel, Y. Chang, M. K. Kelly, L. Dottl, M. Onellion, P. A. Morris, W. A. Bonner, and G. Margaritondo, Phys. Rev. B37, 7952 (1988).

[144]T. Takahashi, F. Maeda, H. Katayama-Yoshida, Y. Okabe, T. Suzuki, A. Fujimori, S. Hosoya, S. Shamoto, and M. Sato, Phys. Rev. B37, 9788 (1988).

[145]M. Tang, N. G. Stoffel, Q.-B. Chen, D. LaGraffe, P. A. Morris, W. A. Bonner, G. Margaritondo, and M. Onellion, Phys. Rev. 38, 897 (1988).

the most important features that will be discussed are intrinsic, bulk properties.

a. *Occupied Valence States*

Angle-integrated valence band photoemission spectra have been obtained for both the 2-1-4 and 1-2-3 systems by using both monochromatic sources and variable energy synchrotron radiation.[123,124,143–166] The latter

[146]P. Steiner, J. Albers, V. Kinsinger, I. Sander, B. Siegwart, S. Hufner, and C. Polititis, *Z. Phys.* B66, 275 (1987); P. Steiner, R. Courths, V. Kinsinger, I. Sander, B. Siegwart, S. Hufner, and C. Politits, *Appl. Phys.* A44, (1987).

[147]T. Takahashi, F. Maeda, S. Hosoya, and M. Sato, *Jpn. J. Appl. Phys.* 26, L349 (1987).

[148]B. Reihl, T. Riesterer, J. G. Bednorz, and K. A. Müller, *Phys. Rev.* B35, 8804 (1987).

[149]N. Nücker, J. Fink, B. Renker, D. Ewert, C. Politis, P. J. Weijs, and J. C. Fuggle, *Z. Phys.* B67, 9 (1987).

[150]D. D. Sarma and C. N. R. Rao, *J. Phys.* C20, L659 (1987).

[151]J. C. Fuggle, P. J. W. Weijs, R. Schoorl, G. A. Sawatzky, J. Fink, N. Nücker, P. J. Durham, and W. M. Temmerman, *Phys. Rev.* B37, 123 (1988).

[152]P. Steiner, V. Kinsinger, I. Sander, B. Siegwart, S. Hufner, and C. Politis, *Z. Phys.* B67, 19 (1987).

[153]F. C. Brown, T. C. Chiang, T. A. Friedman, D. M. Ginsberg, G. N. Kwawer, T. Miller, and M. G. Mason, *J. Low Temp. Phys.* 69, 151 (1987).

[154]P. D. Johnson, S. L. Qiu, L. Jiang, W. M. Ruckman, M. Strongin, S. L. Hulbert, R. F. Garrett, B. Sinkovic, N. V. Smith, R. J. Cava, C. S. Jee, D. Nichols, E. Kaczanowicz, R. E. Salomon, and J. E. Crow, *Phys. Rev.* B35, 8811 (1987).

[155]R. L. Kurtz, R. L. Stockbauer, D. Mueller, A. Shih, L. E. Toth, M. Osofsky, and S. A. Wolf, *Phys. Rev.* B35, 8818 (1987).

[156]M. Onellion, Y. Chang, D. W. Niles, R. Joynt, G. Margaritondo, N. G. Stoffel, and J. M. Tarascon, *Phys. Rev.* B36, 819 (1987).

[157]D. D. Sarma, K. Streedhar, P. Ganguly, and C. N. R. Rao, *Phys. Rev.* B36, 2371 (1987); D. D. Sarma and C. N. R. Rao, *Solid State Commun.* 65, 47 (1988).

[158]J. A. Yarmoff, D. R. Clarke, W. Drube, U. O. Karlsson, A. Taleb-Ibrahami, and F. J. Himpsel, *Phys. Rev.* B36, 3967 (1987).

[159]N. G. Stoffel, J. M. Tarascon, Y. Chang, M. Onellion, D. W. Niles, and G. Margaritondo, *Phys. Rev.* B36, 3986 (1987).

[160]B. Dauth, T. Kachel, P. Sen, K. Fischer, and M. Campagna, *Z. Phys.* B68, 407 (1987); T. Kachel, P. Sen, B. Dauth, and M. Campagna, *Z. Phys.* B70, 137 (1988).

[161]T. Takahashi, F. Maeda, H. Arai, H. Katayama-Yoshida, Y. Okabe, T. Suzuki, S. Hosya, A. Fujimori, T. Shidara, T. Koide, T. Miyahara, M. Onoda, S. Shamoto, and M. Sato, *Phys. Rev.* B36, 5686 (1987).

[162]Z. Iqbal, E. Leone, R. Chin, A. J. Signorelli, A. Bose, and H. Eckhardt, *J. Mater. Res.* 2, 768 (1987).

[163]Y. Petroff, P. Thiry, G. Rossi, A. Revcolevschi, and J. Jegoudez, *Int. J. Mod. Phys.* B1, 183 (1987); P. Thiry, G. Rossi, Y. Petroff, A. Revcolevschi, and J. Jegoudez, *Europhys. Lett.* 5, 55 (1988).

[164]A. Bianconi, A. Clozza, A. Congiu Castellano, S. Della Longa, M. De Santis, A. Di Cicco, K. Garg, P. Delogu, A. Gargano, R. Giorgi, P. Lagarde, A. M. Flank, and A. Marcelli, *Int. J. Mod. Phys.* B1, 205 (1987); A. Bianconi, A. Congiu Castellano, M. De Santis, P. Delogu, A. Gargano, and R. Giorgi, *Solid State Commun.* 63, 1135 (1987).

is particularly useful for determining the orbital composition of various states since the excitation cross sections of different orbitals vary in different manners with photon energy.[141]

The solid curves in Fig. 12 show representative valence band photoemission spectra for $La_{2-x}Sr_xCuO_4$ obtained at different photon energies by different groups.[147,148,149] The Sr concentration also varies from panel (a) through (d), but this has been found to have a negligible effect on the resulting spectrum. The dashed curves in Fig. 12 represent theoretical spectra calculated by using LDA energy bands by Redinger et al.[167] These calculations are more meaningful for comparison than valence band densities of states alone (cf. Fig. 8), since they also include cross-section effects and lifetime broadening. For photon energies of 21.2 eV, the spectra are dominated by the O 2p contribution to the valence bands. The Cu 3d contribution increases in importance with increasing photon energy and is essentially the only emission seen at 1487 eV.

At all photon energies, the main peak in the experimental spectrum occurs at higher binding energy compared to the LDA prediction and yields a much weaker emission within 1 to 2 eV of the Fermi level.[151,167] Similar discrepancies have been observed in 1-2-3 compounds.[168] The larger discrepancies observed at higher photon energies, and complementary x-ray emission results,[169] indicate that the principal difficulty lies in the location of the Cu d states. The LDA predicts that the largest Cu d contribution lies closer to the Fermi level than the O p contribution, whereas the experimental situation is just the opposite. The O p contribution appears to be reasonably described by the LDA. The LDA underestimate of the binding energy of the d-related feature is a common problem in transition-metal compounds[81,82,128] that results from an overscreening of charged excitations.[85]

Further insight into the nature of Cu 3d excitations has been obtained from resonant photoemission studies.[141] For photon energies near 75 eV, which corresponds to the Cu 3p core level absorption edge, the cross section for Cu 3d emission exhibits a Fano-like resonance whenever the ground state contains a partially unoccupied Cu 3d shell. The resonant

[165]D. van der Marel, J. van Elp, G. A. Sawatzky, and D. Heitman, *Phys. Rev.* B**37**, 5136 (1988).

[166]A. Samsavar, T. Miller, T.-C. Chiang, B. G. Pazol, T. A. Friedman, and D. M. Ginsberg, *Phys. Rev.* B**37**, 5164 (1988).

[167]J. Redinger, J. Yu, A. J. Freeman, and P. Weinberger, *Phys. Lett.* A**124**, 463 (1987).

[168]J. Redinger, A. J. Freeman, J. Yu, and S. Massida, *Phys. Lett.* A**124**, 469 (1987).

[169]J.-M. Mariot, V. Barnole, C. F. Hague, V. Geiser, and H.-J. Guntherodt, *Solid State Commun.* **64**, 1203 (1987) and *J. de Physique, Colloque* C**9**, C9-1203 (1987).

La₂CuO₄

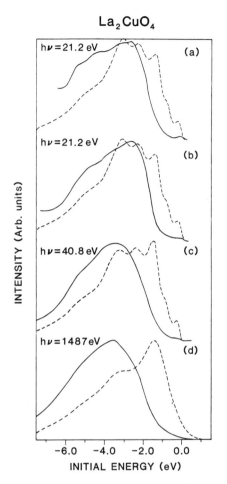

FIG. 12. Comparison of calculated photoemission spectra (solid curves) based on LDA energy bands for $La_{2-x}Sr_xCuO_4$ with experimental spectra (dashed curves) for (a) $x = 0.15$ (Ref. 147), (b, c) $x = 0.2$ (Ref. 148), and (d) $x = 0$ (Ref. 149). Photon energies indicated at left. [After J. Redinger, J. Yu, A. J. Freeman and P. Weinberger, *Phys. Lett.* **A124**, 463 (1987).]

behavior results from a quantum interference between a direct valence band photo-excitation and an indirect process involving a Cu 3p–3d transition followed by a two-electron, super Coster–Kronig decay.[126]

Figure 13 shows the measurements of Shen *et al.*[124] for $La_{1.8}Sr_{0.2}CuO_4$ over a range of photon energies near the Cu 3d resonance. In addition to the main valence band peak discussed above, the figure also shows structure below −15 eV due to La 5p emission and two peaks labelled C

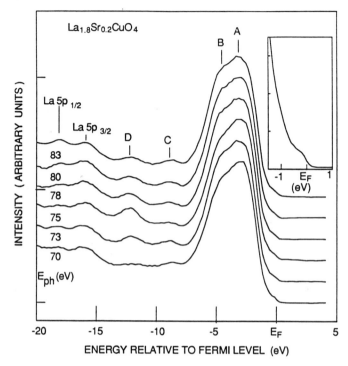

FIG. 13. Valence band photoemission spectra for $La_{1.8}Sr_{0.2}CuO_4$ for photon energies (E_{ph}) in the vicinity of the Cu 3d resonance. Inset shows detail near the Fermi level for a photon energy of 70 eV. [After Z. Shen, J. W. Allen, J. J. Yeh, J.-S. Kang, W. Ellis, W. E. Spicer, I. Landau, M. B. Maple, Y. D. Dalichaouch, M. S. Torikachvili, J. Z. Sun, and T. H. Geballe, *Phys. Rev.* B**36**, 8414 (1987).

and D at -9 eV and -12.5 eV, respectively. The strong enhancement of the intensity of peak D for a photon energy of 75 eV indicates that this peak has a predominantly Cu 3d origin. This peak is usually referred to as a valence band satellite.[126] Its presence cannot be accounted for in conventional band theory and its similar location[90] to that in CuO and $CuCl_2$ suggests that the average Cu valence in the ground state is close to $+2$. Detailed analyses of the intensities of the main valence band features in Fig. 13 as functions of photon energy also indicate resonant behavior,[124] even at binding energies as low as 0.5 eV. The Cu 3d states thus contribute throughout the valence band complex. Whether or not peak C also represents an intrinsic valence band feature has been a subject of much debate.[141] The only single crystal data[143] for $La_{2-x}Sr_xCuO_4$ does not show this peak, but other evidence suggests that peak C has at least some intrinsic character.[124,141]

The above results are consistent with the strong correlation picture described by the cluster CI diagram of Fig. 10. In this scheme, the emission of a photoelectron from the $N = 9$ ground state produces the "two-hole" excitation spectrum shown for $N = 8$. Many authors,[141,142] beginning with Fujimori et al.,[123] have thus attributed the main valence band features in Figs. 12 and 13 to final states of predominatly mixed $3d^9\underline{L}$ and $3d^{10}\underline{L}^2$ (and nonbonding O p) character, and the satellite peak to an almost pure $3d^8$ final state. The >5 eV separation between the satellite and the main peak suggests a large value of U_d. This implies a high energy for the $3d^8$ configuration and suggests that charge fluctuations to a Cu^{3+} state are extremely unfavorable even in doped $La_{2-x}M_xCuO_4$. The main valence band features associated with Cu d emission can be thought of as resulting in part from the removal of a Cu d electron followed by a screening process[125,126,142] in which an electron is transferred from the ligand orbitals to the Cu. It is these screened Cu d hole excitations that are approximated (poorly) in the LDA band calculations since the LDA is based on a ground-state charge distribution at each Cu site.[44]

Quantitative fits to experiment using the cluster CI model[123,124,151] yield $U_d \simeq 6$ eV, Δ values of less than one eV and a hybridization mixing (T in Section 6b) of greater than one eV. Such appreciable hybridization is consistent with the much lower intensity in the satellite compared to the main peak in Fig. 13, since, in the absence of hybridization, all of the intensity at resonance would be in the satellite.[124–126] The extracted parameter values yield an average $N = 9$ ground-state Cu valence of approximately 1.6 to 1.7. Similar conclusions have been drawn from the Anderson impurity model analysis of $N = 8$ final states by McMahan et al.[57] In this latter picture, the low photoemission intensity within one eV of the Fermi level is attributed to two-hole bound states, whereas the rapid rise at slightly larger binding energy represents the onset of the O p continuum. The fact that resonant behavior is still observed[124] near 75 eV for emission at -0.5 eV suggests that the low energy "O p holes" are strongly hybridized with the Cu and are thus more likely of O p_σ character.

Figure 14 shows an angle-integrated x-ray photoemission (XPS) spectrum[165] for $YBa_2Cu_3O_7$. As stated above, the main valence band region again peaks at higher binding energy compared to LDA calculations and exhibits a much weaker emission close to the Fermi level.[168] The large peak at -14 eV in Fig. 14 is due to spin-orbit split Ba 5p levels, which are poorly resolved in XPS. The presence of these states makes it difficult to observe the predominantly d^8 valence band satellite in this compound; careful resonant photoemission studies, however, again place

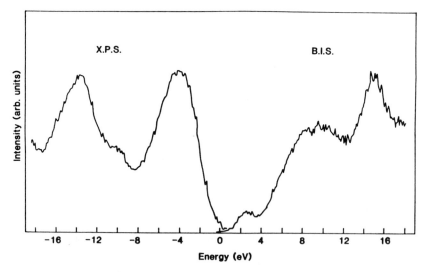

FIG. 14. Valence band XPS spectrum and unoccupied state BIS spectrum for $YBa_2Cu_3O_7$ relative to a Fermi level at 0 eV. [After D. van der Marel, J. van Elp, G. A. Sawatzky, and D. Heitmann, *Phys. Rev. B37*, 5136 (1988).]

this satellite near -12.5 eV.[124,155,156] The main valence band features in $YBa_2Cu_3O_7$ are thus well accounted for by the same CI model parameters as in the 2-1-4 case.[123,124,164,165] An additional small peak occurs between -9 and -10 eV in Fig. 14, similar to peak C in Fig. 13. Most evidence in the 1-2-3 case,[141] including recent single crystal results,[143,145] suggests that this peak is intrinsic. Several groups[145,163] have reported resonant behavior for this peak for photon energies near the O 2s core level excitation threshold at 20 eV. This behavior is consistent with the interpretation of this peak[142,145,163] as an O satellite resulting from a $3d^{10}\underline{L}^2$ final state in which the two ligand holes occupy the same O site and interact with a U_p value of approximately 4 eV.

Photoemission spectra for O-deficient $YBa_2Cu_3O_{7-\delta}$ differ in several respects.[159,164] The valence band features all shift to higher binding energy; the -9 eV peak, the main O p-like region near -2 eV and the -12.5 eV satellite all decrease in intensity; and a new satellite peak appears near -15.3 eV, which also resonates at the Cu 3p threshold. This latter feature is similar to that which occurs[90] in Cu_2O; its appearance with increasing δ signals the conversion of some $Cu2^+$ to Cu^{1+} in the ground state, which is believed to occur primarily in the disrupted Cu1–O1 chains.[28a,45]

In both the 2-1-4 and 1-2-3 systems only a few experiments have revealed a clear Fermi edge in the metallic regime. One example is the weak edge seen in the inset to Fig. 13 for $La_{1.8}Sr_{0.2}CuO_4$, which was not observed in the insulating parent compound.[124] Attempts to observe a superconducting energy gap in the 1-2-3 system at low temperatures by photoemission have so far been unsuccessful.[166,170] This result is inconclusive, since, even if a gap exists large enough to be observable (>10 meV), there is no guarantee that the surface regions in these studies have been superconducting. Some groups[150,157,160] have reported reversible changes in other valence band features upon thermal cycling from 300 K to below 77 K. Most of these changes can probably be attributed to surface contamination.[141,162]

Stoffel et al.[143] have recently reported the first angle-resolved photoemission study of $YBa_2Cu_3O_7$ single crystals. No appreciable dispersion is observed for any of the main valence band features below -1 eV. The weak leading edge of the valence band emission, by contrast, does move upwards by 0.25 to 0.3 eV from the Γ point of the Brillouin zone to about 75% of the way to the S point (π/a, π/b, 0). This result is qualitatively consistent with the one-electron band structure of Fig. 9, although a more quantitative comparison is difficult. The increase in energy from Γ to S is also consistent with a primarily O p-like band within the correlation gap of the Cu d levels. No matter what the ultimate interpretation, the observation of dispersion so close to E_f is significant since it can only be accounted for by the full consideration of lattice effects.

b. Unoccupied States

Inverse photoemission, or bremsstrahlung isochromat spectroscopy (BIS) at x-ray energies, has also been applied extensively to both the 2-1-4 and the 1-2-3 systems.[148,149,151,158,165,171–173] The right side of Fig. 14 shows a typical BIS spectrum[165] for $YBa_2Cu_3O_7$. The weak intensity closest to E_f and the lowest energy peak at 2.5 eV are difficult to reconcile with the unoccupied LDA bands in Fig. 9. In terms of the CI cluster model, the 2.5 eV peak is associated[165] with the single $N = 10$ final state in Fig. 10 corresponding to Cu $3d^{10}$. A similar structure[173] is observed at this energy in CuO. The broader and more intense structure starting at 6 eV in Fig. 14 corresponds to a mixture of O 3s, Cu 4s, Ba 5d,

[170]E. R. Moog, S. D. Bader, A. J. Arko, and B. K. Flandermeyer, *Phys. Rev.* B**36**, 5583 (1987).

[171]Y. Gao, T. J. Wagener, J. H. Weaver, A. J. Arko, B. Flandermeyer, and D. W. Capone, *Phys. Rev.* B**36**, 3971 (1987).

[172]T. J. Wagener, Y. Gao, J. H. Weaver, A. J. Arko, R. Flandermeyer, and D. W. Capone, *Phys. Rev.* B**36**, 3899 (1987).

[173]A. J. Viescas, J. M. Tranquada, A. R. Moodenbaugh, and P. D. Johnson, *Phys. Rev.* B**37**, 3738 (1988).

and Y 4d states and occurs at higher energy than in the LDA. The cross-section dependence of this structure at lower photon energies suggests[172] that the Ba 5d levels contribute to the onset region near 6 eV, while the Y 4d levels contribute to the peak at about 8.5 eV. The narrow peak at 15 eV is attributed to relatively localized Ba 4f states. The large separation of the Ba and Y levels from the Fermi level is consistent with the Ba^{2+} and Y^{3+} ionic character assumed in virtually all of the theoretical models proposed to date.

Inverse photoemission spectra for 2-1-4 compounds are quite similar. A low-intensity region extends up to about 4 eV and is followed by a broad peak that has been attributed to La 5d states at 5.8 eV and La 4f states at 8.7 eV, which also mix with O 3s and Cu 4s orbitals.[171] BIS measurements,[149,151,171] which have the largest cross section for Cu 3d states, also indicate a weak shoulder at 2 eV. Like the 2.5 eV peak in Fig. 14, this structure is most likely due to $Cu 3d^{10}$ final states. The lower intensity of this feature in the 2-1-4 case[173] may simply be due to the much smaller Cu/La ratio in La_2CuO_4 compared to the Cu/Ba ratio in $YBa_2Cu_3O_7$. Other possibilities include difficulties associated with charging effects[124] (which fill the Cu d shell) and surface O depletion.

c. *Core Levels*

XPS core level measurements in the new superconductors have been used primarily to address issues of valence.[146, 174–180] Analyses of core level line shapes and satellite structure also provide information on the strength of Cu–O hybridization and on the magnitude of intraatomic Coulomb interactions.[142]

To test the hypothesis that any holes added to insulating 2-1-4 and 1-2-3 compounds have significant amplitude on the Cu sites (as in band

[174]S. Horn, J. Cai, S. A. Shaheen, Y. Jeon, M. Croft, C. L. Chang, and M. L. den Boer, *Phys. Rev. B***36,** 3895 (1987).

[175]P. Steiner, V. Kinsinger, I. Sander, B. Siegwart, S. Hufner, C. Politis, R. Hoppe, and H. P. Muller, *Z. Phys. B***67,** 497 (1987); P. Steiner, S. Hufner, V. Kinsinger, I. Sander, B. Siegwart, H. Schmitt, R. Schulz, S. Junk, G. Schwitzgebel, A. Gold, C. Politis, H. P. Muller, R. Hoppe, S. Kemmler-Sack, and C. Kunz, *Z. Phys. B***69,** 449 (1988).

[176]D. M. Hill, H. M. Meyer, J. H. Weaver, R. Flandermeyer, and D. W. Capone, *Phys. Rev. B***36,** 3979 (1987).

[177]D. E. Ramakar, N. H. Turner, J. S. Murday, L. E. Toth, M. Osofsky and F. L. Hutson, *Phys. Rev. B***36,** 5672 (1987).

[178]T. A. Sasaki, Y. Baba, N. Masaki, and I. Takano, *Jpn. J. Appl. Phys.* **26,** L1569 (1987).

[179]T. Gourieux, G. Krill, M. Maurer, M. F. Ravet, A. Menny, H. Tolentino, and A. Fontaine, *Phys. Rev. B***37,** 7516 (1988).

[180]D. H. Kim, D. D. Berkley, A. M. Goldman, R. K. Schulze, and M. L. Mecartney, *Phys. Rev. B***37,** 9745 (1988).

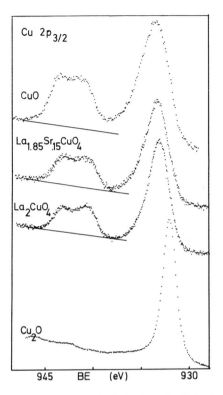

FIG. 15. Cu $2p_{3/2}$ core level XPS spectrum in CuO, La$_{2-x}$Sr$_x$CuO$_4$ ($x = 0.15$ and $x = 0$) and Cu$_2$O. [After N. Nücker, J. Fink, B. Renker, D. Ewert, C. Politis, P. J. W. Weijs, and J. C. Fuggle, *Z. Phys. B***67**, 9 (1987).]

theory), most of the earliest core level photoemission studies were concerned with the measurement of Cu valence as a function of hole concentration. Figure 15 compares typical XPS data[149] for the Cu $2p_{3/2}$ core level in both pure and doped La$_{2-x}$Sr$_x$CuO$_4$ with reference data for CuO (+2) and Cu$_2$O (+1). The most important feature to note is the remarkable similarity between the $x = 0$ and $x = 0.15$ spectra, which is inconsistent with any change in Cu valence. The main peak at 933 eV in both spectra, and the broad satellite centered at 941 eV, are very similar to the CuO features. The slightly lower binding energies of the main 2-1-4 peaks suggest an average Cu valence of slightly less than +2. Virtually identical Cu $2p_{3/2}$ spectra have been observed in 1-2-3 compounds.[123,124] In neither case has any new satellite structure nor any significant shift to higher binding energy been reported that would provide convincing evidence for the presence of Cu^{3+} (d^8).

The higher and lower binding energy features in Fig. 15 are usually attributed to $\underline{2p}3d^9$ and $\underline{2p}3d^{10}\underline{L}$ final states, respectively, where $\underline{2p}$ denotes a $Cu\,2p_{3/2}$ core hole.[123–126] (Structure due to the spin-orbit split-off $Cu\,2p_{1/2}$ core levels lie above 950 eV.) The intensity of the "main" peak results not only from the presence of some $3d^{10}\underline{L}$ component in the ground state, but also from an O to Cu charge transfer induced by the creation of a core hole. Most of the width of this peak is believed to be due to an intrinsic O p ($\underline{L}$) bandwidth.[164] The width of the satellite peak, by contrast, is due primarily to multiplet effects.[126,164] The separation between the satellite and the main peak is given approximately by the intraatomic Coulomb repulsion U_{cd} between a Cu 2p core hole and a 3d hole. Since $U_d/U_{cd} \simeq 0.7$ in most 3d transition-metal compounds,[123] the observed separation of 8 eV in Fig. 14 is in good agreement with the U_d values of 5 to 6 eV determined in Section 8a from the location of the Cu $3d^8$ satellite in the valence band spectra.

Rigorously, the above analysis only applies to the insulating $N = 9$ ground-state configuration in Fig. 10. In the metallic regime, in which the overall ground state contains some additional O p holes, an additional feature is expected in the $Cu\,2p_{3/2}$ core level spectrum due to $\underline{2p}d^{10}\underline{L}^2$ final states.[164,179] This additional structure is believed to occur at about 935 eV, on the high binding energy side of the main $\underline{2p}d^{10}\underline{L}$ peak. The slight increase in the width of the main peak in Fig. 15 from $x = 0$ to $x = 0.15$ is consistent with this effect. A more pronounced emission[164,179] near 935 eV is observed in $YBa_2Cu_3O_{7-\delta}$. With increasing δ, the main $Cu\,2p_{3/2}$ peak in this system narrows considerably and shifts to lower binding energy while its intensity increases relative to that of the satellite.[178,179] All of these effects are consistent with a simultaneous decrease in the number of O p holes and a decrease in the average Cu valence in the ground state.

XPS studies of the O 1s core levels in the new superconductors are much more controversial;[142] little agreement exists at present as to what constitutes an intrinsic, bulk spectrum. Most spectra which have been recorded (e.g., van der Marel et al.[165]) suggest the presence of two peaks at approximately 529 and 531 eV. The latter of these is particularly sensitive to surface treatments[162,181] and is absent in some measurements,[176,182] including the only single crystal data[144] for $La_{2-x}Sr_xCuO_4$. An intrinsic component to this second peak would not be

[181]W. K. Ford, C. T. Chen, J. Anderson, J. Kwo, S. H. Liou, M. Hong, G. V. Rubenacker, and J. E. Drumheller, *Phys. Rev.* **B37**, 7924 (1988).

[182]M. Grioni, J. C. Fuggle, P. J. Weijs, J. B. Goedkoop, G. Rossi, F. Schaefers, J. Fink, and N. Nücker, *J. de Physique*, Colloque **C9**, C9-1189 (1987).

surprising in view of the presence of inequivalent O sites[178] and the likelihood of more than one O valence[164] (O^{2-} and O^-). Some groups[150,160] have also reported the growth of a third feature at 533 eV at low temperatures, which they attribute to the formation of O_2^{2-} dimers. As with the valence band spectra, much further characterization would be required to prove that such temperature-dependent effects are not surface related.[162]

Core level studies of elements other than Cu and O are consistent with the ionic configurations La^{3+}, Ba^{2+}, Sr^{2+}, and Y^{3+} (e.g., Steiner et al.[146] and Iqbal et al.[162]). Many groups have observed a slight splitting of the Ba and Sr peaks which, if not a surface effect,[162] could be due to different environments resulting from the presence of O vacancies and/or O 2p holes.[152]

9. AUGER

Auger electron spectra for the new superconductors provide further confirmation of the presence of strong electron–electron interactions. Cu L_3VV Auger spectra, which have been measured by several groups for both the 2-1-4 and 1-2-3 systems,[150,151,157,165,177,180,183] are dominated by a broad peak centered at a kinetic energy of about 918 eV, just as in CuO. The Cu L_3VV Auger process[126] involves the creation of a Cu $2p_{3/2}$ core hole (L_3) which, when filled by a valence electron (V), causes the ionization of a second valence electron (V). The kinetic energy of the second electron is determined by the difference between the Cu $2p_{3/2}$ binding energy and the energy of the final state, which contains two valence band holes. Such a final state corresponds in the CI model of Section 6b to an $N = 7$ cluster (not shown in Fig. 10) whose spectrum is determined by a mixture of $3d^9\underline{L}^2$, $3d^8\underline{L}$, and $3d^7$ configurations.[151] The peak in the Auger spectrum results from the predominantly $3d^8\underline{L}$ final state. The ~15 eV separation between this peak and the energy of the main Cu $2p_{3/2}$ XPS peak in Fig. 14 is consistent with the ~12.5 eV energy of the $3d^8$ satellite in valence band photoemission, and thus with a U_d value of about 6 eV.

Recently, van der Marel et al.[165] have also reported an O KVV Auger spectrum for $YBa_2Cu_3O_{7-\delta}$. This spectrum results from the creation of an O 1s core hole and is dominated by a final state in which the two valence band holes have predominantly O 2p character. The peak in the measured spectrum occurs at a kinetic energy of approximately 514 eV,

[183]A. Balzarotti, M. De Crescenzi, C. Giovannella, R. Messi, N. Motta, F. Patella, and A. Sgarlata, Phys. Rev. B36, 8285 (1987).

which is again about 15 eV less than the O 1s XPS binding energy. Since the average O 2p binding energy is roughly 5 eV according to valence band photoemission, the energy of the two hole Auger final state again provides evidence for a U_p value of the order of 5 eV.

10. X-RAY ABSORPTION AND HIGH ENERGY ELECTRON ENERGY LOSS

Additional insight into the ground-state charge distribution and the character of states near the Fermi level in 2-1-4 and 1-2-3 systems has been obtained from x-ray absorption spectroscopy (XAS).[158,164,184-193] This technique, together with high-energy electron energy loss spectroscopy (EELS),[149,194,195] provides detailed information on localized transitions from atomiclike core levels to low-lying unoccupied states.

[184]J. M. Tranquada, S. M. Heald, A. R. Moodenbaugh, and M. Suenaga, *Phys. Rev. B***35**, 7187 (1987); J. M. Tranquada, S. M. Heald, and A. R. Moodenbaugh, *Phys. Rev. B***36**, 5263 (1987).

[185]E. E. Alp, G. K. Shenoy, D. G. Hinks, D. W. Capone, L. Soderholm, H.-B. Schuttler, J. Guo, D. E. Ellis, P. A. Montano, and M. Ramnathan, *Phys. Rev. B***35**, 7199 (1987); E. E. Alp, G. K. Shenoy, L. Soderholm, G. L. Goodman, D. G. Hinks, and B. W. Veal, in *Materials Research Society Symposium Proc.* (M. B. Brodsky, R. C. Dynes, K. Kitazawa, and H. L. Tuller, eds.), Vol. 99, p. 177, MRS, Pittsburgh, 1988.

[186]D. Sondericker, Z. Fu, D. C. Johnston, and W. Eberhardt, *Phys. Rev. B***36**, 3983 (1987).

[187]H. Oyanagi, H. Ihara, T. Matsubara, M. Tokumoto, T. Matsushita, M. Hirabayashi, K. Murata, N. Terada, T. Yao, H. Iwasaki, and Y. Kimura, *Jpn. J. Appl. Phys.* **26**, L1561 (1987).

[188]F. Baudelet, G. Collin, E. Dartyge, A. Fontaine, J. P. Kappler, G. Krill, J. P. Itie, J. Jegoudez, M. Maurer, P. Monod, A. Revcolevschi, H. Tolentino, G. Tourillon, and M. Verdaguer, *Z. Phys. B***69**, 141 (1987).

[189]K.-L. Tsang, C. H. Zhang, T. A. Callcott, L. R. Canfield, D. L. Ederer, J. E. Blendell, C. W. Clark, N. Wassdahl, J. E. Rubensson, G. Bray, N. Mortensson, J. Nordgren, R. Nyholm, and S. Cramm, *Phys. Rev. B***37**, 2293 (1988).

[190]T. Iwazumi, I. Nakai, M. Izumi, H. Oyanagi, H. Sawada, H. Ikeda, Y. Saito, Y. Abe, K. Takita, and R. Yoshizaki, *Solid State Commun.* **65**, 213 (1988).

[191]A. Bianconi, J. Budnick, A. M. Flank, A. Fontaine, P. Lagarde, A. Marcelli, H. Tolentino, B. Chamberland, C. Michel, B. Raveau, and G. Demazeau, *Phys. Lett.* *A***127**, 285 (1988).

[192]D. D. Sarma, O. Strebel, C. T. Simmons, U. Neukirch, G. Kaindl, R. Hoppe, and H. P. Muller, *Phys. Rev. B***37**, 9784 (1988); D. D. Sarma, *Phys. Rev. B***37**, 7948 (1988).

[193]S. M. Heald, J. M. Tranquada, A. R. Moodenbaugh, and Y. Xu, *Phys. Rev. B***38**, 761 (1988); J. M. Tranquada, S. M. Heald, A. R. Moodenbaugh, and Y. Xu, submitted for publication.

[194]P. E. Batson and M. F. Chisholm, *Phys. Rev. B***37**, 635 (1988).

[195]N. Nücker, J. Fink, J. C. Fuggle, P. J. Durham, and W. M. Temmerman, *Phys. Rev. B***37**, 5158 (1988) and *Physica C***153–155**, 119 (1988); J. Fink, N. Nücker, H. Romberg, J. C. Fuggle, P. J. W. Weijs, R. Schoorl, P. J. Durham, W. M. Temmerman, and B. Gegenheimer, to be published (*Proc. Schloss Mauterndorf*).

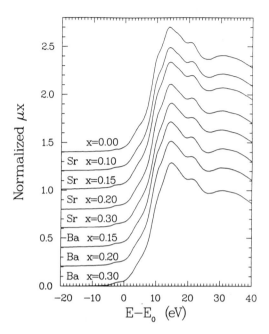

FIG. 16. Normalized Cu K near edge x-ray absorption spectra for $La_{2-x}M_xCuO_4$ with M = Sr, Ba ($E_0 = 8980$ eV). Curves for different x values have been shifted vertically for clarity. [After J. M. Tranquada, S. M. Heald, and A. R. Moodenbaugh, *Phys. Rev.* **B36**, 5263 (1987).]

Figure 16 shows characteristic XAS spectra for $La_{2-x}M_xCuO_4$ (M = Sr, Ba) in the near edge region for Cu K absorption.[184] The main peak in each spectrum, near 13 eV, results from transitions from Cu 1s core levels to a continuum of unoccupied states with predominately Cu 4p character. Quantum chemical calculations indicate that the lower-energy shoulder near 4 eV is essentially an excitonic effect resulting from similar 1s to 4p transitions, followed by a screening of the core hole due to ligand (O) to Cu charge transfer.[196] The weak structure at negative energies has been attributed, on the basis of atomic calculations, to quadrupole-allowed (but dipole-forbidden) Cu 1s to Cu 3d transitions.[184] Note that both of these last two features signal the presence of some Cu d holes in the ground state. Conflicting interpretations[184,185] have been offered for the structure above 15 eV; the most convincing arguments favor multiple scattering effects within the CuO_2 planes.[188]

All of the spectra in Fig. 16 are in excellent agreement with the Cu K

[196]R. A. Bair and W. A. Goddard, *Phys. Rev.* **B22**, 2767 (1980).

XAS spectrum of CuO. By contrast, the leading edge of the Cu K spectrum[184] of Cu_2O lies about 3 eV lower in energy and exhibits a sharp excitonic peak near the zero of energy in Fig. 16. The Cu K edge for a Cu^{3+} ion is expected to lie at several eV above the CuO edge.[184] The observed 2-1-4 spectra are thus consistent with a predominantly Cu^{2+} ground state. More importantly, the striking similarity between the results for different x values in Fig. 16 confirms the conclusion of Section 8c concerning the insensitivity of the Cu valence to the hole concentration. The fact that the intensity in the 4 eV shoulder does decrease slightly with increasing x has been attributed to an increase in the number of O p holes and a corresponding decrease in the number of charge-transfer screened final states.[184]

Cu K XAS spectra of 1-2-3 compounds are basically similar.[190] The main features are insensitive to the particular choice of rare-earth element[190] ($R = Y, Nd, Sm, \ldots$) and again reflect a predominantly Cu^{2+} ground state. Significant changes[185,187,188,193] are observed with increasing δ in $YBa_2Cu_3O_{7-\delta}$ that are consistent with a decrease in the number of O p holes and the conversion of some Cu^{2+} to Cu^+.

Recently, Heald $et\ al.$[193] have observed anisotropies in the Cu K absorption spectra of both 2-1-4 and 1-2-3 powdered samples that have been oriented in a magnetic field. Both the main peak and the leading edge of the absorption in $La_{1.85}Sr_{0.15}CuO_4$ occur at about 3 eV lower energy for final states of p_z symmetry compared to those of p_x or p_y symmetry. This effect is consistent with previous studies of layered Cu^{2+} compounds.[196] An additional anisotropy in the intensity of the low-energy quadrupole feature is consistent with the ground state Cu d hole having x^2-y^2 symmetry, as expected from the theoretical results of Parts III and IV. The anisotropic behavior in $YBa_2Cu_3O_7$ is more difficult to interpret because of the presence of two different Cu environments.[193]

Further information on transitions to unoccupied Cu d states has been obtained from Cu $L_{2,3}$ XAS and EELS spectra which involve the excitation of a Cu $2p_{1/2,3/2}$ core electron.[149,158,164,186,191,194] The L_3 XAS spectrum of pure La_2CuO_4 consists of a single strong absorption ("white") line at 931.3 eV, similar to that of CuO.[191] This line has been attributed to a $\underline{2p}3d^{10}$ final state,[142,191] which again implies a predominantly Cu^{2+} ($3d^9$) ground state. A similar white line has been observed in $YBa_2Cu_3O_7$ along with a higher-energy shoulder at 933 eV.[164] Because this shoulder is also present in doped $La_{2-x}Sr_xCuO_4$, it is believed to be characteristic of the additional ground-state holes present in the metallic regime. The coincidence between the energy of this shoulder and that of the main XPS peak in Fig. 15 suggests that both of these features result from similar $\underline{2p}3d^{10}\underline{L}$ final states.[142,164,191] The Cu L_3 XAS spectra thus provide additional evidence for the presence of a significant $3d^9\underline{L}$ component (O p holes) in the ground state of metallic 2-1-4 and 1-2-3 compounds. Initial results on a weak orientation dependence of the Cu L_3 EELS spectrum in $YBa_2Cu_3O_7$ single crystals are again difficult to interpret because of the different behavior expected at Cu1 and Cu2 sites.[194]

The main peak in La $L_{2,3}$ XAS spectra for $La_{2-x}M_xCuO_4$ has been attributed to the presence of unoccupied O 2p states with d symmetry about both the La and the Cu sites.[184] The monotonic increase in the intensity of this peak[184] with increasing x is consistent with the principal effect of doping being the creation of O 2p holes.

Still more direct evidence in favor of this picture has been obtained from studies of O 1s (K) absorption.[149] In both the 2-1-4 and 1-2-3 systems, the onset of such absorption occurs at about 529 eV, which coincides with the lowest binding energy feature in O 1s XPS measurements. The O 1s absorption increases rapidly at about 3 to 5 eV above the onset, presumably because of transitions to empty conduction states with p-like symmetry about the O sites. The intensity right at 529 eV increases monotonically[149] with x in $La_{2-x}M_xCuO_4$ and grows into a distinct peak which is also present[195] in $YBa_2Cu_3O_{7-\delta}$ for small values of δ. This peak is clearly due to O p hole states just above the Fermi level in the metallic regime. The composition dependence of this peak is inconsistent with LDA band calculations[195] but is qualitatively consistent with the schematic Anderson Hamiltonian picture of Fig. 11. A more detailed analysis of the O 1s absorption spectrum in these materials is complicated by the previously mentioned uncertainties in the O 1s XPS spectrum.[142] An initial study of the orientation dependence of the O 1s EELS spectrum in $YBa_2Cu_3O_7$ single crystals by Batson and Chisholm[194] has suggested that the O p holes have primarily p_x and p_y character, but similar measurements by Fink et al.[195] indicate a much weaker anisotropy.

11. Optical and Low Energy Electron Energy Loss

Lower-energy interband transitions in 2-1-4 and 1-2-3 compounds have also been studied by using EELS,[197–201] as well as by using optical

[197]Y. Chang, M. Onellion, D. W. Niles, R. Joynt, G. Margaritondo, N. G. Stoffel, and J. M. Tarascon, *Solid State Commun.* **63,** 717 (1987); M. Tang, Y. Chang, M. Onellion, J. Seuntjens, D. C. Larbalestier, G. Margaritondo, N. G. Stoffel, and J. M. Tarascon, *Phys. Rev.* B**37,** 1611 (1988).

[198]J. Yuan, L. M. Brown, and W. Y. Liang, *J. Phys.* C**21,** 517 (1988).

[199]A. Ando, K. Saiki, K. Ueno, and A. Koma, *Jpn. J. Appl. Phys.* **27,** L304 (1988).

[200]C. H. Chen, L. F. Schneemeyer, S. H. Liou, M. Hong, J. Kwo, H. S. Chen, and J. V. Waszczak, *Phys. Rev.* B**37,** 9780 (1988).

[201]Y. Chang, Y. Hwu, M. Onellion, G. Margaritondo, N. G. Stoffel, P. A. Morris, and W. A. Bonner, *Phys. Rev.* B**38,** 4996 (1988).

techniques.[101,137,202–210] The results are basically consistent with the electron and hole excitation spectra determined by photoemission and inverse photoemission, respectively, but disagree with theoretical calculations of the complex dielectric function $\varepsilon = \varepsilon_1 + i\varepsilon_2$ based on LDA energy bands.[211]

Etemad et al.[204] have reported ellipsometry measurements of the ε_2 spectra for a series of $La_{2-x}Sr_xCuO_4$ samples in the energy range 1.5 to 5.8 eV. The results indicate a broad absorption peak between 2 and 3 eV and the onset of much stronger absorption above 5 eV. The similarity of these features to those observed in BIS measurements on these materials suggests that the lower energy peak is due primarily to transitions from below the Fermi level to unoccupied $Cu\,d^{10}$ states, while the higher energy absorption results from transitions to the continuum of unoccupied states derived primarily from La 5d and 4f levels. The observed x dependence of the lower energy structure is consistent with the optical gap of the insulating $x = 0$ compound being ~ 2 eV, which is also the value in CuO.[149] Similar gap values have also been obtained from low-temperature photoluminescence measurements[209] and from a Kramers–Kronig analysis of reflectivity.[202,208]

Strong absorption features in $YBa_2Cu_3O_7$ are also well correlated with the energies of unoccupied states determined by BIS measurements (e.g., Fig. 14). A variety of optical techniques[137,202,203,205,210] all indicate a broad absorption peak centered at about 2.5 eV and the onset of much stronger absorption at about 4.5 eV. Low energy EELS measurements[197,200] also

[202]J. Orenstein, G. A. Thomas, D. H. Rapkine, C. G. Bethea, B. F. Levine, R. J. Cava, E. A. Rietman, and D. W. Johnson, Phys. Rev. B36, 729 (1987); J. Orenstein, G. A. Thomas, D. H. Rapkine, C. G. Bethea, B. F. Levine, B. Batlogg, R. J. Cava, D. W. Johnson, and E. A. Rietman, Phys. Rev. B36, 8892 (1987).

[203]K. Kamaras, C. D. Porter, M. G. Doss, S. L. Herr, D. B. Tanner, D. A. Bonn, J. E. Greedan, A. H. O'Reilly, C. V. Stager, and T. Timusk, Phys. Rev. Lett. 59, 919 (1987).

[204]S. Etemad, D. E. Aspnes, M. K. Kelly, R. Thompson, J.-M. Tarascon, and G. W. Hull, Phys. Rev. B37, 3396 (1988).

[205]Ch. B. Lushchik, I. L. Kuusmann, E. Kh. Fel'dbakh, K. E. Vallaste, P. Kh. Liblik, A. A. Maaroos, I. A. Meriloo, and T. L. Savikhina, JETP Lett. 46, 151 (1987).

[206]Z. Schlesinger, R. T. Collins, D. L. Kaiser, and F. Holtzberg, Phys. Rev. Lett. 59, 1958 (1987).

[207]I. Bozovic, D. Kirillov, A. Kapitulnik, K. Char, M. R. Hahn, M. R. Beasley, T. H. Geballe, Y. H. Kim, and A. J. Heeger, Phys. Rev. Lett. 59, 2219 (1987).

[208]H. P. Geserich, D. Sack, G. Scheiber, B. Renker, and D. Ewert, Z. Phys. B69, 465 (1988).

[209]J. M. Ginder, M. G. Roe, Y. Song, R. P. McCall, J. R. Gaines, E. Ehrenfreund, and A. J. Epstein, Phys. Rev. B37, 7506 (1988).

[210]M. K. Kelly, P. Barboux, J.-M. Tarascon, D. E. Aspnes, W. A. Bonner, and P. A. Morris, Phys. Rev. B38, 870 (1988).

[211]G.-L. Zhao, Y. Xu, W. Y. Ching, and K. W. Wong, Phys. Rev. B36, 7203 (1987).

indicate structure at these energies, as well as an additional increase in intensity near 8 eV. From Fig. 14, it would seem that the 2.5 eV and 4.5 eV absorption peaks again result from transitions from just below the Fermi level to unoccupied $Cu\,d^{10}$ states and unoccupied continuum states, respectively. The increased absorption near 8 eV is consistent with transitions across the "pseudo-gap" in Fig. 14 from below -2 eV to above 6 eV.

A sharp new peak in ε_2 at 4.1 eV has been observed in recent ellipsometry measurements[210] on O-deficient $YBa_2Cu_3O_{7-\delta}$. This peak first appears for $\delta \simeq 0.3$–0.4, and its intensity then increases monotonically with δ. The origin of this peak is believed to be a $Cu\,3d^{10}$ to $3d^9 4s^1$ exciton occuring on Cu^{1+} sites within the disrupted Cu1–O1 chains. An increase in the Cu^{1+} concentration in the chains with δ still less than 0.5 is consistent with the material still being metallic (cf. Fig. 4) because of an increase in the hole concentration in the planes. An additional peak also appears at 1.7 eV for $\delta > 0.4$, which the authors suggest[210] could be due to a d–d excitation at a Cu^{2+} site (cf. Section 6d).

A clear picture of the optical properties below 2 eV in 2-1-4 and 1-2-3 compounds has not yet emerged. Reflectivity studies in the metallic regime are dominated by intraband transitions resulting from the presence of free holes in a nominally Cu^{2+}, O^{2-} background.[202,203,208] Most polycrystalline data also exhibit a non-Drude-like contribution that many authors have interpreted[202–204] as evidence for an excitonic transition between 0.3 and 0.6 eV. The reliability of such data has been questioned, however, because of a distortion of anisotropy effects.[212] Recent transmission measurements by Geserich et al.[137] on oriented 1-2-3 thin films do indicate excitonic features at 0.6 eV and 1.5 eV, which were missed in an earlier study on thinner films.[207] Such features are suggestive of either d–d excitations[58] or of the "charge-transfer excitons" proposed by Varma et al.[113] Deviations from pure Drude behavior have also been observed in reflectivity measurements on 1-2-3 single crystals.[101,206] A recent study over a wide range of temperatures[101] has been interpreted in terms of a frequency- and temperature-dependent carrier lifetime and effective mass (with $m^*/m \leq 10$); such behavior is consistent with strong interactions between the carriers and some other excitation of the system, such as spin fluctuations.

VI. Concluding Comments

The theoretical considerations and spectroscopic data reviewed in this chapter provide a reasonably consistent picture of the gross features of

[212]J. Orenstein and D. H. Rapkine, *Phys. Rev. Lett.* **60**, 968 (1988).

the electronic structures of the 2-1-4 and 1-2-3 copper oxides. Two conclusions which now seem well established are that (1) the behavior of both systems is dominated by the same microscopic physics, and (2) the crucial Cu 3d and O 2p electrons are strongly interacting and are not adequately described by conventional one-electron band theory.

The most successful theoretical models that have been proposed to date are the extended Hubbard and Anderson lattice Hamiltonians of Section 6. The key feature of these models is the presence of O p states within the coreelation gap of the Cu d states, as illustrated in Fig. 11. Most existing evidence points to parameter values in the ranges $U_d \simeq 6$–$8\,\mathrm{eV}$, $U_p \simeq W \simeq 4$–$5\,\mathrm{eV}$, $V^0 \simeq \Delta \simeq 1$–$2\,\mathrm{eV}$. A further refinement of these parameters is clearly required, as is additional information on the magnitude of U_{pd}, the relative energies of O p_σ and O p_π states and the possible role of d–d excitations. The relevance of simpler model Hamiltonians, such as the single-band Hubbard Hamiltonian, must also be examined in more detail.

By far the most important unresolved problem in the area of electronic structure is the nature of the low-lying excitations in the "normal" metallic phase of these materials. The primary issue here is whether such excitations can be adequately described by Fermi liquid theory, or whether one needs to consider a more exotic alternative, such as RVB theory. The first Fermi surface studies on the new superconductors, based on two-dimensional, angular correlation of positron annihilation radiation measurements, yield some features that are in reasonable agreement with LDA band theory, but yields other features that disagree.[213] It is often argued that, since the Fermi surface is essentially a ground-state property, it should be well described by the LDA; the LDA Fermi surface for the heavy fermion metal UPt$_3$, in fact, is remarkably accurate.[214] Schönhammer and Gunnarsson[215] have recently shown, however, that even an exact density-functional theory calculation is not guaranteed to give the correct Fermi surface for an interacting system. A detailed understanding of the Fermi surfaces in the 2-1-4 and 1-2-3 copper oxides and, in particular, a knowledge of whether they obey

[213]L. Hoffmann, A. A. Manuel, M. Peter, E. Walker, and M. A. Damento, *Europhys. Lett.* **6,** 61 (1988); A. L. Wachs, P. E. A. Turchi, Y. C. Jean, K. H. Wetzler, R. H. Howell, M. J. Fluss, D. R. Harshman, J. P. Remeika, A. S. Cooper, and R. M. Fleming, *Phys. Rev. B***38,** 913 (1988); L. C. Smedskjaer, J. Z. Liu, R. Benedek, D. G. Legnini, D. J. Lam, M. D. Stahulak, H. Claus, and A. Bansil, *Physica C***156,** 269 (1988).

[214]C. S. Wang, M. R. Norman, R. C. Albers, A. M. Boring, W. E. Pickett, H. Krakauer, and N. E. Christensen, *Phys. Rev. B***35,** 7260 (1987); L. Taillefer and G. G. Lonzarich, *Phys. Rev. Lett.* **60,** 1570 (1988).

[215]K. Schönhammer and O. Gunnarsson, *Phys. Rev. B***37,** 3128 (1988).

Luttinger's theorem,[100] would clearly provide important insight into the possible origins of high-temperature superconductivity.

ACKNOWLEDGMENTS

The author is indebted to his many colleagues who provided figures for this chapter and/or sent preprints prior to publication. He has also benefitted from discussions with A. Bansil, A. D. Brailsford, A. E. Carlsson, G. Graham, B. I. Halperin, E. M. Logothetis, L. F. Mattheiss, A. McMahan, and W. H. Weber and is especially grateful to J. W. Allen and L. C. Davis for sharing their many insights into the physics of transition-metal compounds. Special thanks are also due to L. C. Davis for a critical reading of the manuscript and to H. Ehrenreich for his early stimulation and encouragement.

Notes Added in Proof

Since the completion of this chapter, the author has become aware of several more recent contributions of potential significance.

Experimentally, (1) Torrance et al.[216] have succeeded in inhibiting the formation of oxygen vacancies in $La_{2-x}Sr_xCuO_{4-y}$ samples with large x values. Their results suggest that the superconducting transition temperature in this system is relatively constant for hole concentrations between 0.15 and 0.24 per cell, and then decreases rapidly to zero at 0.32 holes per cell. (2) List et al.[217] have observed the first clear Fermi edge in the photoemission spectrum of a 1-2-3 compound. The edge is present in spectra recorded from clean surfaces of $EuBa_2Cu_3O_{6.7}$ single crystals cleaved in vacuum at 20 K, but not for samples annealed above 55 K. The unannealed spectra also differ in that they do not show the controversial -9-eV peak. The authors contend that the higher temperature spectra, which resemble most previous photoemission data (cf. Section 8), are characteristic of an insulating surface which is probably oxygen deficient. (3) Nücker et al.[218] have extended their analysis[195] of the orientation dependence of the O 1s and Cu 2p EELS spectra of $YBa_2Cu_3O_{7-\delta}$ single crystals. The observed anisotropies are consistent with the lowest lying unoccupied states in the insulating compound having primarily Cu d

[216]J. B. Torrance, Y. Tokura, A. I. Nazzal, A. Bezinge, T. C. Huang, and S. S. P. Parkin, *Phys. Rev.* **61,** 1127 (1988).

[217]R. S. List, A. J. Arko, Z. Fisk, S.-W. Cheong, J. D. Thompson, J. A. O'Rourke, C. G. Olson, A.-B. Yang, T.-W. Pi, J. E. Schirber, and N. D. Shinn, submitted for publication.

[218]N. Nücker, H. Romberg, X. X. Xi, J. Fink, B. Gegenheimer, and Z. X. Zhao, submitted for publication.

$(x^2 - y^2)$ character, and the additional holes for $\delta < 0.5$ having primarily O p_σ or O $p_\pi^{\parallel}$ character.

Theoretical developments include the following: (1) Eskes and Sawatzky[219] have performed Anderson impurity calculations of the electron removal spectra for Cu^{2+} compounds in D_{4h} symmetry. The results are similar to those of McMahan et al.,[57] including the observation that the lowest ionization state is a two-hole singlet bound state. (2) Stechel and Jennison[220] have extended their previous model[116] to conclude O p_π orbitals and have obtained exact numerical results for the ground and low lying excited states of a variety of finite clusters. They include that the O p holes have p_π character only in the smallest clusters, with O p_σ holes favored by more than $0.6\,eV$ in an infinite Cu-O_2 layer due to valence fluctuations. These results point to a possible deficiency in the cluster GVB calculations of Goddard and co-workers.[109] (3) Bansil et al.[221] have presented a detailed comparison of the electron-positron momentum density in $YBa_2Cu_3O_7$ calculated in the LDA with positron annihilation data. The excellent agreement obtained suggests that the ground state and four-sheeted Fermi surface in this system are well described by LDA band theory.

[219]H. Eskes and G. A. Sawatzky, *Phys. Rev. Lett.* **61,** 1415 (1988).

[220]E. B. Stechel and D. B. Jennison, *Phys. Rev.* B38 (1988).

[221]A. Bansil, R. Pankaluoto, R. S. Rao, P. E. Mijnarends, W. Dlugosz, R. Prasad, and L. C. Smedskjaer, *Phys. Rev. Lett.* **61** (1988).

SOLID STATE PHYSICS, VOLUME 42

Electron Correlations in Two Dimensions

A. ISIHARA

Department of Physics
State University of New York at Buffalo
Buffalo, New York

Preface

In the past two decades, remarkable progress has been made in the study of two-dimensional (2-D) electron systems. During this period, clear recognition of the 2-D character of interface electrons, development of different types of 2-D electron systems, and disclosure of very important and unusual properties have taken place. The recent discovery

271

of the quantized Hall effect has brought the study of these electrons to a new stage in which solid state physics is linked with quantum electrodynamics.

In what follows, only a few low-temperature properties of 2-D electron systems with Coulomb interaction, especially those related to electron correlations, will be discussed. Efforts will be made to avoid encylopedic presentation and to achieve coherence and uniformity. Emphasis will be placed on the basics and on two-dimensionality. Consequently, thickness effects, for example, will not be covered in any depth. On the other hand, many references are listed for those interested in expanding their study beyond what is presented.

The author had the privilege of visiting the Department of Physics of Harvard University during the fall of 1987. This provided him the opportunity to write the present article with the encouragement of his colleagues. He is grateful to them, to those who gave valuable suggestions, and to the authors and publishers who granted permission to reproduce their graphs. The publishers include the American Physical Society, American Vacuum Society, Pergamon Journals, Inc., North-Holland Publishing Co., Springer-Verlag, Physical Society of Japan, and IOP Publishing Ltd.

I. Ground-State Energy

1. INTRODUCTION

In 1948 Shockley and Pearson[1] tried to develop a new active circuit element by modifying a parallel plate condenser. They proposed to replace one of the condenser plates by a semiconductor, because charges induced by a bias field could be significant compared with the carrier density in the semiconductor, causing appreciable changes in the electrical properties that might be controlled by the bias field. It took more than a decade, however, to finally develop the condenser device into what is now called *MOSFET*, i.e., metal-oxide-semiconductor-field-effect-transistor. This type of transistor is one of the most important components in modern electronics. The basic concept for this transistor type was actually forwarded by Lillienfeld in the early 1930s.[2]

A typical MOSFET device is a condenser consisting of a silicon crystal as one of the plates and silicon dioxide as a dielectric (Fig. 1). The other

---

[1]W. Shockley and G. Pearson, *Phys. Rev.* **74,** 232 (1948).
[2]J. E. Lillienfeld, US Patent #1,745,175, (1930).

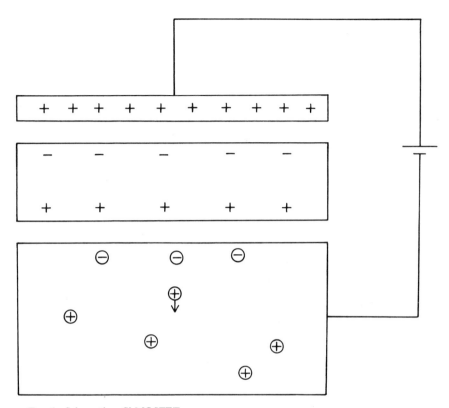

FIG. 1. Schematic p-Si MOSFET.

plate, called the *gate,* is a metal. The oxidation of a silicon surface is easy, and the oxide layer, which reaches up to 2000 Å in thickness, is chemically stable.

If a p-type silicon is used as a plate, the positive carriers near the oxide surface can be driven into the bulk by a voltage onto the gate. This induces negative charges at the interface between silicon dioxide and silicon. Further increase in the voltage will produce conduction electrons. Figure 1 illustrates the basic Si structure without any external lead. These electrons appear in a narrow layer near the interface, its thickness being of the order of 100 Å or less. The sign of these charges is opposite to that of the bulk carriers in silicon. Hence, the layer is called the *inversion layer.* If an n-type silicon is used instead of a p-type for the condenser plate, we will have a so-called *accumulation layer,* because carriers of the same sign as in the bulk are induced.

It has been found that these electrons can move rather freely in the layer, while their motion perpendicular to the condenser plates is restricted due to quantization of the energy levels. Therefore, the electrons constitute a two-dimensional system for low energies.

By connecting similar condensers to each other electrically, it is possible to move such electrons from one condenser plate to another. The existence and nonexistence of such electrons at a particular condenser can be used as a binary code. Thus, a Si-MOSFET has become one of the most important components in electronic devices.

The progress in the study of these 2-D electrons has been phenomenal in the past two decades. It has been revealed that these electrons show very interesting and unusual properties, some of which are unique, whereas others are shared by electrons in bulk.[3] It is the purpose of the present article to review some of the work on their important properties. However, our attention will be focused more or less on electron correlations in two dimensions. Therefore, a Si-MOSFET is not the only system of our interest; at least, two additional 2-D electron systems play equally important roles in our discussion.

Let us now introduce these additional 2-D electron systems. It has been found that electrons can be trapped two-dimensionally by image forces near the surface of liquid helium.[4] The surface of liquid helium presents a barrier V_0 of more than 1 eV to electron transmission into the liquid, while the image potential attracts the electron to the interface. As a result, the wave function is localized, as is illustrated in Fig. 2.

Another 2-D electron system has been realized in the interface between two semiconductors. By computerized molecular-beam epitaxy, alternating thin layers of two semiconductors can form a sandwiched structure, known as a *quantum well,* or a one-dimensional periodic structure, called a *superlattice,*[5] that has periodicity larger than that of the original lattices. The most familiar semiconductor combination is $GaAs/Ga_xAl_{1-x}As$, where x is a fraction. The Al atoms do not alter the lattice structure of the GaAs lattice and yet cause an adequate superlattice potential that modifies the original band structure. The potentials between the layers form a one-dimensional array like the one in the Kronig–Penney model, creating minizones in momentum and subbands in energy. The period ranges from tens to hundreds of angstroms and is shorter than the electron mean free path but longer than the lattice constants of the host crystals.

[3] T. Ando, A. B. Fowler, and F. Stern, *Rev. Mod. Phys.* **54,** 437 (1982).

[4] M. W. Cole and M. H. Cohen, *Phys. Rev. Lett.* **23,** 1238 (1966).

[5] L. Esaki and R. Tsu, IBM, *J. Res. Dev.* **24,** 61 (1970); M. J. Kelly and R. J. Nicholas, *Rep. Prog. Phys.* **48,** 1699 (1985).

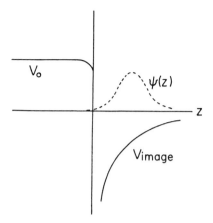

FIG. 2. Schematic potential diagram on the surface of liquid He. $\psi(z)$ represents a charge distribution.

Figure 3(a) shows schematically the enrgy diagram of GaAs/Ga$_{0.7}$-Al$_{0.3}$As, which is undoped, and ε_c and ε_v are the conduction- and valence-band edges, respectively. The Fermi energy ε_F is at about the middle of the band gap. One can dope Ga$_x$Al$_{1-x}$As layers with Si as donor impurity and modulate the energy band as in Fig. 3(b). The Fermi

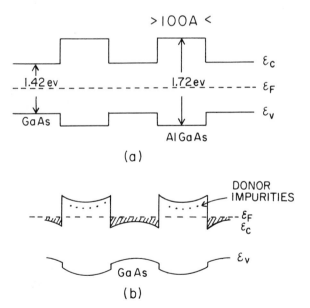

FIG. 3. Energy diagram of GaAs/GaAlAs.

level is now moved near the conduction-band edge. Since the conduction-band edge of GaAs is below the donor states in $Ga_xAl_{1-x}As$, donor electrons move into the GaAs region, causing high mobility in the layer. Other types of superlattices include InAs/GaSb.

The density of electrons differs widely among these 2-D electron systems. Of the three systems, the density of electrons, to be denoted by n, can be changed most easily in Si-MOSFETs in the typical range of 10^{11}–10^{12}/cm^2 by varying the gate voltage. In the case of electrons on the surface of liquid helium, a typical density is of the order of 10^9/cm^2, and in the case of GaAs/GaAlAs it is 10^{12}/cm^2. At these densities, the electrons in Si/SiO$_2$ and GaAs/GaAlAs are quantum-mechanically degenerate at low temperatures, whereas those on the surface of liquid helium constitute classical 2-D systems.

For low temperatures, the areal density n of electrons is expressed by the Fremi wavenumber k_F and by a dimensionless parameter r_s. These are listed below in comparison with those for 3D:

2D	3D
$k_F^2 = 2\pi n$	$k_F^3 = 3\pi^2 n$
$r_s = \dfrac{1}{a_0}\dfrac{1}{\sqrt{(\pi n)}}$	$r_s = \dfrac{1}{a_0}\left(\dfrac{3}{4\pi n}\right)^{1/3}$

The parameter r_s represents the radius of the circle per electron measured in units of the effective Bohr radius a_0. Note that for electrons in semiconductors, the bare electron mass m_0 must be replaced by an effective mass m and also e^2 by e^2/κ, where κ is the average dielectric constant. In Si inversion layers, $m = 0.19m_0$ and $\kappa = 11.5$, and in GaAs superlattices $m = 0.068m_0$, and κ is approximately 12. Besides, for Si [100] inversion layers, the valley degeneracy factor g_v must be introduced such that $k_F = (2\pi n/g_v)^{1/2}$. In the natural unit in which $\hbar = 1$ and $2m = 1$, the Fermi momentum is given by the above expression for the Fermi wave number. In addition, sometimes units in which $e^2 = 2$ are used to set $a_0 = 1$.

Because the available density range in thse 2-D electron systems is wide in contrast to the case of metals, they provide as a whole a very good testing ground for many-body theory. Besides, new and very unusual properties have been discovered, some of which follow the predictions of many-body theory, while others still wait for theoretical explanations. The physics of 2-D electron systems, together with the developments in their important industrial applications, will continue to be one of the most exciting fields in condensed-matter physics.

2. Dielectric Function

For an electron system with neutralizing positive charges in background, the dielectric function is defined by

$$\tilde{D}(\mathbf{q}, \omega) = \varepsilon(\mathbf{q}, \omega)\tilde{E}(\mathbf{q}, \omega), \tag{2.1}$$

where $\mathbf{q}$ is the wave vector and ω is the frequency. $\tilde{D}$ and $\tilde{E}$ are the Fourier transforms of the displacement and electric field vectors, respectively.

It is convenient to adopt the natural unit in which $\hbar = 1$ and $2m = 1$, where m is the electron's effective mass in the system. In this unit, $\mathbf{q}$ and ω represent momentum and energy, respectively. For simplicity, this unit is frequently adopted in many-body theory. Therefore, we shall also use this unit, although explicit manifestation of electron mass or $\hbar$ will be made whenever desirable. The restoration of $\hbar$ or m in equations can easily be made through a dimensional condideration.

The dielectric function plays an important role in discussing the dynamic and static properties of electron systems. The displacement vector can be determined by introducing a test charge into the system and by evaluating the density change due to the charge either by linear-response theory or by first-order perturbation theory based on the assumption that the effect of the test charge is small.

If the eigenvalues of the total Hamiltonian that includes both the electrons and the test charge are given, the dielectric function is expressed in the form

$$1/\varepsilon(q, \omega) = 1 + u(q) \sum_j |(n_q)_{j0}|^2 \left[\frac{1}{\omega - \omega_{j0} + i\eta} - \frac{1}{\omega + \omega_{j0} + i\eta} \right], \tag{2.2}$$

where $\omega_{j0} = E_j - E_0$, with E_j being the eigenvalues, and $(n_q)_{j0}$ is the matrix element of the density fluctuations in momentum space. The symbol η represents a small parameter that is brought to zero after completing the summation.

As shown by Ehrenreich and Cohen,[6] a self-consistent expression for the dielectric function of an electron gas is

$$\varepsilon(q, \omega) = 1 - \frac{2u(q)}{(2\pi)^2} \int \frac{f(\varepsilon_{\mathbf{k}+\mathbf{q}}) - f(\varepsilon_k)}{\varepsilon_{\mathbf{k}+\mathbf{q}} - \varepsilon_k + \omega + i\eta} \, d\mathbf{k}, \tag{2.3}$$

[6]H. Ehrenreich and M. H. Cohen, *Phys. Rev.* **115**, 786 (1959).

where $f(\varepsilon)$ is the Fermi distribution function, and $u(q)$ is the Fourier transform of the Coulomb potential:

$$u(q) = \frac{2\pi e^2}{q}.$$ (2.4)

Note that this is a two-dimensional Fourier transform of the ordinary 3-D Coulomb potential, since our 2-D systems always have certain widths, and the electrons interact with each other through the ordinary Coulomb potential instead of through the logarithmic one. In this sense, our systems are *quasi-two-dimensional*. However, for most of the discussions in this article, the width will not be explicitly considered because of our interest in properties intrinsic to 2-D.

Equation (2.3) is based on replacing $u(q)$ by $u(q)/\varepsilon(q, \omega)$. For convenience, we introduce an equivalent form:

$$\varepsilon(q, \omega) = 1 + u(q)\chi(s, y),$$ (2.5)

where $\chi(s, y)$, called the *polarization function*, is expressed as

$$\chi(s, y) = F(s, y)/2\pi$$ (2.6)

$$F(s, y) = 1 - \frac{1}{2s}[g_+ + g_-]$$ (2.7)

$$g_\pm = [(s \pm iy/s)^2 - 4]^{1/2}.$$

In these equations, we have used the reduced variables $s = q/p_F$ and $y = -i\omega/p_F^2$, where p_F is the Fermi momentum ($\hbar = 1$, $2m = 1$).

Note the following limiting forms of the function $F(s, y)$:[7]

$$F(s, y) = \begin{cases} 2\left(\frac{s}{y}\right)^2\left[1 - 3\left(\frac{s}{y}\right)^2\right] \\ \dfrac{2}{s^2}. \end{cases}$$ (2.8)

Note also that in the static limit, the dielectric function is given by

$$\varepsilon(q, 0) = 1 + \frac{e^2}{q}\begin{cases} 1 & s \le 2 \\ [1 - (1 - 4/s^2)^{1/2}], & s \ge 2. \end{cases}$$ (2.9)

[7]F. Stern, *Phys. Rev. Lett.* **30**, 278 (1973); A. Isihara and T. Toyoda, *Z. Phys.* B**23**, 389 (1976).

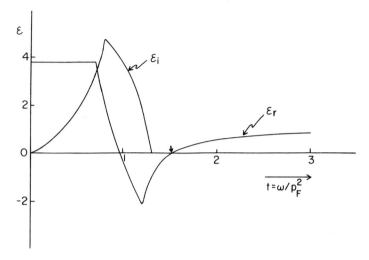

FIG. 4. The RPA dielectric function for $r_s = 1$ and $q = 0.5p_F$. ε_r and ε_i refer respectively to the real and imaginary parts. The abscissa is dimensionless ($\hbar = 1$, $2m = 1$). The arrow indicates ω_p.

The real and imaginary parts of the (RPA) dielectric function are illustrated in Fig. 4 for $r_s = 1.0$ and $q = 0.5p_F$ as functions of ω/p_F^2. The real part ε_r is constant for small frequency, vanishes at $\omega_p = 1.5p_F^2$, and gradually approaches 1 for large ω. The relatively large values of ε_r for small ω mans that static screening is effective. The imaginary part ε_i is zero at the origin and is finite only up to around $\omega = 1.3p_F^2$. At large q, the real part becomes smaller, as shown in Fig. 5, although it is still finite at the origin. This graph corresponds to the case of $q = 2p_F$. The singularity of the real part is due to the *plasmon*, which we shall discuss in the next section. The imginary part is now smooth and parabolic below the plasmon frequency, where it vanishes.

The dielectric function can be expressed in several forms. It can be obtained from the eigenvalues of the effective propagator representing the unit in the ring diagrams.[6] The eigenvalues are given by

$$\lambda_j(q) = \frac{1}{\pi^2} \int d\mathbf{p} f(p) \frac{q^2 + 2\mathbf{p} \cdot \mathbf{q}}{(q^2 + 2\mathbf{p} \cdot \mathbf{q})^2 + \left(\dfrac{2\pi j}{\beta}\right)^2}, \qquad (2.10)$$

where $\beta = 1/kT$ and j is an integer. By replacing $2\pi j/\beta$ by $-i\omega + \eta$, the eigenvalue function is converted into the polarization function.

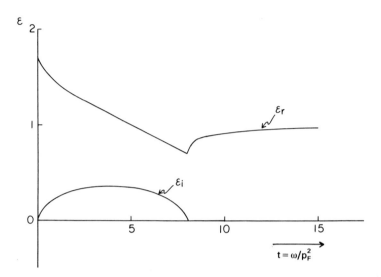

FIG. 5. The RPA dielectric function at $q = 2p_F$.

Figure 6 illustrates $F(s, y)$ as a function of s.[7] Note that all the curves are finite and that the curve corresponding to $y = 0$ or $j = 0$ has a singularity at $s = 2$, i.e., $q = 2p_F$. These behaviors are characteristic of two dimensions. In fact, the corresponding function for 3-D increases from $s = 0$, diverges at $s = 2$, and then decreases very fast for $s > 2$. The 1-D polarization function decreases smoothly from 1 without a kink or a divergence.

The eigenvalues have an interesting electrostatic analogue.[8] In terms of new variables defined by

$$\rho = 2p_F/q, \qquad z = 2\pi j/\beta q^2, \qquad (2.11)$$

and after the p-integration, it is transformed into

$$\lambda(\rho, z) = \frac{\rho}{2\pi} \int dx e^{-|z|x} \frac{\sin x}{x} J_1(\rho, x) \, dx. \qquad (2.12)$$

This form suggests that the eigenvalue function represents the electric flux passing through a disc of radius ρ at a distance z from a conducting disc of radius 1 and charge $Q = 1/(2\pi)$, as illustrated in Fig. 7.

[8]A. Isihara and L. C. Ioriatti, Jr., *Physica A***103**, 621 (1980); L. C. Ioriatti, Jr. and A. Isihara, *Z. Phys. B***44**, 1 (1981).

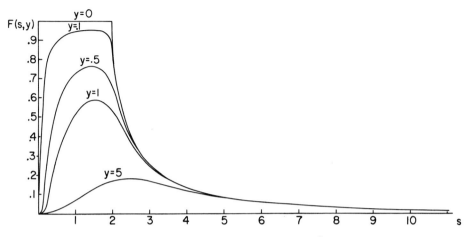

FIG. 6. Polarization function $F(s, y)$ with $s = q/p_F$, $y = -i\omega/p_F^2$, $(\hbar = 1, 2m = 1)$.

In view of the geometry of the electric flux, let us introduce an oblate-spheroidal coordinate system given by

$$\rho = \cosh \xi \sin \phi; \qquad z = \sinh \xi \sin \phi, \qquad (2.13)$$

where

$$0 \le \phi \le \pi/2; \qquad 0 \le \xi < \infty.$$

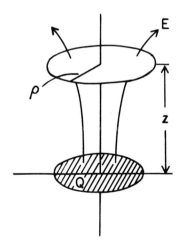

FIG. 7. Electrostatic analogue of the RPA dielectric function.

In this coordinate system, the eigenvalue function is simply

$$\lambda(\phi) = \frac{1}{2\pi}(1 - \cos \phi).$$ (2.14)

We shall show later that this simple form is advantageous for treating electron correlations.

The dielectric function given by Eqs. (2.5) and (2.6) is based on the RPA. For low densities, or large wave number, it is necessary to introduce corrections. For example, it has been found that improvements can be achieved by adopting the following form:

$$\chi(q, \omega) = \frac{\chi^0(q, \omega)}{1 - u(q)[1 - G(q, \omega)]\chi^0(\mathbf{q}, \omega)},$$ (2.15)

where χ^0 is the response function of an ideal gas and $G(q, \omega)$ represents a local-field correction. Johnson[9] used the approximation originally due to Singwi et al.,[9] in which a frequency-independent correction $G(q)$ was determined so as to reproduce the structure factor $S(q)$ in a self-consistent way:

$$S(q) = -\frac{\hbar}{\pi n u(q)} \int_0^\infty d\omega \, \mathrm{Im}[\varepsilon^{-1}(q, \omega) - 1],$$ (2.16)

$$G(q) = -\frac{1}{(2\pi)^2 n} \int d\mathbf{q}' \frac{\mathbf{q}}{|\mathbf{q}|} \cdot \frac{\mathbf{q}'}{|\mathbf{q}'|}[S(\mathbf{q} - \mathbf{q}') - 1].$$ (2.17)

In this approach, $\varepsilon(q, \omega)$ is evaluated from Eq. (2.15) for a given $G(q, \omega)$. The result used to determine $S(q)$ in accordance with Eq. (2.16) and Eq. (2.17) should yield the same $G(q)$ as that given by the original $G(q, \omega)$. One way of determining $G(q)$ is to use the RPA structure factor. Jonson showed that $G(q)$ is then given approximately by

$$G(q) = \frac{1}{2} \frac{q}{(q^2 + k_F^2)^{1/2}},$$ (2.18)

where k_F is the Fermi wave number. As we see, $G(q)$ is small for small q but becomes significant as q increases. That is, in strongly interacting

[9]M. Jonson, J. Phys. C9, 3055 (1976); K. S. Singwi, M. P. Tosi, R. H. Land, and A. Sjölander, Phys. Rev. 176, 589 (1968); S. Nagano and K. S. Singwi, Phys. Rev. B28, 6286 (1983).

cases the role played by such a correction is important. Of course, the function G may depend on ω. For a repulsive δ-function potential, Nagano and Singwi evaluated such a correction function numerically.[8] A self-consistent approach to the dynamical density–density correlation function will be discussed in Section 12.

3. PLASMON DISPERSION

One of the most characteristic differences between 2-D and 3-D electrons appears in their plasmon dispersion relations. Such relations are obtained by equating the dielectric function to zero:

$$\varepsilon(q, \omega) = 0. \tag{3.1}$$

This determines the *plasmon* frequency ω_p as a function of the *wave number q*.

The dielectric function depends on the geometry and other conditions of a given system. For a classical system Caplik[10] and Nakayama[10] derived a formula

$$\omega_p^2 = \frac{ne^2}{m} \frac{q}{\kappa_s + \kappa_{ox} \coth(qd)}, \tag{3.2}$$

where κ_s and κ_{ox} are the dielectric constants of semiconductor and oxide, respectively, and d is the thickness of the oxide layer.

The above plasmon frequency vanishes in the long wavelength limit. This is an important 2-D characteristic. In fact, in the long wavelength limit, the 3-D plasmon frequency approaches a constant given by $(4\pi ne^2/m)^{1/2}$. In 1977 the plasmon frequency for inversion layers was confirmed by Allen et al.[11] at a fixed wave vector as a function of electron density. The position, width, and strength of the resonance were found to agree with theoretical expectations for densities above 10^{12} cm^{-2}. At lower densities, the resonance position was found to be below the predicted value, implying an increase in the effective mass. In 1988, Theis et al.[12] determined a plasmon dispersion relation in the inversion layer of

[10]A. V. Chaplik, *Zh. Eksp. i. Teo. Fiz.* **62,** 746 (1972) [Sov. Phys. *JETP* **35,** 395 1972)]; M. Nakayama, *J. Phys. Soc. Jpn.* **36,** 393 (1974).

[11]S. J. Allen, Jr., D. C. Tsui, and R. A. Logan, *Phys. Rev. Lett.* **38,** 980 (1977). See also D. Heitman, *Surf. Sci.* **170,** 332 (1986).

[12]T. N. Theis, J. P. Kotthous, and P. J. Stiles, *Solid State Commun.* **26,** 803 (1978); T. N. Theis, *Surf. Sci.* **98,** 515 (1980).

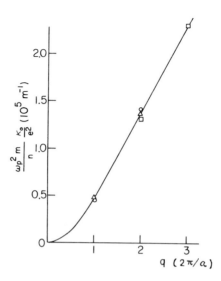

FIG. 8. Experimental plasmon dispersion. From T. N. Theis, J. P. Kotthous, and P. J. Stiles, *Solid State Comm.* **26,** 803 (1978).

[100] p-Si as illustrated in Fig. 8. Here, the ordinate represents the resonance value $(\omega_p^2 m/n)(\kappa_0/e^2)$ in 10^5 m^{-1}, where κ_0 is the vacuum dielectric constant, m is the effective mass, and the abscissa is the wave number in units of $2\pi/a$ with $a = 5\,\mu$m. The width $d = 0.224\,\mu$m, $m = 0.2m_0$, $\kappa_s = 11.4\kappa_0$, and $\kappa_{ox} = 3.8\kappa_0$. The effective mass is determined from cyclotron resonance. The solid curve represents Eq. (3.2). The agreement between the theory and experiment is satisfactory.

For small q, Eq. (3.2) yields[13]

$$\omega_p^2 = \frac{ne^2}{m\kappa_{ox}}\, q^2 d\left\{1 - qd\left(\frac{\kappa_s}{\kappa_{ox}}\right) + (qd)^2\left[\left(\frac{\kappa_s}{\kappa_{ox}}\right)^2 - \frac{1}{3}\right]\right\}. \qquad (3.4)$$

Note that the second term is proportional to qd and has a minus sign. On the other hand, the plasmon dispersion of a purely 2-D degenerate electron gas is given by

$$\omega_p = \left(\frac{2\pi ne^2}{m}\, q\right)^{1/2}[1 + \tfrac{3}{8}a_0 q], \qquad (3.5)$$

[13]A. Isihara and S. V. Godoy, *Kinam* A3, 25 (1981).

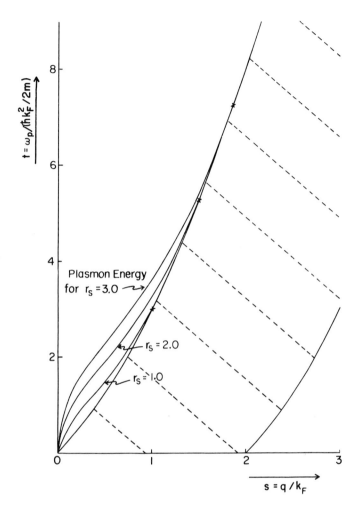

FIG. 9. Quantum plasmon dispersion.

where a_0 is the Bohr radius. Exchange and correlation corrections to this relation have been obtained.[14]

Figure 9 illustrates three plasmon dispersion curves based on the RPA dielectric function for degenerate electrons. At $r_s = 3$, the top curve is strongest in Coulomb interaction. As r_s decreases, or when the electron density increases, the initial convexity is reduced and the curves approach

[14]D. E. Beck and P. Kumar, *Phys. Rev.* **B13**, 2859 (1976); A. K. Rajagopal, *Phys. Rev.* **B15**, 4264 (1977); M. Jonson, *J. Phys.* **C9**, 3055 (1976); A. Gold, *Z. Phys.* **B63**, 1 (1986).

the classical curve, which is concave upward.[14] The shaded area in the graph represents particle–hole creation. The scattering process of an electron from momentum $\hbar\mathbf{k}$ to $\hbar(\mathbf{k} + \mathbf{q})$ is equivalent to a creation of a particle–hole pair with total momentum $\hbar\mathbf{q}$. The excitation frequency is determined from $\omega = \hbar[2\mathbf{k} \cdot \mathbf{q} + q^2]/2m$. This spectrum forms a continuum that lies in a region determined by the conditions:

$$0 \le \omega \le \hbar(2k_Fq + q^2)/2m$$

$$\hbar(-2k_Fq + q^2)/2m < \omega < \hbar(2k_Fq + q^2)/2m.$$

The area with dashed lines represents this region.

The 2-D character of the plasmon dispersion relation has also been observed for electrons on liquid helium by Grimes and Adams.[15] They used standing-wave resonances in a small cell containing helium and electrons. The cell was assembled from metal plates that were electrically isolated from each other. By giving a positive bias potential of a few volts to the bottom submerged electrode, electrons could easily be held onto the surface of liquid helium, which half filled the cell. The finite size of the cell causes a small modification of the dispersion relation as follows:

$$\omega_p^2 = \frac{2\pi n e^2}{m} k_i F(k_i)$$

$$F(k_i) = 2 \sinh(k_i d) \sinh[k_i(h - d)]/\sinh(k_i h),$$

(3.6)

where

$$k_i^2 = k_x^2 + k_y^2; \qquad k_x = \frac{m\pi}{W}; \qquad k_y = \frac{n\pi}{L}.$$

Here, W, L, and h are the width, length, and height of the cell, respectively; d is the helium depth; and m and n are integers. Equation (3.6) shows that ω_p is characterized by a set of integers m, n. Figure 10 illustrates their results, in which the observed standing-wave resonance frequencies are plotted against the charging potential. The electron density n is proportional to this voltage: $n = 8.6 \times 10^{10} V_0$ in accordance with V_0 in volts. Their data points are right on the theoretical lines, both in density and wave vector dependences, confirming the 2-D character of the dispersion relation.

[15]C. C. Grimes and G. Adams, *Phys. Rev. Lett.* **36**, 145 (1976); *Surf. Sci.* **58**, 292 (1976).

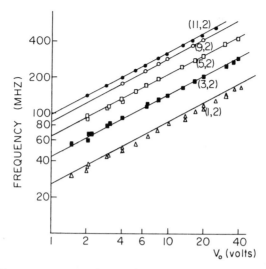

FIG. 10. Standing-wave resonance frequencies vs. charging potential V_0 for $n = 2$, odd m and $d/h = 0.375$. From C. C. Grimes and G. Adams, *Phys. Rev. Lett.* **36,** 145 (1976).

Plasma oscillations of electrons in layered structures have been investigated theoretically since around 1973.[16] For instance, concerning superlattices, Giuliani and Quinn[16] discussed that the surface plasmons are free from Landau damping because quantization of electron motion along the superlattice axis prevents their decay into electron–hole pairs. However, Pinczuk *et al.*[17] reported that they did not observe surface plasmons of the type discussed by Giuliani and Quinn, presumably because of the too-thick depletion layer.

Magnetoplasmon oscillations have also been studied theoretically since around 1974.[18] In the presence of a perpendicular magnetic field, the

[16]D. Grecu, *Phys. Rev.* **B8,** 1958 (1973); A. L. Fetter, *Ann. Phys.* (N.Y.) **88,** 1 (1974); M. Apostol, *Z. Phys.* **B22,** 13 (1975); O. D. Roberts, *Z. Phys.* **B22,** 21 (1975); G. H. Shmelov, I. A. Chaikovskii, V. V. Pavlovich, and E. M. Epstein, *Phys. Stat. Solidi* **B80,** 697 (1977); Das Sarma and J. J. Quinn, *Phys. Rev.* **B25,** 7603 (1982); G. F. Giuliani and J. J. Quinn, *Phys. Rev. Lett.* **51,** 919 (1983); J. W. Wu, P. Hawrylak, and J. J. Quinn, *Phys. Rev. Lett.* **55,** 879 (1985); J. K. Jain and P. B. Allen, *Phys. Rev. Lett.* **54,** 2437 (1985); R. F. Wallis and J. J. Quinn, *Phys. Rev.* **B38,** 4205 (1988-II); P. Hawrylak, *Phys. Rev. Lett.* **59,** 485 (1987).

[17]P. Pinczuk, M. G. Lamont, and A. C. Gossard, *Phys. Rev. Lett.* **56,** 2092 (1986). See also G. Fasol, N. Mestres, H. P. Hughs, A. Fischer, and K. Ploog, *Phys. Rev. Lett.* **56,** 2517 (1986).

[18]M. Nakayama, *Jpn. J. Appl. Phys. Suppl.* **2,** Pt. 2, 901 (1974); K. W. Chiu and J. J. Quinn, *Phys. Rev.* **B9,** 4724 (1974); M. Kobayashi, J. Mizuno, and I. Yokota, *J. Phys. Soc. Jpn.* **39,** 18 (1975); T. K. Lee and J. J. Quinn, *Phys. Rev.* **B11,** 2144 (1975); N. J. M. Horing and M. M. Yildiz, *Ann. Phys.* **97,** 216 (1976); N. J. M. Horing, G. Fiorenza, and H. Chui, *Phys. Rev.* **B31,** 6349 (1985).

magnetoplasmon frequency is expected to shift. The quantum correction, which is proportional to q as in Eq. (3.4), causes an interaction of the plasmon resonance with harmonics $n\omega_c$, where $n = 2, 3, 4, \ldots$, as observed by Batke et al.[19] They found also that in the absence of a magnetic field, the increase in the plasmon frequency due to the q term is canceled by the effect of the finite spatial extent.

4. GROUND-STATE ENERGY AND SPECIFIC HEAT

The ground-state energy per electron of a 2-D electron system can be expressed by

$$\varepsilon_g = \frac{1}{r_s^2} + \varepsilon_x + \varepsilon_c, \qquad \text{(Rydbergs per electron)} \qquad (4.1)$$

where the first term is the *kinetic energy*, the second term ε_x is the *exchange energy*, and the last term ε_c is called the *correlation energy*, which represents the difference between the true ground-state energy and the one evaluated in the Hartree–Fock approximation. The correlation energy is very important because it is negative and contributes to the binding of the electrons. Its high-density series has been obtained by Isihara and Toyoda[20] and Rajagopal and Kimball.[20] The exact treatments of the second-order exchange energy[21] and ring energy[22] have been given by Isihara and Ioriatti.[21,22] Considerations of ladder diagrams, which are important for low densities, have been made by Isihara and Um,[23] Freeman,[24] and Nagano et al.[24] The correlation energy has also been evaluated numerically by Jonson[9] and Ceperley.[25]

a. Exchange Energy

The exchange energy can be evaluated in several different ways. It was first evaluated by Stern.[26] Isihara and Toyoda[27] obtained Stern's result

[19]E. Batke, D. Heitmann, E. Gornik, G. Strangle, A. C. Gossard, and W. Wiegmann, *Surf. Sci.* **113**, 118 (1982); E. Batke, D. Heitmann, and C. W. Wu, *Phys. Rev.* **B34**, 6951 (1986).

[20]A. Isihara and T. Toyoda, *Ann. Phys.* (N.Y.) **106**, 394 (1977); A. Isihara and T. Toyoda, *Ann. Phys.* **114**, 497 (1978); A. K. Rajagopal and J. C. Kimball, *Phys. Rev.* **B15**, 2819 (1977).

[21]A. Isihara and L. Ioriatti, Jr., *Phys. Rev.* **B22**, 214 (1980).

[22]L. C. Ioriatti, Jr. and A. Isihara, *Z. Phys.* **B44**, 1 (1981).

[23]C. I. Um and A. Isihara, *J. Kor. Phys. Soc.* **11**, 60 (1978).

[24]D. L. Freeman, *J. Phys.* **C16**, 711 (1983); S. Nagano, K. S. Singwi, and S. Ohnishi, *Phys. Rev.* **B29**, 1209 (1984).

[25]D. M. Ceperley, *Phys. Rev.* **B18**, 3126 (1978).

[26]F. Stern, *Phys. Rev. Lett.* **30**, 278 (1973).

[27]A. Isihara and T. Toyoda, *Z. Phys.* **B23**, 389 (1978).

with a different method, and Stern[28] and Vinter[29] investigated a thickness effect on this energy. It can also be obtained by a method that uses a sum rule and the electrostatic analogue that was discussed in Section 2. A similar method can be adopted effectively for the correlation energy.

For low but finite temperatures, the exchange energy cannot be evaluated by a simple application of the Sommerfeld method, because the coefficients of the expansion give rise to increasingly divergent integrals. An exact evaluation was made by Isihara and Ioriatti[30] based on a Mellin transform and the electrostatic analogue discussed in Section 2. Their result is expressed by

$$\varepsilon_x = -\frac{8\sqrt{2}}{3\pi r_s} - (\tfrac{1}{6}\ln \eta_0 + 4c - \tfrac{1}{6})\frac{\pi\sqrt{2}}{\eta_0^2 r_s}, \tag{4.2}$$

where $c = 0.02553$, $\eta_0 = \beta\varepsilon_F$, $\beta = 1/kT$, and $\varepsilon_F = (\hbar k_F)^2/2m$ is the ideal Fermi energy, with k_F being $(2\pi n)^{1/2}$.

By combining this result with the ideal gas energy, we find the electronic specific heat to be given by[30]

$$c_A = c_A^0[1 + r_s(0.1977 - 0.2251 \ln \eta_0)], \quad (2\text{-D}), \tag{4.3}$$

where $c_A^0 = \pi k^2 T/6$ is the ideal specific heat per area. Due to the term $\ln \eta_0$ the ratio c_A/c_A^0 becomes larger and larger as the temperature decreases towards absolute zero—although the specific heat itself approaches zero. However, as in the 3-D case, it is quite likely that the sum of all such terms form a convergent series. The 3-D specific heat corresponding to the above expression is

$$c_v = c_v^0[1 + r_s(0.1617 - 0.1659 \ln \eta_0)]. \tag{4.4}$$

Comparing these two expression, we find that the interaction effect is larger in 2-D than in 3-D.

b. Ring Energy

The correlation energy is usually evaluated approximately in consideration of the ring diagrams and second-order exchange diagrams. The ring-diagram contribution to the grand partition function is of order e^4,

[28]F. Stern, *Phys. Rev. Lett.* **35**, 598 (1975).
[29]B. Vinter, *Phys. Rev. Lett.* **35**, 1044 (1975).
[30]A. Isihara and L. C. Ioriatti, Jr., *Physica* A**103**, 621 (1980); L. C. Ioriatti, Jr. and A. Isihara, *Phys. Stat. Sol.* (b)**97**, K65 (1980).

and is given exactly by the following integral:

$$
\varepsilon_r = \frac{8}{\pi} \int_0^\pi d\phi (\cos\phi - \cos 2\phi)(1 - \cos\phi)\sin\phi \left\{ \left[\frac{\pi}{2} - xF(x) \right] I_5(\phi) \right.
$$
$$
\left. - \left[\frac{\pi}{4} - x + \frac{\pi}{2}x^2 - x^3 F(x) \right] I_3(\phi) \right\}.
\tag{4.5}
$$

This formula is based on the electrostatic analogue discussed in Section 2. Here, the function $F(x)$ is defined by the integral:

$$
F(x) = \int_0^\infty \frac{dt}{1 + x\cosh t}
\tag{4.6}
$$

and

$$
I_3 = -\frac{\cos\phi}{2\sin^2\phi} + \frac{1}{2}\ln\left(\tan\frac{\phi}{2}\right); \qquad I_5 = -\frac{\cos\phi}{4\sin^4\phi} + \frac{3}{8}\left[\ln\left(\tan\frac{\phi}{2}\right) - \frac{\cos\phi}{\sin^2\phi}\right],
$$

where x is a variable given by

$$
x = \frac{r_s}{2^{1/2}}\sin\phi(1 - \cos\phi).
\tag{4.7}
$$

It is very important to note that the integral $F(x)$ takes different forms in accordance with $x \le 1$ or $x \ge 1$. This difference corresponds to high and low densities, respectively, as one can judge from the form of x. We find then that the boundary between these two regions takes place at $r_s = 2^{1/2}$.

For high densities, $F(x)$ can be expanded in powers of x without difficulty. For low densities, a Mellin transform can be used to generate an expansion in $1/x$. Finally, we arrive at the ring energy in the following form:

$$
\varepsilon_r = \begin{cases}
-0.6137 - 0.1726 r_s \ln r_s + 0.8653 r_s + O(r_s^2 \ln r_s), & (r_s \le \sqrt{2}) \\[2ex]
-\dfrac{1.2935}{r_s^{2/3}} + \dfrac{1.2004}{r_s} - \dfrac{0.14018}{r_s^{4/3}} + O\left(\left(\dfrac{\ln r_s}{r_s}\right)^2\right), & (r_s \ge \sqrt{2}).
\end{cases}
\tag{4.8}
$$

Note that the low-density series is convergent for $r_s \to \infty$.

The termperature dependence of the ring energy has been approximately evaluated by Isihara and Toyoda[31] in the following form:

$$\varepsilon_r = \frac{4}{\eta_0^2}\left[-0.1824 - 0.02968\ln\eta_0 + \frac{1}{24\pi}(\ln\eta_0)^2\right]. \tag{4.9}$$

The logarithmic terms diverge at absolute zero. Since the last term is exactly the square of the corresponding exchange term, they might be part of a geometric series.

c. Second-Order Exchange Energy

To order e^4, there is yet another class of diagrams that contribute to the grand-partition function as the ring diagrams do. They are the second-order exchange graphs. These graphs are classified into two groups: anomalous and regular second-order exchange graphs, in accordance with whether or not the two Coulomb interaction lines in a graph cross each other. The exact second-order exchange energy has been obtained by Isihara and Ioriatti[21] as follows:

$$\varepsilon_{2x} = \frac{28.363}{4\pi^3} = 0.2287 \text{ Ryd.}, \tag{4.10}$$

which is constant. This energy is considerably larger than the corresponding 3-D energy of 0.04836 evaluated by Onsager et al.[32]

The temperature dependence of the second-order exchange energy has been determined only approximately. Isihara and Toyoda[31] found that the contribution from the anomalous second-order exchange graphs includes $\ln\eta_0$ and $(\ln\eta_0)^2$.

Summarizing the results in this section, we have obtained the ground state energy in the following form:

$$\varepsilon_g = \frac{1}{r_s^2} - \frac{1.2004}{r_s} + 0.2287 + \varepsilon_r, \tag{4.11}$$

where ε_r is given by Eq. (4.8). In comparison, the corresponding high-density series for 3-D is

$$\varepsilon_g = \frac{2.21}{r_s^2} - \frac{0.916}{r_s} + 0.0622\ln r_s - 0.096 + r_s(0.0049\ln r_s - 0.02) + \ldots . \tag{4.12}$$

[31]A. Isihara and T. Toyoda, *Phys. Rev.* **B21**, 3358 (1980).
[32]L. Onsager, L. Mittag, and M. Stephen, *Ann. Phys.* (N.Y.) **18**, 71 (1966).

Hence, the kinetic energy of 2-D systems is one-half of the 3-D value at the same r_s, whereas the exchange energy is approximately 40% higher and the correlation energy is ten times larger in magnitude. Thus, we can conclude that electron correlations are much stronger in 2-D than in 3-D. Note also that in the low-density series, the exchange energy is exactly canceled out by a ring contribution. This means that for $r_s > \sqrt{2}$, the Hartree–Fock approximation is not valid.

The ring diagram contribution is important in the sense that it is obtained by summing over all the ring diagrams. It is important even in the classical limit in which it produces the Debye–Hückel result. Nevertheless, the true ground-state energy is expected to depend on many other types of graphs. Although the true ground-state energy is difficult to obtain, a lower bound has been given in consideration of a charged Bose gas. This is a fictitious system consisting of bosons that interact with Coulomb forces. Its ground state has been obtained[33] in the form

$$\varepsilon_g^B = -2^{4/3}\pi^2[\Gamma(1/3)]^{-3}r_s^{-2/3} = -1.2918r_s^{-2/3} \quad \text{(2-D)}. \quad (4.13)$$

The corresponding value for the 3-D case is[34]

$$\varepsilon_g^B = -0.803r_s^{-3/4} \quad \text{(3-D)}. \quad (4.14)$$

Note that the 2-D energy is lower.

Figure 11 illustrates the ground-state energy of a 2-D electron system, which has been obtained by including the ladder diagram contribution. For comparison, a corresponding 3-D curve is also given. As we see, the 2-D energy becomes negative at $r_s = 0.67$ and reaches a minimum of -0.55 Ryd. at around $r_s = 1.6$. In contrast, the 3-D curve becomes negative more slowly. Its minimum is not shown because it appears at around $r_s = 5$, with the minimum value being -0.15 Ryd., which is much higher than in the 2-D case. Therefore, we can conclude again that electron correlations are much stronger in 2-D.

The correlation energy has been evaluated by several authors in various ways with some differences in their results, particularly for three dimensions.[35] Nevertheless, apart from minor details, there is general

[33]C. I. Um, A. Isihara, and S. T. Choh, *J. Kor. Phys. Soc.* **14**, 39 (1981).

[34]J. L. Foldy, *Phys. Rev.* **124**, 649 (1961); *Phys. Rev.* **125**, 2208 (1962); T. Samulski and A. Isihara, *Physica A82*, 294 (1976).

[35]A. P. Isihara and E. W. Montroll, *Proc. Acad. Sci.* **68**, 3111, (1971); A. Isihara and D. V. Kojima, *Z. Phys.* *B21*, 33, (1975); *B25*, 167 (1979); D. M. Ceperley and B. J. Alder, *Phys. Rev. Lett.* **45**, 566 (1980).

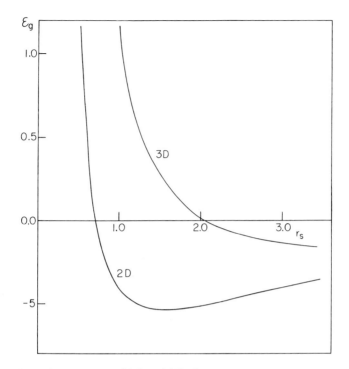

FIG. 11. Ground-state energy of 2-D and 3-D electron systems.

agreement. For example, Freeman's[24] direct and exchange energies in the limit $r_s \rightarrow 0$ are off from the exact values, but his final results for large r_s are in good agreement with those given by Jonson.[9]

II. Phase Transitions

5. VALLEY OCCUPANCY PHASE TRANSITION

The equal energy surfaces of the conduction band of Si consist of equivalent ellipsoids of evolution in the six crystal directions, as illustrated in Fig. 12. The ground state of the inversion layer electrons in the [100] direction of Si is represented by the two overlapping circles obtained by projecting the two ellipsoids onto the [100] plane. These represent *valleys* in which the electrons nest. As in the case of spin, these valleys are completely equivalent, so that a valley degeneracy factor of 2 is necessary for the ground state. Indeed, if the kinetic energy dominates,

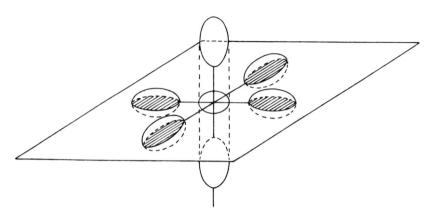

FIG. 12. Equal energy surfaces of the conduction band of electrons in Si inversion layers.

an equal population of the valleys is energetically favorable. However, at low densities, Bloss *et al.*[36] discussed how intravalley exchange and correlations cause a first-order phase transition into a one-valley state at around $(2.3-2.9) \times 10^{11} \, \text{cm}^{-2}$, depending on the density of electrons in the depletion layer that is adjacent to the inversion layer.

An experimental observation of the density dependence of the valley degeneracy of inversion layers on [100] silicon was performed by Cole *et al.*[37] They used a far-infrared Fourier transform spectrometer with light coupled into the inversion layers via a 45° Si prism and measured the intersubband transition energies as a function of electron density. Of the two samples of different mobility they studied, Sample 2 was found to show a peak effective mobility at 4.2 K and pronounced changes in the slopes of transition energies for both $0 \to 1$ and $0 \to 2$ transitions at around $5 \times 10^{11} \, \text{cm}^{-2}$. However, the energy values were continuous. At low densities, the transition energies were found to be higher than those that would result by extrapolating from high densities. That is, they observed a change in slope and shift to higher energies of the intersubband transition energies as a function of carrier density at low electron density. Sample 1 was found to show a similar behavior, although the change in slope was not as pronounced, presumably because of its lower mobility. The conductivity of Sample 2 showed thermally activated behavior at densities below $2 \times 10^{11} \, \text{cm}^{-2}$. The activation energy was around 0.3 meV at $8 \times 10 \, \text{cm}^{-2}$. In the density range between 2.5 and $5 \times 10^{11} \, \text{cm}^{-2}$, the conductivity was not constant but increased slightly

[36]W. J. Bloss, L. J. Sham, and B. Vinter, *Phys. Rev. Lett.* **43**, 1529 (1979).
[37]T. Cole, B. D. McCombe, J. J. Quinn, and R. K. Kalia, *Phys. Rev. Lett.* **46**, 1096 (1981).

with decreasing temperature, as was observed also by Chang and Wheeler.[38]

Neglecting intervalley exchanges, the valley occupancy transition at a low temperature has been determined by Isihara and Ioriatti[39] from the difference in the ground-state energies of the single- and double-valley states. This energy difference can be found by noting that the Fermi energy depends on the valley degeneracy factor g_v.

For low densities, one can make use of the low-density series given by Eq. (4.8) of Section 4. We find that the first two terms are canceled out, indicating that the transition is not due to the exchange energy but rather to the correlation energy. The difference in the ground-state energies between the two states is given by

$$\frac{\Delta \varepsilon}{\varepsilon_0} = 1 - 0.14 r_s^{2/3} + \ldots , \qquad (5.1)$$

where ε_0 is the kinetic energy of the double-valley, equal-occupancy state. This equation shows clearly that the energy difference decreases as r_s increases. A more accurate result, obtained numerically, is illustrated in Fig. 13. The ordinate represents the energy for the single-valley ground state minus that for the double-valley state. The differnce is positive for small r_s, in agreement with the above equation, and vanishes at $r_s = 8.011$. By assuming an average dielectric constant of 7.8 and an effective mass of $0.2m_0$, we find a critical density to be $1.2 \times 10^{11}\,\mathrm{cm}^{-2}$. This critical density is somewhat smaller than the experimental value of around $5 \times 10^{11}\,\mathrm{cm}^{-2}$. However, it is the result of a first-principle calculation for an idealized 2-D model without any artificial adjustment. If a finite thickness is introduced, the Coulomb interaction will be more rapidly weakened for large momenta, resulting in the shift of the energy curve toward a smaller critical r_s.

It is interesting to observe that under essentially the same mechanism, a ferromagnetic phase transition can be expected.[39] The critical density was estimated to be at $r_s = 13$. For a Si inversion layer, this translates into $4.5 \times 10^{10}\,\mathrm{cm}^{-2}$.

In directions other than [100], the theoretical degeneracy is $g_v = 4$ in [110] and $g_v = 6$ in [111]. Experimentally, however, it has been found that sample preparation may cause inhomogeneity, stress, or interface degradation, which in turn results in unequal valley occupancies.[40] This also

[38]K. M. Chang and R. G. Wheeler, *Phys. Rev. Lett.* **44**, 1472 (1980).
[39]A. Isihara and L. C. Ioriatti, Jr., *Phys. Rev.* B**25**, 5534 (1982).
[40]T. Cole and B. D. McCombe, *Phys. Rev.* B**29**, 3180 (1984).

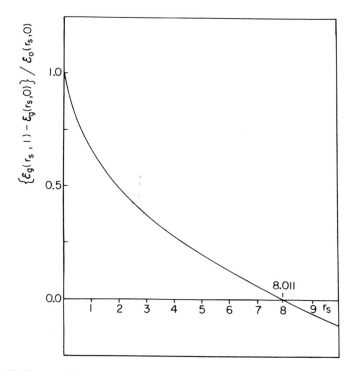

FIG. 13. Energy difference between the single- and double-valley states of Si [100] inversion layers as a function of r_s. A. Isihara and L. C. Ioriatti, Jr., *Phys. Rev.* **B25**, 5534 (1982). The difference vanishes at $r_s = 8.011$; thereafter a single-valley occupancy is preferred.

seems to be the case theoretically; a consideration of lattice distortion and random surface strains favored $g_v = 2$ in the [111] direction.[41] Indeed, $g_v = 4$ for the [110] direction of Si and $g_v = 6$ for the [111] direction have actually been observed.[42]

6. GAS AND LIQUID PHASES

The electron systems introduced in Section 1 differ from each other in detail and require separate theoretical approaches. Nevertheless, they share the basic 2-D character. Therefore, their characterization in terms of an ideal 2-D electron gas becomes fundamentally important. This is

[41]W. L. Bloss, S. C. Ying, and J. J. Quinn, *Phys. Rev.* **B23**, 1839 (1981).
[42]K. C. Woo and P. J. Stiles, *Surf. Sci.* **113**, 515 (1982); D. C. Tsui and G. Kaminsky, *Solid State Commun.* **20**, 93 (1976).

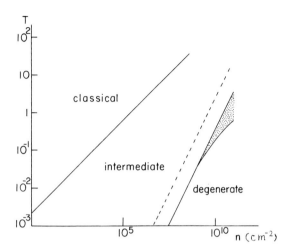

FIG. 14. Classification of 2-D electron systems.

achieved by using various dimensionless combinations of the electron's number density n, temperature T, and charge e.

First, the ratio of the average Coulomb energy to the thermal energy is an important theoretical parameter. Since the average distance between the electrons is given by $(\pi n)^{-1/2}$, this ratio is given by

$$\Gamma = \pi^{1/2} e^2 n^{1/2} / kT. \tag{6.1}$$

If Γ is less than 1, that is, above the top straight line in Fig. 14, the system is classical below 10^5 K. There is another slightly more sophisticated parameter. For nearly classical systems, the so-called plasma parameter plays a natural role. This is defined by the ratio of the Coulomb energy at the Debye length against the thermal energy:

$$\varepsilon = e^2 k_D / kT, \qquad k_D = 2\pi n e^2 / kT. \tag{6.2}$$

Second, at low temperatures and high densities, the system can be quantum mechanically degenerate. The criterion for this region is given by the relation equating the thermal energy to the Fermi energy:

$$kT = \hbar^2 k_F^2 / 2m. \tag{6.3}$$

The bottom straight line represents this relation. However, if Coulomb interaction exists, the line curves down. The shaded area represents the depletion of the quantum region due to Coulomb interaction.[43]

[43]A. Isihara, *Surf. Sci.* **98**, 31 (1980).

Note that in actual systems in semiconductors, the bare electron mass and charge must be replaced by the effective quantities. For example, the effective mass in Si inversion layers is $0.19m_0$ for motion parallel to the interface, and the effective charge is given by e^2/κ, where $\kappa = 11.5$ is the average dielectric constant. Hence, the effective Bohr radius becomes

$$a_0^* = \hbar^2 \kappa / me^2 = 32.57 \times 10^{-8}\,\mathrm{cm}.$$

This is approximately 61 times the true Bohr radius, which is $0.529 \times 10^{-8}\,\mathrm{cm}$, causing a smaller effective r_s. For a typical density of $n = 10^{12}/\mathrm{cm}^2$, $r_s = 1.73$, and the corresponding Fermi energy is 12.6 meV. Also, in the case of Si inversion layers, it is known that the valley degeneracy must be taken into consideration in definining the effective Fermi energy. This depends on the crystal directions. The valley degeneracy g_v for Si inversion layers in the [100] direction of Si is 2. In the [110] and [111] directions, it is 4 and 6, respectively. Thus, the Fermi energy is defined by

$$\hbar^2 k_F^2 / 2m = \hbar^2 (2\pi n / g_v) / 2m. \tag{6.4}$$

The dashed line in Fig. 14 indicates the direction of the shifting of an actual case from the ideal case.

For very low temperatures and for small r_s, a given electron system can be treated by considering Coulomb interaction as a small perturbation. This can be seen by expressing the Hamiltonian in the Rydberg unit as follows:

$$\mathscr{H} = \frac{me^4}{\hbar^2} \frac{1}{r_s^2} \left[\sum_i \frac{p_i'^2}{2} + r_s \sum_{(ij)} \frac{1}{r_{ij}'} \right]. \tag{6.5}$$

In the opposite case of low density at high temperature, the thermodynamic quantities may be obtained in powers of ε or Γ when these parameters are small. However, the vast intermediate region in Fig. 14 is very difficult to treat theoretically, and, in fact, no rigorous analytical theory is available.

Of the two dimensionless parameters, ε bears a many-body character through the Debye constant. It can be shown that the classical Debye–Hückel theory is correct to $O(\varepsilon)$. Improvements on the Deybe–Hückel results can be made by considering higher-order terms in ε. For example, to order ε^2 in 3-D, the internal energy was evaluated by Abe[44] and the pair distribution function by Isihara and Wadati.[45] By using the pair

[44]R. Abe, *Prog. Theo. Phys.* **22**, 213 (1959).
[45]A. Isihara and M. Wadati, *Phys. Rev.* **183**, 312 (1969).

distribution function, one can derive the energy to the next order. For 2-D, Chalupa[46] derived the equation of state as follows:

$$\frac{p}{nkT} - 1 = \tfrac{1}{4}\varepsilon \ln(2\varepsilon) + \tfrac{1}{2}(\gamma - \tfrac{1}{2})\varepsilon - \tfrac{1}{4}\varepsilon^2 \ln^2\left(\frac{\varepsilon}{2}\right) + (\tfrac{1}{2} - \gamma)\varepsilon^2 \ln(2\varepsilon) + O(\varepsilon^2),$$

(6.6)

where $\gamma = 0.5772$ is Euler's constant. However, his second-order term does not seem to be correct in view of Totsuji's later result.[47]

$$\frac{p}{nkT} - 1 = \tfrac{1}{4}\varepsilon \ln(2\varepsilon) + \tfrac{1}{2}(\gamma - \tfrac{1}{2})\varepsilon - \tfrac{1}{4}\varepsilon^2 \ln^2\left(\frac{\varepsilon}{2}\right) + (\tfrac{1}{2} - \gamma)\varepsilon^2 \ln \varepsilon + O(\varepsilon^2).$$

(6.7)

In any case, quantum corrections become increasingly important for higher-order terms.

For classical systems, the hypernetted-chain integral equation and Monte Carlo simulations have been used extensively. Efforts have also been made to combine these results with the the lattice energy, which for the triangular lattice is

$$\frac{U}{nkT} = -1.106\Gamma.$$

(6.8)

For example, Gann et al.[48] adopted Hansen's form[49] for 3-D:

$$\frac{U^{\text{ex}}}{nkT} = \Gamma^2\left[\frac{a_1}{a_2 + \Gamma} + \frac{a_3}{(a_4 + \Gamma)^2} + \frac{a_5}{(a_6 + \Gamma)^3} + \frac{a_7}{(a_8 + \Gamma)^4}\right].$$

(6.9)

Note that in the limit $\Gamma \to \infty$, U^{ex} is proportional to Γ as in Eq. (6.8).

Although our interest in this chapter is in quasi-two-dimensional systems with the ordinary Coulomb potential, there are many articles dealing with the logarithmic potential that characterizes true 2-D. For instance, Caillol and Levesque[50] used Monte Carlo simulations to find a

[46]J. Chalupa, Phys. Rev. B12, 4 (1975); Phys. Rev. B13, 2243 (1976) [Erratum].

[47]H. Totsuji, J. Phys. Soc. Jpn. 40, 857 (1976); Phys. Rev. A19, 889 (1979).

[48]R. C. Gann, S. Chakravarty, and G. V. Chester, Phys. Rev. B20, 326 (1979). See also N. Itoh and S. Ichimaru, Phys. Rev. A22, 1318 (1980); F. Lado, Phys. Rev. B17, 2827 (1978).

[49]J. P. Hansen, Phys. Rev. A8, 3110 (1973).

[50]J. M. Caillol and D. Levesque, Phys. Rev. B33, 499 (1986). See also, C. Deutch and M. Lavaus, Phys. Lett. A40, 349 (1974); R. Calinon et al., Phys. Rev. A20, 329 (1979); P. Bakashi et al., Phys. Rev. 20, 336 (1979).

Kosterlitz–Thouless transition at low densities and a first-order transition between a dielectric gas and a conduction liquid at moderate densities.

Near absolute zero and at high densities, the system is of course in a gaseous state. As the density decreases, Coulomb interaction becomes increasingly important. Although it is difficult to determine where the system becomes liquidlike, Fig. 11 shows that the ground-state energy becomes negative at around $r_s = 0.7$, and the system is more strongly correlated when r_s increases beyond this value. After passing through the minimum, the energy curve approaches zero, as it should. For this to take place smoothly, however, the curve must change its curvature. At this point, which depends on theoretical approximations, the isothermal compressiblility diverges, if evaluated formally. This can be seen from the bulk modulus obtained from.

$$B = \frac{1}{4\pi} r_s \frac{d}{dr_s} \left(\frac{1}{r_s} \frac{d\varepsilon_g}{dr_s} \right). \tag{6.10}$$

If the results in Section 4 are used, the modulus is expressed by[51]

$$B = \frac{2}{\pi^2 r_s} - \frac{2\sqrt{2}}{\pi^2 r_s^3} + B_r, \tag{6.11}$$

where the first and second terms are due to the kinetic and exchange energies, and the last term B_r, due to the ring contribution, is given by

$$B_r = \frac{\sqrt{2}}{\pi r_s^2} \int_0^\pi \frac{(1 - \cos \phi) \sin^2 \phi}{(1 + 2 \cos \phi)^2} [2 \cos \phi + 2 \cos^2 \phi - 1] F(x) \, d\phi, \tag{6.12}$$

where $F(x)$ has been defined in Section 4.

The inverse of the modulus yields the isothermal compressibility κ_T. It has been found that the compressibility is small when r_s is small, indicating that, even though the system is gaseous, its compression requires energy because at absolute zero the electrons cannot make transitions into new states corresponding to new r_s unless the Fermi distribution is changed. As r_s increases, the Coulomb interaction softens the quantum degeneracy of an ideal gas, so that the compressibility becomes larger. Increasing r_s means increasing volume, but at the same time this results in a stronger Coulomb interaction. The compressibility evaluated from the above formula diverges at $r_s = 1.989$, which is rather

---

[51]A. Isihara and L. C. Ioriatti, Jr., Solid State Commun. **41**, 867 (1982).

small and close to the minimum point of the energy curve. Beyond this divergent point, the compressibility becomes negative, although the magnitude becomes increasingly small. This negativeness is due to the neglect of background positive charges. When changing the volume of the system, the role played by the positive background charges must be explicitly taken into consideration.

Although electronic properties differ at high and low densities and the terms of electron gas and liquid are commonly used, the boundary is somewhat unclear. The treatment of the correlation energy in Section 4 shows that $r_s = 2^{1/2}$ is a boundary between the high and low density regions. However, the correlation energy is continuous at this point. On the other hand, the boundary between liquid and solid phases is somewhat clearer, as we shall discuss in Section 8.

7. WIGNER LATTICE

In a study of electron interactions in a metal, Wigner[52] pointed out in 1934 and in 1938 the possibility that the electrons may form a lattice at a low density in order to lower the energy. This possibility arises because the high- and low-density behavior of electron systems is somewhat different from that observed in ordinary molecular systems. In their high-density regime, atoms and molecules become localized in a lattice structure whose zero-point vibrations are quite small. In the case of very light electrons at high densities, localization causes large-momentum fluctuations due to the uncertainty principle, leading to a gaseous state. In the opposite limit of low densities, the Coulomb interaction dominates, so that the electrons try to minimize the electrostatic energy by forming a crystal lattice.

Wigner's suggestion was followed by many theoretical studies to find the crystallization point. However, the theoretical estimates varied quite widely due to the difficulties in evaluating the ground state energy for low densities.[53] Experimentally, no Wigner lattice has been found in three dimensions. It has been suggested that in a star that is cooling toward the white-dwarf state, the ions might form a Wigner lattice.[54] Since this phase change is expected to be in the first order, there would be a release of latent heat at the crystallization, which would temporarily decrease the rate of cooling. This would result in crystallization sequences. However, no conclusive evidence has been found.

[52]E. P. Wigner, *Phys. Rev.* **46,** 1002 (934), *Trans. Farad. Soc.* **34,** 678 (1938).
[53]A. Isihara and E. W. Montroll, *Proc. Nat. Acad. Sci.* **68,** 3111 (1971).
[54]M. H. Van Horn, *Astrophys. J.* **151,** 227 (1968).

The search for a Wigner lattice has attracted renewed interest along with the developments in the study of 2-D electron systems whose density can be varied widely. The possibility of crystallization was suggested by Crandall and Williams[55] for electrons on the surface of liquid helium and by Chaplik[56] for inversion layers.

It is very difficult to treat the crystallization of electrons, and, in fact, no rigorous approach has been developed. However, it is rather straightforward to calculate thermodynamic quantities for a given crystal structure.

The evaluation of the ground-state energy was made by Meissner et $al.$[57] for triangular and square lattices and by Bonsall and Maradudin[58] for arbitrary lattices. The ground-state energy of an electron crystal is obtained by performing a lattice sum. Since electrons repel each other, it is customary to assume a cloud of background positive charges for charge neutrality and to avoid divergence in such an evaluation.

Let us assume that electrons form a two-dimensional triangular lattice. It turns out that this lattice structure gives the lowest energy. When a corner of the unit triangle of side length a is placed at the origin of a rectangular coordinate system, all the lattice points can be generated by the two unit vectors $a(1, 0)$ and $a(1/2, \sqrt{3}/2)$. The electrons can vibrate about their lattice points. In the harmonic approximation for absolute zero, their kinetic and potnetial energies, expanded about equilibrium, can be replaced by the zero-point vibrational energy. The total energy is then given by

$$\varepsilon_g = -\frac{2.21}{r_s} + \frac{1.63}{r_s^{3/2}}, \qquad (7.1)$$

where the second term represents the vibrational energy.

Near absolute zero, the lattice-specific heat of the electron lattice is expected to be proportional to T^2. The proportionality constant has been obtained by Bonsall and Maradudin. For instance, for a square lattice

$$c_A = \frac{3k}{(0.1905)^2\pi} \left(\frac{kT}{\hbar\omega_p}\right)^2 \zeta(3), \qquad (7.2)$$

where $\zeta(3)$ is the Riemann zeta function, the subscript A stands for "per

[55]R. S. Crandall and R. S. Williams, $Phys.$ $Lett.$ A**34**, 404 (1971); R. S. Crandall, $Phys.$ $Rev.$ A**8**, 2136 (1973).
[56]A. V. Chaplik, $Zh.$ $Eksp.$ $Teo.$ $Fiz.$ **62**, 746 (1972) [$Sov.$ $Phys.$-$JETP$ **35**, 395 (1972)].
[57]G. Meissner, H. Namizawa, and M. Voss, $Phys.$ $Rev.$ B**13**, 1370 (1976).
[58]Lynn Bonsall and A. A. Maradudin, $Phys.$ $Rev.$ B**15**, 1959 (1977).

area," and

$$\omega_p^2 = \frac{2\pi ne^2}{ma}, \tag{7.3}$$

a is the unit length of the Bravais lattice. For the hexagonal lattice, a factor $\sqrt{3}/2$ should be multiplied by the right side of Eq. (7.2).

The normal modes of vibrations have also been evaluated for the longitudinal and transverse modes. These modes obey the following dispersion relations:

$$\omega_l^2 = \omega_p^2 aq - 0.1815(\omega_p aq)^2,$$
$$\omega_t^2 = 0.0363(\omega_p aq)^2. \tag{7.4}$$

As we see, $\omega_l \propto q^{1/2}$ and $\omega_t \propto q$. The former dependence resembles the plasmon dispersion in a degenerate electron gas and the latter is similar to that for the classical case. Note in ω_l that the first term yields

$$\omega_l = \left[\frac{2\pi ne^2}{m} q\right]^{1/2}, \tag{7.5}$$

which is independent of the lattice structure and is equal to the plasmon frequency of an electron gas. The above dispersion relations are obtained for an isolated 2-D electron lattice. On the other hand, the electrons on the surface of liquid helium can couple with the ripplon (surface wave) modes.[59]

Let us first discuss Shikin's modes. In 1974, Shikin[60] suggested that if surface electrons crystallize, a series of resonances due to excitation of standing capillary-waves (ripplons) will appear when the electrons are driven up and down against the He surface by a uniform perpendicular rf electric field. The resonances occur when an integral number of wave-lengths of the capillary-wave matches the spacing between rows of electrons in the crystal and the driving frequency satisfies the ripplon dispersion relation at the wavelengths. For the triangular lattice, the resonance frequency is

$$\nu_i = 2^{5/4}3^{-3/8}\pi^{1/2}(\alpha/\rho)^{1/2}n^{3/4}i^{3/4}, \tag{7.6}$$

[59]L. P. Gor'kov and D. M. Chernikova, *Zh. Eksp. Teo. Fiz. Pis'ma Red.* **18**, 119 (1973) [*JETP Lett.* **18**, 68 (1973)]; K. Mima, H. Ikezi, and A. Hasegawa, *Phys. Rev.* **14**, 3953 (1976).
[60]V. B. Shikin, *Zh. Eksp. Teo. Fiz. Pis'ma Red.* **19**, 647 (1974) [*Sov. Phys.-JETP Lett.* **19**, 647 (1974)].

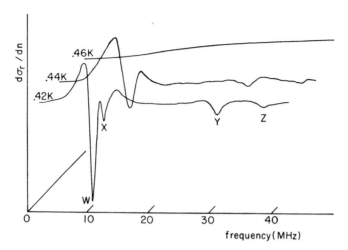

FIG. 15. Disappearance of ripplon–plasmon coupling. C. C. Grimes and G. Adams, *Phys. Rev. Lett.* **42**, 795 (1979).

where α is the surface tension coefficient, ρ is the density of liquid He, and

$$i = k^2 + l^2 + kl,$$

with k and l being zero or integers, so that $i = 1, 3, 4, 7, \ldots$. The integers i appear because the frequency is determined by the reciprocal lattice vector of the electron lattice.

Shikin's modes correspond to purely vertical motion of the electrons in the bare ripplon modes. In 1979 Fisher *et al.*[61] proposed coupled ripplon–plasmon modes in which nonuniform horizontal motions of the electrons take place. The oscillatory motion of the electrons in the coupled modes can have an amplitude of hundreds of Angstroms, in contrast to an Angstrom in Shikin's modes. Hence, the coupled modes give rise to stronger absorption strengths.

In 1979, Grimes and Adams[62] observed such modes, rather than Shikin's modes, thus confirming the existence of a Wigner lattice. Their resonance data are shown in Fig. 15. In their experiment, they used a broadband-swept frequency spectrometer whose output is proportional to the derivative of the radial component of the frequency-dependent conductivity of the electron layer. The ordinate represents this derivative.

[61] D. S. Fisher, B. I. Halperin, and P. M. Platzman, *Phys. Rev. Lett.* **42**, 798 (1979).
[62] C. C. Grimes and G. Adams, *Phys. Rev. Lett.* **42**, 795 (1979).

The electron density is $4.4 \times 10^8/\text{cm}^2$. Resonances are clear for 0.42 K and 0.44 K and disappear at 0.46 K (more precisely, at 0.475 K).

The amplitude of the resonances in Fig. 15 was found to remain almost constant up to a temperature of 0.42 K and to decrease sharply thereafter. Since this decrease is continuous as a function of T, Grimes and Adams determined the melting temperature T_m by extrapolating to zero amplitude of the resonance amplitudes. By repeating the determination of T_m for different values of n, they obtained the melting curve illustrated in Fig. 16. The straight line represents a proportionality to $n^{1/2}$, as expected from the definition of the constant Γ given by Eq. (6.1):

$$\Gamma = \pi^{1/2}e^2n^{1/2}/kT. \tag{7.7}$$

For classical electron systems, this dimensionless parameter characterizes the energy due to Coulomb interaction at finite temperatures. Grimes and Adams found that melting takes place when Γ is equal to

$$\Gamma_m = 131 \pm 7. \tag{7.8}$$

The phase transition was studied for areal electron densities from 3×10^8 to $9 \times 10^8/\text{cm}^{-2}$, with melting temperatures between 0.35 and 0.65 K.

Fisher et al.[61] showed that the resonances observed by Grimes and Adams represent the coupled longitudinal-phonon–ripplon modes. The frequency spectra for these modes were obtained from the secular equation

$$\omega^2 - \omega_l^2(q) - \tfrac{1}{2}\sum_G V_G^2 \frac{\omega^2}{\omega^2 - \Omega^2(G)} = 0, \tag{7.9}$$

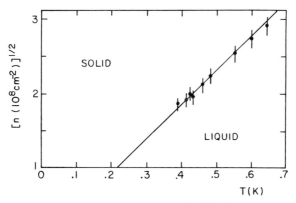

FIG. 16. Melting curve of an electron crystal on liquid helium. C. C. Grimes and G. Adams, *Phys. Rev. Lett.* **42**, 795 (1979).

where **G** is a reciprocal-lattice vector, $\Omega = 2\pi\nu_i$, and V_G, which is the coupling constant between electron and ripplons, depends on the applied vertical electric field and also on the variation in the attraction of the image charge due to curvature of the helium surface. Fisher *et al.* identified the three resonances W, Y, and Z in Fig. 15 as the lowest-wave-vector ($\mathbf{q}_1$) resonances near the ripplon frequencies Ω_1, Ω_2, and Ω_3, respectively, and the fourth resonance X as the second spatial mode ($\mathbf{q}_2$) with frequency near Ω_1. A schematic dispersion relation of the longitudinal-coupled modes is illustrated in Fig. 17 by solid curves. The dashed lines show the uncoupled modes, and the vertical lines represent the wave vectors excited in the experiment. The longitudinal-mode spectrum of the ideal electron crystal, representing a 2-D plasmon, is also illustrated by the dashed curve.

The coupling parameter has the form $V_{Gi} = V_{Gi}^0 \exp(-iW_1)$, where the

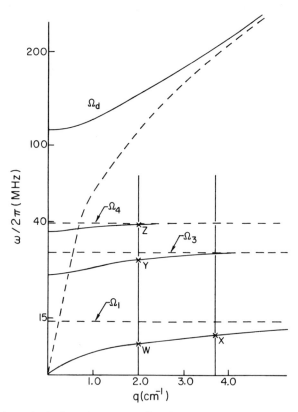

FIG. 17. Schematic ripplon–plasmon coupling. D. S. Fisher, B. I. Halperin, and P. M. Platzman, *Phys. Rev. Lett.* **42**, 798, (1979).

second factor is the Debye–Waller factor given by $W_1 = G_1^2 \langle \mathbf{u}_f \rangle^2 / 4$ in which $G_1^2 = 8\pi^2 n / \sqrt{3}$, and $\mathbf{u}_f$ is part of the displacement vector $\mathbf{u}$, which moves fast. That is, the displacement was split into a slow part, $\mathbf{u}_s$, and a fast part, $\mathbf{u}_f$. Fisher $et~al.$ took $\exp(-2W_1) = 0.23$ and obtained resonance frequencies $W = 10.2$, $X = 12.3$, $Y = 31.4$, and $Z = 39.4$ MHz, in excellent agreement with the experiment. The electron density $n = 4.55 \times 10^8$ cm^{-2}, and the electric field $E = 415$ V/cm. At high frequencies, there is a mode with $\omega^2 = \omega_d^2 + \omega_l^2$, where $\omega_d^2 = \frac{1}{2} \Sigma_G V_G^2$ is the characteristic frequency of a single electron in a static dimple.

An interesting aspect of the electron lattice is that the average square displacement from equilibrium is rather large. An experimental probe of this displacement is the measurement of lines widths of the lattice resonances. Mehrotra and Dahm[63] found that the dominant scattering for the line width is umklapp scattering involving ripplons; they predicted a maximum in the line width as a function of temperature and a large change in the width near the melting temperature.

The scattering of phonons of the 2-D electron lattice by ripplons can be investigated by measuring the mobility of an electron lattice, as Mehrotra $et~al.$ have discussed.[63] Their data showed a minimum in the mobility as a function of temperature. They also found that the melting curve deviates from that predicted by a constant Γ_m. The deviation is of the order of 10 mK and increases with electron density n. They compared their data with the single-phase Fermi liquid theory of Platzman and Fukuyama,[64] which gave an $n^{1/2}$ correction of the classical Γ_m:

$$\Gamma_m = \Gamma_0 (1 + 9 \times 10^{-7} n^{1/2}).$$

Their experimental value for Γ_0 was 124 ± 4.

Grimes and Adams detected the appearance of the acoustical modes. On the other hand, Deville $et~al.$[65] worked on the optical branch to find that the melting electron density is proportional to $T^{1.8\pm0.2}$ rather than T^2; but, if the T^2 dependence is forced, $\Gamma_m = 118 \pm 10$.

Further, with respect to electron–ripplon coupling, we remark that u_s in the zeroth order leads to an equilibrium displacement of the surface. This represents the dimple suggested by Shikin.[66] He pointed out that in

[63]R. Mehrotra and A. J. Dahm, $Phys.~Rev.~Lett.$ **43**, 467 (1979); R. Mehrotra, B. M. Guenin, and A. J. Dahm, $Phys.~Rev.~Lett.$ **48**, 641 (1982).

[64]P. M. Platzman and H. Fukuyama, $Phys.~Rev.$ **B10**, 3150 (1974).

[65]G. Deville, F. Gallet, D. Marty, J. Poitrenaud, A. Valdes, and F. I. B. Williams in "Ordering in Two Dimensions," (S. K. Sinha, ed.), p. 317. North-Holland, 1980.

[66]V. B. Shikin, $Zh.~Eksp.~Teo.~Fiz.$ **58**, 1748 (1970); $Zh.~Eksp.~Teo.~Fiz.$ **60**, 713 (1971) [$Sov.~Phys.-JETP$ **31**, 936 (1970); $Sov.~Phys.-JETP$ **33**, 387 (1971)]; V. B. Shikin and Yu. P. Monarkha, $J.~Low~Temp.~Phys.$ **16**, 193 (1974).

the presence of a pressing electric field, a depression tends to form on the He surface under each electron, which in turn localizes the electrons in the plane of the surface. According to the analysis of Shikin and Monarkha,[66] an electric field of 3000 V/cm localizes an electron over a distance of 1000 A, and the surface is depressed by 1 A.

In addition to the microscopic dimple under each electron, there is a macroscopic deformation of the liquid due to the integrated effect of many other electrons. By placing a positive metal plate under the surface of liquid helium, Crandall[67] observed a macroscopic dimple with about 0.1 mm depression at a field of 2000 V/cm. More recently, Leiderer *et al.*[68] took pictures of isolated dimples as well as of a hexagonal assembly of these dimples. In the former case, each dimple contained approximately 10^7 electrons, whereas in the latter case each dimple contained about 7×10^6 electrons.

The coupling of electrons with ripplons leads to an instability that limits electron density to about $2 \times 10^9 \, \text{cm}^{-2}$, according to the estimate of Gorkov and Chernikova.[69] Ikezi showed that above a crtical electric field, periodic deformations occur.[69] This may be considered a shallow dimple lattice, in contrast to the deep dimple lattice observed by Leiderer *et al.* for electric fields above the critical field of 2600 V/cm.

8. MELTING OF A WIGNER CRYSTAL

Melting and crystallization refer to the same phase change. However, it is more difficult to treat crystallization because the theoretical description of the reference liquid state itself is not easy. Among several numerical approaches, it might be mentioned that Isihara and Montroll[53] treated the quantum-mechanically degenerate case in 3-D. Adopting a Padé approximation, they interpolated the ground-state energy of a high-density electron gas and that of a Wigner crystal and pointed out the possibility of a first-order phase transition at $r_s = 14.4$. Gann *et al.*[48] used Monte Carlo simulations of the classical case and found a directional long-range order below $\Gamma_m = 125 \pm 15$. On the basis of molecular dynamics, Kalia *et al.*[70] arrived at the conclusion that the system undergoes a first-order phase transition in the range $\Gamma_m = 118$–130. They

[67]R. S. Crandall, *Surf. Sci.* **58**, 266 (1976).
[68]P. Leiderer and M. Wanner, *Phys. Lett.* **42**, 315 (1975); W. Ebner and P. Leiderer, *Phys. Lett. A***80**, 277 (1980); P. Leiderer, W. Ebner, and V. B. Shikin, *Physica* **107**, 217 (1981).
[69]L. P. Gorkov and D. M. Chernikova, *Zh. Eksp. Teo. Fiz. Pis'ma* **18**, 119 (1973) [*Sov. Phys.-JETP Lett.* **18**, 68 (1973)]; H. Ikezi, *Phys. Rev. Lett.* **42**, 1688 (1979).
[70]R. K. Kalia, P. Vashishta, and S. W. de Leeuw, *Phys. Rev. B***23**, 4794 (1981).

did not find a lambda-type specific heat anomaly. Instead, a hysteresis, which is characteristic of a first-order transition, was observed. These estimates of Γ at melting are consistent with the experiment of Grimes and Adams. Monte Carlo simulations of the equation of state and of the electron distribution were also reported by Calinon et al.[71]

In addition to these numerical approaches, an important theory of phase transitions in 2-D was forwarded by Kosterlitz and Thouless.[72] Two-dimensional systems do not have true long-range order, as in the case of 3-D, but various phase transitions do occur. Kosterlitz and Thouless proposed a new type of order, called *topological order*, to describe these phase transitions. Thouless applied this theory to the melting of a Wigner lattice.[72]

In a continuum description, the elastic energy of a 2-D isotropic crystal is given by

$$\mathcal{H}_0 = \tfrac{1}{2} \int d\mathbf{r} [2\mu u_{ij} u_{ij} + \lambda u_{ii}^2], \tag{8.1}$$

where μ and λ are Lamé's constants, and

$$u_{ij} = \frac{1}{2} \left(\frac{\partial u_j}{\partial r_i} + \frac{\partial u_i}{\partial r_j} \right) \tag{8.2}$$

represents a displacement tensor. A transrational order parameter may be defined for each reciprocal lattice vector $\mathbf{G}$ by

$$\rho_{\mathbf{G}}(r) = \exp[i\mathbf{G} \cdot \mathbf{u}(r)]. \tag{8.4}$$

One can then show that the correlation function of the order parameter decreases algebraically as follows:

$$\langle \rho_{\mathbf{G}}(r) \rho_{\mathbf{G}}(0) \rangle \sim \left(\frac{1}{r} \right)^{\eta_G}, \tag{8.4}$$

where the exponent η_G is associated with the elastic constants and with $G = |\mathbf{G}|$ through

$$\eta_G = \frac{|G|^2 \, kT (3\mu + \lambda)}{4\pi\mu(2\mu + \lambda)}. \tag{8.5}$$

[71]R. Calinon, Ph. Choquard, E. Jain, and M. Navet, *in* "Ordering in Two Dimensions," (S. K. Sinha, ed.) p. 317. North-Holland, 1980.
[72]J. M. Kosterlitz and D. J. Thouless, *J. Phys.* **C6,** 1181 (1973); D. J. Thouless, *J. Phys.* **C11,** L189 (1978).

Hence, the exponent varies continuously with the temperature T, and the 2-D system has a quasi-long-range order. In the Kosterlitz–Thouless theory, the melting point is determined as the temperature at which the algebraic power decay changes its character due to singularities called dislocations that exist in an elastic body at finite temperatures. These dislocations require extra energy, which can be obtained by decomposing the displacement vector $\mathbf{u}$ into smoothly varying and singular parts. The dynamical energy is found to consist of the harmonic part and a singular part. When a loop integral

$$\int_c \frac{\partial u_i}{\partial \mathbf{r}} \cdot d\mathbf{r} = b_i, \qquad i = (x, y), \tag{8.6}$$

about a lattice point is not zero but is equal to b_i, we say that the point represents an isolated dislocation. The vector $\mathbf{b}$, known as a Burgers vector, represents the strength of the dislocation. It is of the order of a lattice vector in magnitude. The energy associated with an isolated dislocation of strength b is given by

$$E_{\text{disloc}} = \pi J \ln(L/a_c) + E_c, \tag{8.7}$$

where E_c is a core energy, a_c is the core size, and

$$J = \frac{1}{2\pi^2} \frac{\mu(\mu + \lambda)}{2\mu + \lambda} b^2.$$

The energy of isolated dislocations increases logarithmically. However, the energy of opposite dislocation pairs is found to depend on their distance. Therefore, dislocation pairs can be energetically favorable and may exist at low temperatures. At sufficiently high temperatures, these pairs may break up into isolated dislocations.

The probability of finding isolated dislocations is proportional to b^2/L^2, where L is a characteristic length of the system. Since the entropy is proportional to the logarithm of the size of the system, the free energy is given by

$$F_{\text{disloc}} = (\pi J - 2kT) \ln(L/a_c). \tag{8.8}$$

We find then that above the temperature given by

$$T_m = \frac{\pi J}{2k}, \tag{8.9}$$

the free energy is negative, favoring the appearance of isolated dislocations.

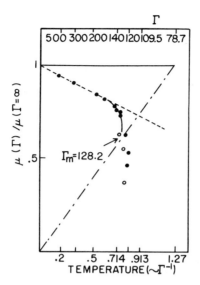

FIG. 18. Theoretical melting curve of a Wigner lattice. R. Morf, *Phys. Rev. Lett.* **43**, 931 (1979). The intersection of the top solid line and the chain straight line represents the melting point obtained by Thouless, while the dependence of the elastic constant on temperature reduces the melting point. As the melting point is approached, heating-up (black circles) and cooling-down (while circles) processes show differences.

The bare value of λ is infinite for a Wigner lattice. Thouless used the modulus μ for absolute zero instead. In a simplified form, the melting temperature was given by

$$T_m = \frac{\rho b^2 c^2}{4\pi k},$$
(8.10)

where c is the phonon velocity, and ρ is the mass density. This T_m yields $\Gamma_m = 78.7$, which is rather small compared with the experimental value of 131 of Grimes and Adams or with the Monte Carlo result of 125 of Gann *et al.*[48] Therefore, Morf[73] suggested that the elastic constant should be renormalized in consideration of its temperature dependence. For this purpose, the renormalization scheme of Kosterlitz and Thouless[72] could be used starting with the initial value μ_0. Morf adopted the linearly extrapolated value μ_0 from absolute zero. The change of the shear modulus is illustrated in Fig. 18 by the dashed line. The renormalization process also requires the core energy of a dislocation. Morf adopted the

[73]R. Morf, *Phys. Rev. Lett.* **43**, 931 (1979). See also D. S. Fisher, B. I. Halperin, and R. H. Morf, *Phys. Rev.* **B20**, 4692 (1979).

crude assumption that the energy at finite temperature is reduced from
the zero temperature value in proportion to the reduction of the bare
shear modulus μ_0 because of phonon interaction. His results for melting
are given by the solid curve in Fig. 18, which yields $\Gamma_m = 128.2$. This
value is in good agreement with the above-mentioned experimental and
Monte Carlo results. The chain line in Fig. 18 represents the original
Thouless criterion that intersects at Thouless' value the horizontal line
corresponding to $\mu = $ constant. The linear variation of the shear modulus
has been obtained on the basis of renormalization due to phonon–
phonon interactions[74] and has been confirmed by experiments.[75]

In addition to the value of Γ_m, it is important to determine the order of
the phase transition. If it is in first order, the change in entropy is
expected to be on the order of 0.3 k per electron. The specific-heat
measurements of Glattli et al.[76] lead to an upper limit of 0.02 k per
electron for the entropy change. Thus, the transition appears to be the
Kosterlitz–Thouless continuous type rather than first order. The temperature variation of the specific heat follows a pure phonon heat
capacity with a renormalized transverse phonon velocity. At 0.2 K, the
specific heat is approximately 1.8 k per electron. This value is close to the
classical value of 2 k. In fact, the Debye temperature is $T_D =$
$0.935 T_m (n/10^8 \, \text{cm}^{-2})$ in which n is of the order $10^8 \, \text{cm}^{-2}$, and the melting
temperature is

$$T_m = 149.5 \pm 1 \, \text{mK}. \tag{8.11}$$

The melting criterion based only on Γ is applicable to classical cases. In
order to describe the entire phase curve down to the lowest temperature,
another parameter must be introduced. For this purpose, the parameter
r_s, which is independent of temperature, may be used because it can be
varied as a density parameter for a given Γ. Figure 19 illustrates the
theoretical curves obtained by Imada and Takahashi[77] by extrapolating
their Monte Carlo results for spin-polarized $12 \sim 30$ electrons to the
thermodynamic limit. They used the assumption that the melting point
decreases in inverse proportion to the total number of electrons and
obtained two curves that represent a kind of hysteresis in their calcula-

[74]D. S. Fisher, B. I. Halperin, and P. M. Platzman, *Phys. Rev. Lett.* **42**, 798 (1979); D. S.
  Fisher, *Phys. Rev.* **B26**, 5009 (1982).
[75]F. Gallet, G. Deville, A. Valdés, and F. I. B. Williams, *Phys. Rev. Lett.* **49**, 212 (1982);
  G. Deville, A. Valdes, E. Y. Andrei, and F. I. B. Williams, *Phys. Rev. Lett.* **53**, 588
  (1984).
[76]D. C. Glattli, E. Y. Andrei, and F. I. B. Williams, *Phys. Lett.* **60**, 420 (1988).
[77]M. Imada and M. Takahashi, *J. Phys. Soc. Jpn.* **53**, 3770 (1984).

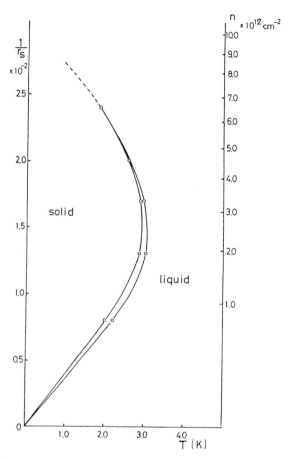

FIG. 19. Solid–liquid boundary line of a 2-D electron system obtained by extrapolation to the thermodynamic limit. M. Imada and M. Takahashi, *J. Phys. Soc. Jpn.* **53**, 3770 (1984).

tion. The maximum melting point is around 3 K for an electron density of 2.2×10^{10} cm^{-2}. Above this point, the melting point decreases monotonically. The solid phase seems to disappear above densities of order 10^{11} cm^{-2} or $r_s \sim 34$.

There is a related and equally interesting subject concerning the electron phase: Aoki[78] studied numerically 2-D disordered electron systems in a strong magnetic field and observed changes in the electron distribution ranging from a crystal to Anderson localization.

[78]H. Aoki, *J. Phys. C***12**, 633 (1979).

III. Correlations in a Magnetic Field

9. EFFECTIVE g-FACTOR

In the absence of Coulomb interaction, the solutions of the Schrödinger problem of one electron in a perpendicular magnetic field are known; the energy levels are given by the Landau levels

$$\varepsilon_n = (n + \tfrac{1}{2})\hbar\omega_c \pm \frac{g}{4}\hbar\omega_c, \tag{9.1}$$

where g is Lande's factor and ω_c is the cyclotron frequency. If the system has a finite length L_x (or L_y) in the x (or y) direction, the wave vector $\mathbf{k}$, which characterizes the center coordinate of the harmonic motion of the electron, can take values between 0 and eHL_xL_y/ch, leading to degeneracy of eH/ch per unit area. The total magnetic flux Φ can be written as

$$\Phi = HL_xL_y = \left(\frac{eH}{ch}\right)\phi_0 L_x L_y. \tag{9.2}$$

The quantity ϕ_0 defined by

$$\phi_0 = \frac{ch}{e} \tag{9.3}$$

is the *flux quantum*. The equation $(eH \cdot ch)L_xL_y = \Phi/\phi_0$ represents the total degeneracy of each Landau level.

For a many-electron system with Coulomb interaction, the problem of finding its susceptibility is very difficult. Nevertheless, the magnetic susceptibility can be obtained by generalizing the evaluation of the correlation energy to the case with a magnetic field. However, in order to eliminate divergences, it is necessary to combine the first-order exchange contribution with the ring contribution. The susceptibility evaluated for high densities by Isihara and Toyoda[79] reads as follows:

$$\chi = \chi_0 \left[\frac{1}{12\pi} (\tfrac{3}{48}g^2 - 1) + 8.956 \times 10^{-3}g^2 r_s + 0.441 \times 10^{-4} r_s \right.$$
$$- 2.985 \times 10^{-3} r_s \ln r_s + (-0.943g^2 + 8.525) \times 10^{-3} r_s^2$$
$$\left. - 1.344 \times 10^{-3} r_s^2 \ln r_s \right]. \tag{9.4}$$

[79] A. Isihara and T. Toyoda, *Phys. Rev.* **19**, 831 (1979).

Note that

$$\chi_0 = \frac{m\mu_B^2}{4\pi\hbar^2} g^2 \tag{9.5}$$

is the ideal susceptibility. Unlike the 3-D case, the susceptibility has a dimension. Equation (9.4) includes both the paramagnetic and diamagnetic susceptibility, the former being given by the terms with the g-factor. For $g = 2$ the ratio of the ideal paramagnetic susceptibility to the diamagnetic susceptibility is 3 to 1, as in the case of 3-D. For small r_s the deviations of both susceptibilities from the ideal gas values are almost linear when the susceptibilities are plotted against r_s.

It is interesting to find out how the spin paramagnetic susceptibility changes with r_s for a wide range. In order to find this variation, we note that the theory of the correlation energy discussed in Section 4 can be applied by modifying the Fermi energy in accordance with the spin directions. This application is convenient because formally it enables the evaluation of the ring diagram contribution for all densities, as shown by Isihara and Ioriatti.[80]

Let us introduce a polarization factor P such that it represents the difference in the fractions of spin-up and -down electrons. In terms of the Fermi momentum p_F of the unpolarized state, the Fermi momenta of these electrons are represented by the following expressions:

$$\begin{aligned} p_F^+ &= (1 + P)^{1/2} p_F, \\ p_F^- &= (1 - P)^{1/2} p_F. \end{aligned} \tag{9.6}$$

To second order in P, we find the shift in the ground-state energy of the total system due to the polarization as follows:

$$E_g(r_s, P) - E_g(r_s, 0) = \frac{Ne^2}{2\kappa^2 a_0} \frac{m}{m_0} A(r_s) P^2 - N\mu_B HP. \tag{9.7}$$

Here, for a later application to a Si inversion layer, the effective mass m has been introduced such that

$$\pi r_s^2 \left(\frac{m_0 a_0 \kappa}{m} \right)^2 = \frac{1}{n}, \tag{9.8}$$

where κ is the average dielectric constant and m_0 is the bare electron

[80]A. Isihara and L. C. Ioriatti, Jr., *Physica B***113,** 42 (1982).

mass. The factor $A(r_s)$ is a function of r_s and valley degeneracy g_v, consisting of the following:

$$A(r_s) = A^0(r_s) + A^x(r_s) + A^c(r_s), \tag{9.9}$$

where the first and second terms are the kinetic and exchange contributions given by

$$A^0(r_s) = \frac{1}{g_v r_s^2}; \qquad A^x(r_s) = -\frac{\sqrt{2}}{\pi \sqrt{g_v}\, r_s}. \tag{9.10}$$

The last term represents the correlation contribution.

The polarization parameter can be determined from the condition to minimize the energy. We arrive at

$$P = \mu_B H a_0 \frac{m_0}{m} \frac{\kappa^2}{e^2 A(r_s)}. \tag{9.11}$$

On the other hand, the magnetic moment per unit area is given by

$$M = n\mu_B P. \tag{9.12}$$

Since the susceptibility is

$$\chi = \left(\frac{\partial M}{\partial H}\right)_n, \tag{9.13}$$

we obtain the ratio of the actual susceptibility to the ideal susceptibility as follows:

$$\frac{\chi_0}{\chi} = 1 - \frac{\sqrt{(2g_v)}}{\pi} r_s + g_v r_s^2 A(r_s), \tag{9.14}$$

$$\chi_0 = \frac{g_v m}{\pi \hbar^2} \mu_B^2. \tag{9.15}$$

This ratio may be used to derive an effective g-factor.

In order to find the correlation contribution A_c, we first note that the second-order exchange energy in the unpolarized system is given by

$$E_{2x} = (0.2287 \ldots) \frac{Ne^2 m}{2\kappa^2 a_0 m_0}, \tag{9.16}$$

which is constant and is unaltered by the spin polarization. Hence, within the neglect of higher-order exchanges, the entire correlation contribution to the susceptibility comes from the ring diagrams. This contribution can be expressed in a simple form by adopting the variables $z = 2\pi j/\beta q^2$ and $\rho = 2P_F/q$. These are essentially the same variables as those adopted in Section 2 in consideration of the electrostatic analogue of the RPA dielectric function. The energy difference between the polarized and unpolarized cases is found to be given by

$$E_r(r_s, P) - E_r(r_s, 0)$$
$$= -\frac{Ne^2 m g_v P^2}{2\pi\kappa^2 m_0 a_0} \int_0^\infty dz \int_0^\infty d\rho \left[4\pi^2 \lambda \frac{\partial}{\partial\rho} \left(\frac{\partial\lambda}{\rho\,\partial\rho} \right) \right] \left[1 + \frac{2\pi r_s g_v^{3/2}\rho\lambda}{\sqrt{2}} \right]^{-1}, \quad (9.17)$$

where

$$\lambda(\rho, z) = \frac{\rho}{2\pi} \int_0^\infty dx \frac{\sin x}{x} e^{-|z||x|} J_1(\rho x). \quad (9.18)$$

From these results, we find

$$A^c(r_s) = \frac{\sqrt{2}}{\pi r_s \sqrt{g_v}} + \frac{2g_v}{\pi r_s g_v^{3/2}} \int_0^\infty \int_0^\infty \frac{dz\,d\rho}{\rho} \frac{\partial}{\partial\rho} \left(\frac{\partial\lambda}{\rho\,\partial\rho} \right) \frac{2\pi}{1 + \frac{2\pi r_s g_v^{3/2}\rho\lambda}{\sqrt{2}}}. \quad (9.19)$$

Note that the first term exactly cancels the first-order exchange contribution $A^x(r_s)$ in Eq. (9.10) so that there will be no exchange effect. In other words, the collective couplings of the system serve to increase the correlation energy and to decrease the susceptibility. Furthermore, by analyzing Eq. (9.19), we find that $A^c(r_s, g_v)$ for arbitrary valley degeneracy g_v can be obtained from the case of $g_v = 1$ if we replace r_s by $g_v^{3/2}r_s$ and multiply the results by g_v:

$$A^c(r_s, g_v) = g_v A^c(g_v^{3/2}r_s, 1). \quad (9.20)$$

Further simplification of Eq. (9.19) is possible, and the quantity $A^c(r_s)$ can be expressed by a single finite integral over the angle variable, as in the case of the ring energy.

Figure 20 illustrates the paramagnetic susceptibility thus obtained without effective quantities as a function of r_s. The ordinate is a dimensionless ratio χ/χ_0. The curve is fairly straight in the range $0.5 < r_s < 2.5$, even though the curvature changes in this region. The

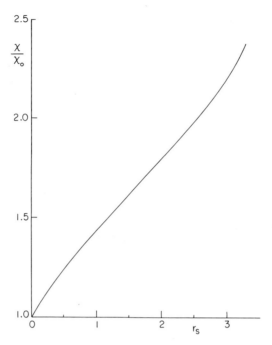

FIG. 20. Density dependence of the paramagnetic susceptibility for $g_v = 2$. A. Isihara and L. C. Ioriatti, Jr., *Physica* **B113**, 42 (1982).

curve starts to increase nonlinearly after the inflection point, indicating increasing Coulomb effects.

The 2-D character of the electrons in n-type Si inversion layers and their effective g-factor were observed by Fang and Stiles[81] in 1968 by an ingenious method in which a tilted magnetic field was used. If a magnetic field is applied at an angle θ from the direction perpendicular to the surface of electrons, the orbital energy will depend only on the vertical component $H \cos \theta$, but the spin-field coupling will be determined by the total field H. The Landau levels are given by

$$\varepsilon_n = (n + \tfrac{1}{2})\hbar\omega_c \cos\theta \pm \tfrac{1}{2}g\mu_B H,$$

where μ_B is the Bohr magneton. Hence, if the energy is plotted against $H \cos \theta$, a series of parallel lines appears in correspondence to the two spin directions. At a particular angle θ_1, the energy difference due to spin in the $n = 0$ Landau levels become equal to the energy difference

[81]F. F. Fang and P. J. Stiles, *Phys. Rev.* **B28**, 6992 (1983).

between the lower $n = 1$ Landau level and the upper $n = 0$ level. Below this angle, the orbital energy difference is larger than the spin splitting, and above, the orbital energy separation will be less than the spin separation. Fang and Stiles utilized the fact that at this angle the phase of the oscillating conductance changes and determined the effective g-factor through the relation $\hbar\omega_c \cos \theta_1 = 2g\mu_B H$.

In actuality, the effective mass in Si inversion layers depends on the electron density n due to Coulomb interaction. From the diamagnetic susceptibility, Isihara and Toyoda derived the following effective mass formula for Si [100] inversion layers:

$$m^* = m\left[1 + \frac{3.21}{n^{1/2}} \times 10^4\right].$$

(9.21)

On the other hand, Smith and Stiles obtained an empirical effective mass

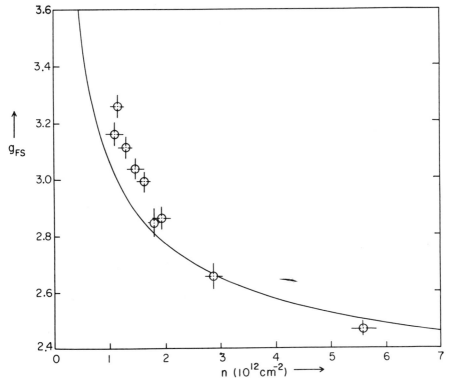

FIG. 21. Density dependence of the effective g-factor. Circles: F. F. Fang and P. J. Stiles, *Phys. Rev. B28*, 6992 (1983). Solid curve: A. Isihara and L. C. Ioriatti, Jr., *Physica B113*, 42 (1982).

formula:

$$m/m^* = 1 - 0.071 r_s^*, \qquad (9.22)$$

where $r_s^* = me^2\hbar^{-2}\kappa_{sc}^{-1}(\pi n)^{1/2}$ with $\kappa_{sc} = 12$ and $m = 0.195 m_0$.

The effective g-factor obtained by Isihara and Ioriatti[80] with Smith–Stiles' effective mass correction is compared in Fig. 21 to the Fang and Stiles data. As we see, the agreement is excellent. Historically, Janak[82] was the first to show that the effective g-factor is enhanced by electron correlations by evaluating the difference in self-energies of spin-up and -down electrons. Since then, several theoretical calculations of the effective g-factor have been reported.[83] However, most of the results showed much weaker density dependencies of the effective g-factor than the data of Fang and Stiles, mainly because of inadequate treatments of the correlation energy. Although the effective g-factor evaluated from the susceptibility and the one determined from the conductance may not be exactly the same, we can conclude from Fig. 21 that the electron correlations enhance the g-factor towards low densities more than twice its bulk value of 2.

10. DE HAAS–VAN ALPHEN AND MAGNETOTHERMAL EFFECTS

In 1930, de Haas and van Alphen observed that the susceptibility of Bi was not constant but oscillatory in strong magnetic fields. For a while, this oscillatory behavior was considered to be a peculiarity of Bi, but soon its generality began to emerge when the same oscillatory behavior was revealed in many other metals. A concrete theoretical account of this phenomenon, now called the *de Haas–van Alphen* (dHvA) *effect,* took more than twenty years to develop: it was 1952 when Lifshitz and Kosevich derived an ideal gas formula and Onsager independently showed how the oscillations were related to the cross section of the Fermi surface. Since then, the dHvA effect has been used very effectively for the determination of the Fermi surfaces of metals. These works were all concerned with bulk metals. However, the basic oscillatory feature comes from the quantization of an electron's energy in the plane perpendicular to the magnetic field. Since 2-D electrons do not have any other degree of freedom than that in the plane, the energy quantization should in

[82]J. F. Janak, *Phys. Rev.* **178**, 1416 (1969).

[83]K. Suzuki and Y. Kawamoto, *J. Phys. Soc. Jpn.* **35**, 1456 (1973); T. Ando and Y. Uemura, *J. Phys. Soc. Jpn.* **37**, 622 (1974); C. S. Ting, T. K. Lee, and J. J. Quinn, *Phys. Rev. Lett.* **34**, 870 (1975); B. Vinter, *Phys. Rev. Lett.* **35**, 1044 (1975); *Phys. Rev. B***13**, 4447 (1976); F. J. Ohkawa, *Surf. Sci.* **58**, 326 (1976).

principle be perfect, making the oscillatory phenomenon even more interesting.

Generally, the effect is observed under the conditions that the field energy is less than the Fermi energy but is no less than the thermal energy. The conditions are

$$kT \leq \mu_B H < (\hbar k_F)^2/2m.$$

That is, the field must be strong enough to suppress the thermal broadening of Landau levels but not so strong as to destroy the sharpness of the Fermi surface.

A many-body approach to the dHvA effect in 2D was made by Isihara and Kojima.[84] They followed a high-density approach by taking into consideration the exchange and ring-diagram contributions and evaluated the energy, specific heat, susceptibility, etc. For instance, their formula for the oscillating specific heat is given by

$$c_A^{\text{osc}} = -\frac{2\pi^4}{3} k\gamma_0 \sum_l (-)^{l+1} l \frac{\cos(\frac{1}{2}g\pi l)}{\sinh(\pi^2 l/\alpha)} \cos(\pi l/\gamma_0), \tag{10.1}$$

and the susceptibility is

$$\chi^{\text{osc}} = \chi_0 \frac{2}{\alpha} \sum_l (-)^{l+1} \frac{\cos(\frac{1}{2}g\pi l)}{\sinh(\pi^2 l/\alpha)}$$
$$\times \left\{ \left[1 + \left(\frac{\pi^2 l}{\alpha} \right) \coth\left(\frac{\pi^2 l}{\alpha} \right) \right] \cos\left(\frac{\pi l}{\gamma_0} \right) + \left(\frac{\pi l}{\gamma_0} \right) \sin\left(\frac{\pi l}{\gamma_0} \right) \right\}, \tag{10.2}$$

where $\gamma_0 = \mu_B H/[(\hbar k_F)^2/2m]$ and $\alpha = kT/(\mu_B H)$ are dimensionless variables. In the oscillating susceptibility, note that the constant phase $\pi/4$, which characterizes the 3-D oscillations, does not appear in 2D. This is a 2-D peculiarity. Moreover, the period of oscillations is not affected by electron correlations. Many years ago, Kohn[85] discussed that a short-range interaction does not change the period of the dHvA oscillations.

The chemical potential is changed from its ideal gas value by magnetic field as well as by Coulomb interaction, even for absolute zero. The reason is simply that it is the chemical potential as the actual Fermi energy that is relevant to low-temperature properties. As Sommerfeld's theory of an ideal gas specific heat shows, the chemical potential depends

[84]A. Isihara and D. Y. Kojima, *Phys. Rev.* **B19**, 846 (1979).
[85]W. Kohn, *Phys. Rev.* **123**, 1242 (1961).

also on temperature. Hence, in the presence of a strong magnetic field or Coulomb interaction, the ideal Fermi energy $p_0^2 = 2\pi n$ and the actual Fermi energy can be different from each other. The actual Fermi energy is a function of the ideal gas Fermi energy, which represents the density of electrons.

The chemical potential for the case where Lande's factor $g = 2$ is plotted against electron density in Fig. 22. The ordinate $1/\gamma = \mu/\mu_B H$ is the ratio of the chemical potential μ to the field energy $\mu_B H$, and the abscissa $1/\gamma_0 = \varepsilon_F/\mu_B H$ is a dimensionless ratio of the ideal Fermi energy against the field energy. The staircase change shows that at absolute zero and in the ideal case, the chemical potential jumps to the next Landau level whenever a Landau level is filled. However, at finite temperatures and in the presence of Landau-level broadening, the sudden changes are eased, as shown by the two curves: one with $\pi^2/\alpha = 1.5$, which corresponds roughly to 1 K and $2T$, and the other with $\pi^2/\alpha = 5.0$, where $\alpha = \mu_B H/kT$. The latter curve is very close to the straight line representing the limiting case in which $1/\gamma = 1/\gamma_0$. The broadening of Landau levels is assumed to be Lorentzian with a dimensionless parameter $\bar{\Gamma} = \pi\hbar/(\tau\mu_B H)$ in Eq. (10.15) below. The two smooth curves correspond to the case in which $\Gamma = 0.3$. Similar softening of sharp changes is caused by the Coulomb interaction and the magnetic field. Nevertheless, the

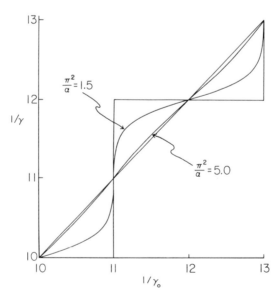

FIG. 22. Chemical potential in units of $\mu_B H$ as a function of $\gamma_0^{-1} = (\hbar k_F)^2/(2m\mu_B H)$, where $k_F^2 = 2\pi n$, n being the electron density for the case $g = 2$.

chemical potential increases fast when the abscissa consists of odd intergers (because $g = 2$ in this graph) if the temperature is not very high. That is, the ideal gas behavior is preserved to some extent. Since similar behavior holds for interaction effects, the period of dHvA oscillations is expected not to change, as Kohn discussed. However, the amplitude may be affected by Coulomb interaction.

In order to find correlation effects, Shiwa and Isihara[86] extended the correlation energy theory discussed in Section 4 to the case of a strong magnetic field. The eigenvalues of the effective electron propagator are given by

$$\lambda_j(q) = \frac{1}{2\pi} \sum_l (-)^{l+1} z^l \frac{\cosh(\frac{1}{2}g\alpha l)}{\sinh l\alpha} \int_{-\infty}^{\infty} dx \, \exp(-\pi |j| x/\alpha) \sin(Q \sin x)$$
$$\times \exp[-Q(1 - \cos x) \coth \alpha l], \quad (10.3)$$

where $Q = \hbar^2 q^2/(2m\hbar\omega_c)$ and $z = \exp(\mu/kT)$. The hyperbolic cosine consists of the contributions from the spin-up and -down electrons. By separating these from each other, the eigenvalues are given by

$$\lambda_j(q) = \frac{1}{2}\left[\lambda_j^*\left(q, \mu + \frac{g}{4}\hbar\omega_c\right) + \lambda_j^*\left(q, \mu - \frac{g}{4}\hbar\omega\right)\right], \quad (10.4)$$

$$\lambda_j^*(q) = \frac{1}{2\pi} \int_0^{\infty} \exp\left(-\frac{\pi |j|}{\alpha} x\right) \sin(Q \sin x) \sum_0^{\infty} 2\theta(k) \exp[-Q(1 - \cos x)]$$
$$\times L_k[2Q(1 - \cos x)] \, dx. \quad (10.5)$$

$\theta(k)$ is a step function. The k sum can be replaced by a contour integral:

$$\sum_{k=0}^{\infty} 2\theta(k) \exp(-\tfrac{1}{2}y) L_k(y) = \frac{1}{2\pi i} \int_c dt \, \exp(-\tfrac{1}{2}yt) g(t),$$

where the path c encircles $1, y = 2Q(1 - \cos x)$, L_k is Laguerre's function. If $2i - 1 < 1/\gamma < 2i + 1$ for large integer i and for $0 \, \text{K}, g(t) = -1 + [(t + 1)/(t - 1)]^i$.

The eigenvalues are found to be given by

$$\lambda_j^*(q) = \frac{1}{2\pi}(1 - r) + \dots \quad (10.6)$$

[86]Y. Shiwa and A. Isihara, Phys. Rev. B27, 4743 (1983).

for large integers i such that $2i - 1 < 1/\gamma < 2i + 1$. Here, r is the solution of

$$s^4 = \frac{8i\gamma s^2}{1 - r^2} - \frac{J^2}{r^2},$$

where $s = \hbar q/p_F$, $J = 2\pi j/\eta$, with $\eta = \mu/kT = p_F^2/(2mkT)$.

Once the eigenvalues are found, the exchange and correlation contributions can be obtained. For small γ or large i, the internal energy per electron in Rydbergs is

$$\frac{\varepsilon_g(H)}{(1/r_s^2)} = 4[1 + r_s^2\varepsilon(r_s)]i\gamma_0 - 4\left[1 + \frac{(2i-1)(2i+1)}{(2i)^2}r_s^2\varepsilon(r_s)\right]i^2\gamma_0^2. \quad (10.7)$$

Here, $\gamma_0 = \mu_B H/[(\hbar k_F)^2/2m]$ as before, and $\varepsilon(r_s)$ defined below represents (the ground-state energy − the kinetic energy) with the ground-state energy given by Eq. (4.12) in Part I. The magnetic susceptibility is given then by

$$\frac{\chi}{\chi_0} = \frac{4}{\gamma_0}\left\{\left[1 + \frac{(2i-1)(2i+1)}{(2i)^2}r_s^2\varepsilon(r_s)\right]i^2\gamma_0 - [1 + r_s^2\varepsilon(r_s)]i\right\}. \quad (10.8)$$

These formulae are restricted to the region:

$$2i - 1 < 1/\gamma_0 < 2i + 1.$$

The function $\varepsilon(r_s)$ is defined by

$$\varepsilon(r_s) = -\frac{8\sqrt{2}}{3\pi}\frac{1}{r_s} + \frac{8}{\pi}\Phi, \quad (10.9)$$

where Φ is the ϕ-integral in Eq. (4.11) of Part I. That is, Φ is the sum of the first-order exchange and ring-diagram contributions to the ground-state energy in the absence of the magnetic field.

The internal energy is parabolic as a function of electron density, as shown in Fig. 23, in which the abscissa is a dimensionless variable $1/\gamma_0 = \varepsilon_F/\mu_B H$. The ordinate represents the energy in units of the average kinetic energy. The left (resp. right) set of curves corresponds to relatively small (resp. large) values of the abscissa. The curves show stronger variations on the right side because the magnetic field is stronger. From top to bottom, r_s increases. The curves are first convex-up, but this convexity is reduced as r_s increases, and finally, at the

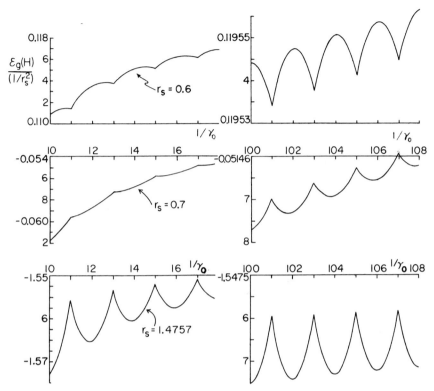

FIG. 23. Magnetic field dependence of the energy per electron in units of the kinetic energy for $g = 2$. Y. Shiwa and A. Isihara, *Phys. Rev.* **B27**, 4743 (1983).

point where the ground state becomes negative, the curvature changes. The reason for this change is simply that the magnetic moment is proportional to the ground-state energy.

The magnetic susceptibility is plotted against $1/\gamma_0$ in Fig. 24. The left and right sets have the same meaning as they have in Fig. 23. In the top two graphs, the dashed lines represent the case without Coulomb interaction. That is, electron correlations tend to suppress the oscillations. The middle set shows that the phase of oscillations changes at around $r_s = 0.67$, at which point the ground-state energy becomes negative. After this point, sawtoothlike oscillations take place with some enhancement with r_s and $1/\gamma_0$. These sharp oscillations are of course characteristic of absolute zero; at finite temperatures the changes are softened. Also, as r_s increases, the contributions from higher-order exchange interactions will become increasingly important. Nevertheless,

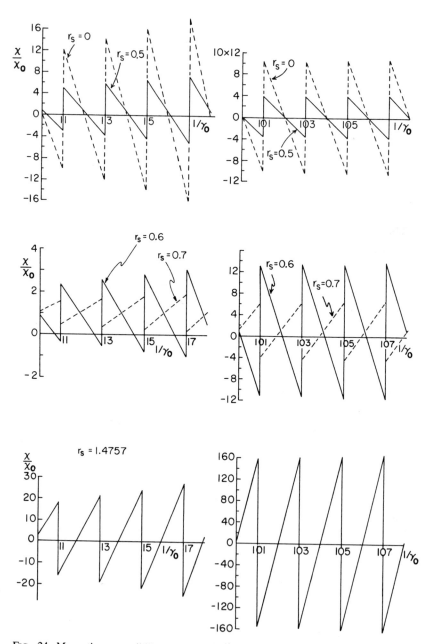

FIG. 24. Magnetic susceptibility at absolute zero as a function of the dimensionless variable γ_0^{-1} for $g = 2$. Y. Shiwa and A. Isihara, *Phys. Rev. B27*, 4743 (1983). The right-hand set corresponds to larger values of the abscissa.

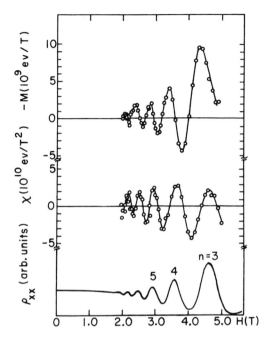

FIG. 25. Experimental dHvA oscillations in GaAs/GaAlAs. H. L. Störmer, T. Haavasoja, V. Narayanamurti, A. C. Gossard, and Wiegmann, *J. Vac. Sci. Tech. B1*, 423 (1983). From top to bottom: magnetization, susceptibility, and resistivity.

it is very interesting to observe that the gaseous and liquidlike phases show somewhat different behaviors of the susceptibility.

On the experimental side, Störmer et al.[87] observed that the oscillatory magnetic moment in GaAs/GaAlAs superlattices is a function of the magnetic field. Since the magnetic moment is proportional to the surface area, the signal was enhanced by stacking 4000 layers equivalent to a total area of 240 cm^2. Figure 25 represents their data. From top to bottom the magnetic moment, susceptibility, and longitudinal resistivity are shown. As we see, the oscillations of the magnetic moment and of the resistivity match well. However, Störmer et al. found that the amplitude of the dHvA oscillations was smaller by a factor of 30 than the theoretical expectation, presumably because of inhomogeneity, and that the density of states never vanishes between neighboring Landau levels. In a related experiment, Haavasoja et al.[87] found a difference between the resistivity

[87]H. L. Störmer, T. Haavasoja, V. Narayanamurti, A. C. Gossard, and W. Wiegmann, *J. Vac. Sci. Tech. B1*, 423 (1983); T. Haavasoja, et al., private commun.; J. P. Eisenstein, H. L. Störmer, V. Narayanamurti, A. Y. Cho, and A. C. Gossard, *Surf. Sci.* **170**, 271 (1986).

and magnetic moment curves: While the resistivity stays flat at the center between adjacent Landau levels, the magnetic moment is parabolic. Although both dHvA and Shubnikov–de Haas (SdH) oscillations originate from an electron's orbital motion, the latter is apparently very sensitive to impurity scattering near the center.

There is yet another experiment to be mentioned. Fang and Stiles[81] carried out an experiment on Si inversion layers by using twenty $25 \times 500 \ \mu m$ finger-tip gates that provided a total area of 25 cm². At $12T$ and 1.7 K, they observed a few peaks representing dM/dn, but the signal-to-noise ratio was not adequate to resolve the quantum step, which is twice the effective Bohr magneton. Störmer et al. pointed out that the dHvA effect, although difficult to observe in 2D, can be adopted for a more effective determination of the density of states than the transport phenomena, because no knowledge of scattering processes is needed for interpretation.

The dHvA effect is observed under the isothermal condition. If, instead, the magnetic field is changed adiabatically, the *magneto-thermal effect*,[88,89] in which the temperature oscillates, occurs. Isihara and Shiwa[88] have shown that the two dimensionality of electrons is reflected upon the unusually large amplitude and also on the constancy of the period. The large amplitude in 2D is due to the complete conversion of the field energy into the thermal energy. The largeness of the temperature oscillations has been observed by Gornik et al.[90] When plotted against $1/\gamma_0$, the theoretical expression of Isihara and Shiwa for the period is

$$\Delta(1/\gamma_0) = \Delta(\varepsilon_F/\mu_B H) = 2, \qquad (10.10)$$

where ε_F is the ideal Fermi energy. Hence, it is possible to determine the effective Bohr magneton. Moreover, the nodes appear at particular values of $1/\gamma_0$.

Figure 26 illustrates the theoretical magnetothermal oscillations in the case when $\pi^2/\alpha = 1.5$, which corresponds roughly to 1 K and $2T$ and when the Landau levels are elliptically broadened for approximately 0.2 K. The corresponding 3-D oscillations are much smaller.

The temperature variation T under the adiabatic condition is given by

[88]A. Isihara and Y. Shiwa, *Solid State Commun.* **48**, 1081 (1983), *Solid State Commun.* **50**, 35 (1984); *J. Phys.* C**17**, 5075 (1984), *J. Phys.* **18**, 4703 (1985).

[89]W. Zawadzki and R. Lassnig, *Surf. Sci.* **142**, 225 (1984).

[90]E. Gornik, R. Lassnig, H. L. Störmer, W. Seidenbusch, A. C. Gossard, W. Wiegmann, and M. V. Ortenberg, *in Proc. Internat. Conf. Phys. Semicond*, (J. D. Chadi and W. A. Harrison, eds.) p. 303. Springer, 1984.

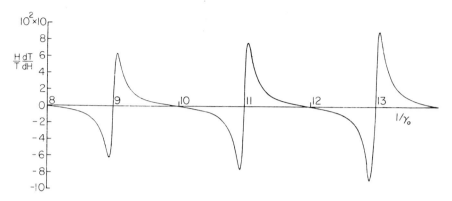

FIG. 26. Magnetothermal oscillations for elliptically broadened Landau levels at $\pi^2/(kT/\mu_B H) = 1.5$ (Isihara and Shiwa, Ref. 88). The curve corresponds roughly to a system at 1 K in an average field of 2T.

the ratio of two entropy derivatives:

$$\Delta T = -\frac{(\partial S/\partial H)_{T,n}}{(\partial S/\partial T)_{H,n}}\,dH. \tag{10.11}$$

The denominator is proportional to the specific heat, which is per unit area given by

$$c_{A,H} = \frac{1}{A}\,T\left(\frac{\partial S}{\partial T}\right)_H. \tag{10.12}$$

If Landau levels are not broadened, the specific heat becomes exponentially small because the linear specific heat is canceled out, causing large ΔT. In the presence of broadening, the linear specific heat is restored. Near absolute zero and for Lorentzian broadening represented by the density of states per unit energy interval and area,

$$g(\varepsilon) = \frac{1}{\pi^2 l^2}\sum_{i=0}^{\infty}\frac{\hbar/\tau}{(\varepsilon - \varepsilon_i^0)^2 + (\hbar/\tau)^2}; \qquad \left(l^2 = \frac{c\hbar}{eH}\right), \tag{10.13}$$

and the specific heat at constant area is given by

$$c_{A,H} = \frac{\pi k^2 T}{(6\hbar^2/2m)}\left\{1 + 2W\left[\cos\left(\frac{\pi}{\gamma}\right) - W\right]\Big/\left[1 - 2W\cos\left(\frac{\pi}{\gamma}\right) + W^2\right]\right\} \tag{10.14}$$

where

$$W(\Gamma) = \exp(-\pi\hbar/\mu_B H\tau) = \exp(-\tilde{\Gamma}) \tag{10.15}$$

is a broadening parameter in which $\hbar/\tau$ represents the half-width of the Lorentzian function. The function $W(\Gamma)$ appears when Eq. (10.13) is Fourier transformed. As each energy level is broadened, its contribution to the grand partition function is reduced by the amplitude function $W(\Gamma)$ such that the higher-harmonic contributions are progressively reduced.

In Fig. 26 we have adopted an elliptic density of states given by

$$g(\varepsilon) = \frac{1}{\pi^2 l^2} \sum_{i=0}^{\infty} \frac{\tau}{\hbar} \left[1 - \frac{(\varepsilon - \varepsilon_i)^2}{(\hbar/\tau)^2} \right]^{1/2}. \tag{10.16}$$

This results in thermodynamic functions in which the damping factor $W^n(\tilde{\Gamma}) = \exp(-n\tilde{\Gamma})$ for the nth harmonic of the Fourier transform of the Lorentzian case is replaced by $2J_1(\tilde{\Gamma}n)/\tilde{\Gamma}n$, where J_1 is the Bessel function of the first kind and $\tilde{\Gamma} = \pi\hbar/(\tau\mu_B H)$. We note that $\tau/2$ represents the scattering time and that for Si [100] inversion layers with mobility μ and effective mass m,

$$\frac{\hbar}{\tau} = 0.3047 \left[\frac{0.19m_0}{m} \right] \left[\frac{10^4}{\mu(\mathrm{cm}^2\,\mathrm{V}^{-1}\,\mathrm{s}^{-1})} \right] \mathrm{meV},$$

$$\hbar\omega_c = 0.6093H \left[\frac{0.19m_0}{m} \right] \mathrm{meV}, \tag{10.17}$$

$$\tilde{\Gamma} = \frac{\pi}{H} \left[\frac{10^4}{\mu(\mathrm{cm}^2\,\mathrm{V}^{-1}\,\mathrm{s}^{-1})} \right].$$

Here, H is given in Tesla.

11. MAGNETOCONDUCTIVITY

The *Shubnikov–de Haas effect,* discovered in the same year as the dHvA effect, refers to the oscillation of the magnetoconductivity as a function of electron density or a magnetic field. For 2D, this effect was first observed as a function of gate voltage by Fowler et al. in 1966 in an n-channel Si inversion layer.[91] By analyzing the oscillations, they were led to the conclusion that the electron system was two-dimensional, thus

[91]A. B. Fowler, F. F. Fang, W. E. Howard, and P. J. Stiles, *J. Phys. Soc. Jpn. Suppl.* **21,** 331 (1966); *Phys. Rev. Lett.* **16,** 901 (1966).

opening the new field of 2-D electron systems. Subsequently, the effect was observed by many investigators.[92]

On the theoretical side, Ohta[93] was the first to treat 2-D magnetoconductivity in consideration of level broadening. He used the coherent potential approximation (CPA), in which the single-particle Green's function is averaged over the configurations of randomly distributed impurities. Following the prescription of Duke[94] for the self-energy, he then obtained self-consistent coupled equations and derived a conductivity formula. In the same year, Bastin *et al.*[95] obtained a formula for the magnetoconductivity. Although their work is primarily for 3D, its application to 2D is straightforward. Both Ohta and Bastin *et al.* found that the conductivity is proportional to the Landau-level index. In 1974, Ando[96] arrived at the same conclusion. He adopted a simpler self-energy equation than Ohta, which produced more reasonable conductivity curves.

The magnetoconductivity has been observed to reach a maximum as a function of gate voltage. This maximum depends on the field strength, so that the higher the field the lower the maximum and larger the maximum point in electron density. Such a behavior was recently explained by Isihara and Smrčka.[97]

a. *Classical Description*

The conductivity σ and resistivity ρ tensors are defined by

$$\mathbf{J} = \sigma \cdot \mathbf{E}, \qquad \mathbf{E} = \rho \cdot \mathbf{J}, \qquad (11.1)$$

where $\mathbf{J}$ is the current and $\mathbf{E}$ is the electric field. In the presence of a crossing magnetic field $\mathbf{H}$ and in a single relaxation time approximation, the motion of an electron may be determined by the Langevin equation:

$$\dot{\mathbf{v}} = -\frac{e}{m}\left(\mathbf{E} + \frac{\mathbf{v}}{c} \times \mathbf{H}\right) - \frac{\mathbf{v}}{\tau}, \qquad (11.2)$$

[92]K. F. Komatsubara, K. Narita, Y. Katayama, N. Kotera, and M. Kobayashi, *J. Phys. Chem. Solids* **35**, 723 (1974); S. Kawaji, T. Igarashi, and J. Wakabayashi, *Prog. Theo. Phys. Suppl.* **57**, 176 (1975).

[93]K. Ohta, *Jpn. J. Appl. Phys.* **10**, 850 (1971); *J. Phys. Soc. Jpn.* **31**, 1627 (1971).

[94]C. B. Duke, *Phys. Rev.* **168**, 816 (1968).

[95]A. Bastin, C. Lewiner, O. Betbeder-Matibet, and P. Nozières, *J. Phys. Chem. Solids* **32**, 1811 (1971); P. Středa and L. Smrčka, *Phys. Stat. Sol.* (*b*)**70**, 537 (1975).

[96]T. Ando, *J. Phys. Soc. Jpn.* **37**, 1233 (1974).

[97]A. Isihara and L. Smrčka, *J. Phys.* C**19**, 6777 (1986); L. Smrčka and A. Isihara, *Solid State Commun.* **57**, 259 (1986).

where **v** is the average velocity and τ is the relaxation time. The current is given by $\mathbf{J} = -ne\mathbf{v}$ so that in a steady state, the above equation becomes

$$\sigma_0 E_x = \omega_c \tau j_y + j_x,$$
$$\sigma_0 E_y = -\omega_c \tau j_x + j_y, \qquad (11.3)$$

where

$$\sigma_0 = \frac{ne^2\tau}{m} \qquad (11.4)$$

is the Drude conductivity. We arrive at the following resistivity and conductivity formulae:

$$\rho_{xx} = \rho_{yx} = 1/\sigma_0, \qquad \rho_{xy} = -\rho_{yx} = \omega_c\tau/\sigma_0, \qquad (11.5)$$

$$\sigma_{xx} = \sigma_{yy} = \frac{\sigma_0}{1 + (\omega_c\tau)^2}, \qquad \sigma_{xy} = -\sigma_{yx} = -\frac{\sigma_0\omega_c\tau}{1 + (\omega_c\tau)^2}, \qquad (11.6)$$

where σ_{xx} is the longitudinal component and σ_{xy} is the transverse component of the conductivity tensor. These are related to the resistivity through

$$\sigma_{xx} = \frac{\rho_{xx}}{\rho_{xx}^2 + \rho_{xy}^2}, \qquad \sigma_{xy} = -\frac{\rho_{xy}}{\rho_{xx}^2 + \rho_{xy}^2}. \qquad (11.7)$$

Hence, if the resistivity vanishes, the conductivity also vanishes.
On the other hand, Eqs. (11.6) lead to

$$-\sigma_{xy} = \frac{nec}{H} + \frac{\sigma_{xx}}{\omega_c\tau}. \qquad (11.8)$$

We observe that if σ_{xx} vanishes, σ_{xy} is given by the first term. This is a purely classical result. However, if quantum mechanically the electrons fill up to the ith Landau level, $n = ieH/ch$ so that at such a particular *point* (that is, density or magnetic field),

$$\sigma_{xy} = -\frac{e^2}{h}i. \qquad (11.9)$$

b. *Coherent Potential Approximation*

According to linear response theory, the conductivity components are given in terms of the velocity–velocity correlation function:

$$\sigma_{ij}(0) = \frac{\pi \hbar e^2}{A} \int d\varepsilon \left(-\frac{\partial f}{\partial \varepsilon} \right) \langle \mathrm{tr}\, \delta(\varepsilon - \mathcal{H}) v_i \delta(\varepsilon - \mathcal{H}) v_j \rangle, \quad (11.10)$$

where A is the surface area, the average is taken over the configurations of impurities, $f(\varepsilon)$ is the Fermi distribution, and $\mathcal{H}$ is the Hamiltonian. We can evaluate the integrand by using one-electron Green's functions:

$$G_0(z) = \frac{1}{z - \mathcal{H}_0}, \qquad G(z) = \frac{1}{z - \mathcal{H}}, \quad (11.11)$$

where $\mathcal{H}_0$ is the ideal Hamiltonian without impurities. For the complex z, $\delta(\varepsilon - \mathcal{H})$ in Eq. (11.10) can be expressed by $\mathrm{Im}\, G(z)$.

In multiple-scattering theory, the T matrix is introduced such that[98]

$$G = G_0 + G_0 T G_0. \quad (11.12)$$

Averaging over impurity configurations is expressed by

$$\langle G \rangle = G_0 + G_0 \langle T \rangle G_0. \quad (11.13)$$

The averaged Green's function $\langle G \rangle$ is represented by the effective Hamiltonian $\mathcal{H}_{\mathrm{eff}}$:

$$\langle G \rangle = \frac{1}{z - \mathcal{H}_{\mathrm{eff}}}. \quad (11.14)$$

In the presence of impurity scattering, the electron energy may be shifted and its motion damped. Hence, we introduce a complex self-energy such that

$$\mathcal{H}_{\mathrm{eff}} = \mathcal{H}_0 + \Sigma. \quad (11.15)$$

We find immediately that

$$\langle G \rangle = G_0 + G_0 \Sigma \langle G \rangle. \quad (11.16)$$

The effective Hamiltonian is determined by averaging the T matrix. We

[98] B. Velický, S. Kirkpatrick, and H. Ehrenreich, *Phys. Rev.* **175**, 747 (1968); P. Soven, *Phys. Rev.* **156**, 809 (1967).

first note the identity $G = G_0 + G_0(\mathcal{H} - \mathcal{H}_0)G$. In combination with Eq. (11.12), we arrive at

$$T = \frac{1}{1 - (\mathcal{H} - \mathcal{H}_0)G_0}(\mathcal{H} - \mathcal{H}_0). \qquad (11.17)$$

On the other hand, from Eqs. (11.13) and (11.16), we find

$$\mathcal{H}_{\text{eff}} = \mathcal{H}_0 + \langle T \rangle(1 + G_0\langle T \rangle)^{-1}. \qquad (11.18)$$

For a short-range potential of strength V_0 and a small impurity fraction c, the following self-consistent equation for Σ may be adopted:[98]

$$\Sigma = cV_0 + \frac{c(1 - c)V_0^2 K(z - \Sigma)}{1 + [\Sigma + (c - 1)V_0]K(z - \Sigma)}, \qquad (11.19)$$

where

$$K(z - \Sigma) = \text{tr}\langle G(z) \rangle. \qquad (11.20)$$

Note that $K(z) = \text{tr } K_0(z)$. A combination of Eqs. (11.16), (11.19), and (11.20) represents a self-consistent set of equations for $\langle G \rangle$.

Ohta's theory is equivalent to neglecting Σ on the right side of Eq. (11.19). For strong magnetic fields, mixing of Landau levels may be neglected. Hence, we may have K for each Landau level. Neglecting in Eq. (11.19) the first term, the c^2 term, and the denominator representing multiple scattering, we write for the nth Landau level:

$$\Sigma_n = \tfrac{1}{4}\Gamma^2\langle G_n \rangle, \qquad (11.21)$$

where Γ is a constant. The density of states, which is given in terms of the imaginary part of $\langle G_n \rangle$, is then expressed by

$$g_n(\varepsilon) = \frac{1}{\pi^2 l^2 \Gamma}\left[1 - \left(\frac{\varepsilon - \varepsilon_n}{\Gamma}\right)^2\right]^{1/2}. \qquad (11.22)$$

This agrees with what Ando et al.[99] obtained. Here, $l^2 = c\hbar/eH$, the factor $1/(2\pi l^2)$ has been introduced so as to yield eH/ch, which is the natural width, when the density of states is integrated over ε, and

$$\Gamma^2 = \frac{4}{2\pi l^2}n_i V_0^2, \qquad (11.23)$$

[99]T. Ando, Y. Matsumoto, Y. Uemura, M. Kobayashi, and K. F. Komatsubara, *J. Phys. Soc. Jpn.* **32**, 859 (1972).

where n_i is the impurity concentration, and V_0 is the strength of short-range delta-function-type potential. The conductivity can be shown to be given by[100]

$$\sigma_{xx}(T) = \frac{e^2}{\pi^2 \hbar} \sum_n (n + \tfrac{1}{2}) \int_{-\infty}^{\infty} d\varepsilon \left(-\frac{\partial f}{\partial \varepsilon}\right)\left[1 - \left(\frac{\varepsilon - \varepsilon_n}{\Gamma}\right)^2\right]. \quad (11.24)$$

Kawabata[101] derived a low-temperature series for σ_{xx}.

c. Shubnikov–de Haas Effect

For discussing the gate voltage and magnetic field dependencies of the static conductivity, we note

$$K(z - \Sigma) = \frac{\hbar \omega_c}{\varepsilon_M} \sum_0^{n_M} \frac{1}{z - \hbar \omega_c (n + \tfrac{1}{2}) - \Sigma}, \quad (11.25)$$

where n_M is a cutoff associated with the maximum band energy by $\varepsilon_M/\hbar\omega_c - \tfrac{1}{2}$. Assuming that the oscillating part of K is small in comparison with its nonoscillating part, Isihara and Smrčka[97] arrived at the conductivity given by

$$\sigma_{xx} = \frac{\sigma_0}{1 + (\omega_c \tau)^2}\left[1 + \frac{2\omega_c^2\tau^2}{1 + (\omega_c\tau)^2}\frac{\Delta g}{g_0}\right], \quad (11.26)$$

$$\sigma_{xy} = -\frac{\sigma_0 \omega_c \tau}{1 + (\omega_c \tau)^2}\left[1 - \frac{3\omega_c^2\tau^2 + 1}{\omega_c^2\tau^2(\omega_c^2\tau^2 + 1)}\frac{\Delta g}{g_0}\right], \quad (11.27)$$

where σ_0 is the Drude conductivity given by Eq. (11.4), τ is determined by Σ, and

$$\frac{\Delta g}{g_0} = 2 \sum_{s=1}^{\infty} \exp(-\pi s/\omega_c\tau)\frac{2\pi^2 kT/(\hbar\omega_c)}{\sinh(2\pi^2 skT/(\hbar\omega_c))}\left[\cos\left(\frac{2\pi s \varepsilon_F}{\hbar\omega_c} - \pi s\right)\right]$$

$$(11.28)$$

represents the ratio of the oscillating part of the density of states to the nonoscillating part. The resistivity components are

$$\rho_{xx} = \frac{1}{\sigma_0}\left(1 + 2\frac{\Delta g}{g_0}\right),$$

$$(11.29)$$

$$\rho_{xy} = \frac{1}{\sigma_0}\left(1 - \frac{1}{(\omega_c\tau)^2}\frac{\Delta g}{g_0}\right)\omega_c\tau.$$

[100]T. Ando and Y. Uemura, *J. Phys. Soc. Jpn.* **36**, 959 (1974).
[101]A. Kawabata, *Surf. Sci.* **98**, 276 (1980).

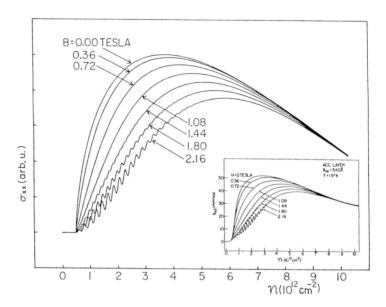

FIG. 27. Density dependence of the magnetoconductivity in arbitrary units. Solid curves: Reprinted with permission from A. Isihara and L. Smrčka, *J. Phys.* *C*19, 6777 (1986), Pergamon Journals, Ltd. Inset: Data of F. F. Fang, A. B. Fowler, and A. Hartstein, *Phys. Rev. B*16, 4446 (1977).

The solution to the self-consistent equation (11.19) can be found easily for a delta-function potential. The magnetoconductivity thus evaluated is illustrated in Fig. 27 and compared with the data of Fang *et al.*[102] on [100] Si accumulation layers. The theoretical parameters have been chosen such that $\varepsilon_M = 190$ meV, $c = 0.06$, and $V_0 = \infty$. As in the experiment, the effective mass is $m = 0.21m_0$ and the temperature is 1.9 K. The inset represents the data of Fang *et al.* The theoretical curves reproduced the data well, but even more refined theoretical curves can be obtained by using a better impurity potential.[97] The conductivity increases first rather rapidly, and reaches a maximum. The maximum decreases with the magnetic field, and the maximum point is shifted toward a larger concentraion when the field is increased. The maximum appears because the conductivity is effectively determined by the product of the electron density and the relaxation time. Thus, the conductivity increases first as the gate voltage increases, but then, after a maximum starts decreasing because impurity scattering becomes more and more effective and the relaxation time decreases.

[102]F. F. Fang, A. B. Fowler, and A. Hartstein, *Phys. Rev. B*16, 4446 (1977).

The conductivity depends on the effective mass. However, the effective mass varies in a complicated way. Fang *et al.* found that with substrate bias, a reverse bias between the inversion layer and substrate reduced the apparent effective mass for a given carrier density, and an open channel geometry gave markedly different results. Also, various degrees of substrate doping caused wide spreading of the effective mass. Moreover, artificially introduced Na^+ ions were found to cause a striking effect on the effective mass where even the apparent effective mass increased with electron concentration when Na^+ ions were above a certain concentration.

The line shape of SdH oscillations in GaAs/GaAlAs heterostructures[103] or Si MOSFETs[104] has been found to be asymmetric. For high-mobility cases, Haug *et al.*[105] explained such asymmetry on the basis of an asymmetric shape of the density of states that is expected if either attractive or repulsive scatterers dominate and if the number of scattering centers within one Landau orbit is small. By applying a back-gate voltage, they inverted the oscillations. This voltage shifts the wave function relative to the differently charged impurities and explains the observed line shape.

d. *Narrow 2-D Systems*

Recently, narrow 2-D systems have attracted considerable attention because they show a variety of new phenomena ranging from 1-D to 2-D.[106] In very narrow wires, the conductance shows irregular but reproducible spikes. Such structures were numerically simulated by Lee and Stone.[107] They also predicted universal fluctuations of order e^2/h in the conductance of small systems. In wider but finite-width systems,

[103] K. von Klitzing, *Physica B***126**, 242 (1984); H. Z. Zhen, K. K. Choi, D. C. Tsui, and G. Weimann, *Phys. Rev. Lett.* **55,** 1144 (1985); R. Woltjer, J. Mooren, J. Wolter, J.-P. André, and G. Weimann, *Physica B***134**, 352 (1985).

[104] K. von Klitzing, G. Ebert, N. Kleinmichel, H. Obloh, G. Dorda, and G. Weimann, *in Proc. Internat. Conf. on Phys. Semicond.* (J. D. Chadi and W. A. Harrison, eds.) p. 271. Springer, 1985.

[105] R. J. Haug, K. von Klitzing, and K. Ploog, *Phys. Rev. B***35,** 5933 (1987).

[106] R. G. Wheeler, K. K. Choi, A. Goel, R. Wisnieff, and D. E. Prober, *Phys. Rev. Lett.* **49,** 1674 (1982). See also *Surf. Sci.* **142,** 19 (1984); A. B. Fowler, A. Hartstein, and R. A. Webb, *Phys. Rev. Lett.* **48,** 196 (1982); *Physica,* *B***117/118,** 661 (1983); A. Hartstein, R. A. Webb, A. B. Fowler, and J. J. Wainer, *Surf. Sci.* **142,** 1 (1984); W. J. Skocpol, L. D. Jackel, R. E. Howard, H. G. Craighead, L. A. Fetter, P. M. Mankiewich, P. Grabbe, and D. M. Tennant, *Surf. Sci.* **142,** 14 (1984); R. A. Webb, A. B. Fowler, A. Hartstein, and J. J. Wainer, *Surf. Sci.* **170,** 14 (1986); A. D. C. Grassie, K. M. Hutchings, M. Larkimi, C. T. Foxon, and J. J. Harris, *Phys. Rev. B***36,** 4551 (1987).

[107] P. A. Lee, *Phys. Rev. Lett.* **53,** 2042 (1984); P. A. Lee and A. D. Stone, *Phys. Rev. Lett.* **55,** 1622 (1985); A. D. Stone, *Phys. Rev. Lett.* **54,** 2692 (1985).

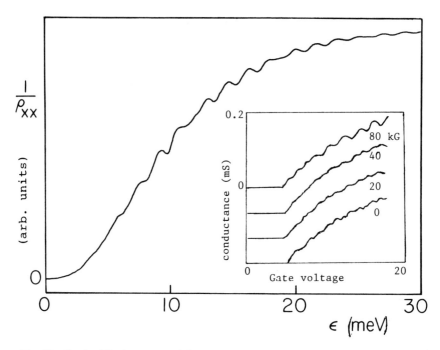

FIG. 28. Conductivity in a narrow 2-D system. Top curve: A. Isihara, H. Havlova, and L. Smrčka, *J. Phys.* **C21,** 645 (1988). Inset: Data by W. J. Skocpol *et al., Surf. Sci.* **142,** 14 (1984).

irregular oscillations have still been observed in certain ranges of electron density and the magnetic field. This has been explained by Isihara *et al.*[108] in terms of a parabolic potential model with width fluctuations. In this model, the system is considered to consist of several segments with some fluctuations in the confining potential $V(y)$:

$$V(y) = V_0 + \frac{m\Omega^2}{2} y^2.$$

Their typical result for $H = 0$ and with three segments is illustrated in Fig. 28 and compared with the data of Skocpol *et al.*[106] (inset). Note that irregular oscillations are strong, particularly when the conductivity increases fast and the magnetic field is weak. The width is 110 nm and $T = 2$ K. When the field becomes sufficiently strong, regular SdH

---

[108]L. Smrčka, H. Havlova, and A. Isihara, *Proc. Euro. Phys.,* to be published; A. Isihara, H. Havlova, and L. Smrčka, *J. Phys.* **C21,** 645 (1988).

oscillations replace the irregular oscillations. These theoretical results agree generally well with the data of Skocpol et al. on Si inversion layers at 110 nm. The appearance of irregular oscillations due to width fluctuations is interesting, especially in view of the recent experiment of Grassie et al.[106] on submicron-width GaAs/GaAlAs heterostructures. They found that the narrower the wires, the higher the scattering rate. Their experiment shows that the finiteness of the sample width is not negligible at around $0.4 \mu m$.

In the above parabolic potential model, the "width" for electron energy ε is represented by $w(\varepsilon) = (2/\Omega)(2\varepsilon/m)^{1/2}$. Figure 28 was obtained for 1 K and $m = 0.19 m_0$ and $\Omega_1 = 1.8 \times 10^{12} \, s^{-1}$, $V_{01} = 4\hbar\Omega_1/5$, $\Omega_2 = 2.7\pi^{-1/2} \times 10^{12} \, s^{-1}$, $V_{02} = \hbar\Omega_2/3$, $\Omega_3 = 2.7 \times 10^{12} \, s^{-1}$, $V_{03} = 0$ for the three segments.

Transport in narrow 2-D systems depends not only on the width but also on several characteristic lengths such as the Fermi wavelength, inelastic diffusion length, phase coherence length, mean free path, screening length, etc. At low electron densities, transport is generally dominated by tunneling and hopping through localized states. As the electron density increases, diffusion becomes important. Electron interactions reduce the interference, which is crucial to localization, and also change the Fermi energy, which determines the diffusion. These effects are expected especially in low-dimensional systems where the diffusion is constrained. Interaction effects have been reported to cause important contributions to the temperature dependence in narrow systems. However, the temperature dependence must be unambiguously dissociated from a similar dependence due to the zero-field localization. We shall discuss some aspects of interaction effects on localization in Section 13.

12. CYCLOTRON RESONANCE

Since around 1974, cyclotron resonance has been used extensively for the investigation of 2-D electron systems.[109] Cyclotron resonance provides two important quantities—effective mass and relaxation time—and is a valuable tool for probing electron systems. Experiments, especially, have shown that these quantities vary with electron density, indicating a possible role played by electron correlations.

Theoretically, cyclotron resonance is determined by the real part of the dynamic conductivity. Hence, let us first discuss some theoretical aspects

[109]G. Abstreiter, P. Kneschaurek, J. P. Kotthaus, J. F. Koch, and G. Dorda, *Phys. Rev. Lett.* **32,** 104 (1974); S. J. Allen, Jr., D. C. Tsui, and J. V. Dalton, *Phys. Rev. Lett.* **32,** 107 (1974).

of the dynamic conductivity. In 1972, Götze and Wölfle[110] developed a memory functional formalism for 3-D and expressed the dynamic conductivity in the following form:

$$\sigma(\omega) = \frac{\omega_p^2}{4\pi[\omega + M(\omega)]}, \qquad (12.1)$$

where ω_p is the 3-D plasmon frequency, and $M(\omega)$ is called the *memory function*. The memory function consists of the real part M_1 and the imaginary part M_2. Götze and Wölfle derived an explicit formula for M_2 to lowest order in impurity scattering. The form of Eq. (12.1) is convenient because both static and dynamic cases can be discussed. For 2-D, the same form was used by Fukuyama et al.[111] for a dimensional consideration of electron correlation effects.

The dynamic conductivity of a 2-D system in a magnetic field was treated approximately by Ting et al.[112] and Götze and Hajdu.[113] An exact theory within the framework of linear response theory has been given by Shiwa and Isihara.[114] They found the dynamic conductivity to be

$$\sigma_\pm(\omega) = \sigma_{xx} \pm i\sigma_{xy}$$

$$= \frac{ine^2/m}{\omega \mp \omega_c + M(\omega)\{1 - M(\omega)[\omega \mp \omega_c]^{-1}\}^{-1}}, \qquad (12.2)$$

where the memory function was expressed in terms of a correlation function. For small impurity effects, all these theories lead to

$$\sigma(\omega) = \frac{ine^2/m}{\omega \mp \omega_c + M(\omega)}, \qquad (12.3)$$

and

$$M(\omega) = \frac{1}{4\pi\omega m}\frac{n_i}{n}\int_0^\infty dq q^3 |v(q)|^2 u^{-1}(q)[\varepsilon^{-1}(q, \omega) - \varepsilon^{-1}(q, \omega)], \qquad (12.4)$$

where n_i is the impurity concentration, $v(q)$ is the impurity potential, $u(\mathbf{q}) = 2\pi e^2/q$, and $\varepsilon(q, \omega)$ is the dielectric function. In the static limit, the integrand may be expressed by the derivative $\lim_{\omega \to 0} d(1/\varepsilon)/d\omega$.

[110]W. Götze and P. Wölfle, *Phys. Rev.* **B6**, 1226 (1972).

[111]H. Fukuyama, Y. Kuramoto, and P. M. Platzman, *Phys. Rev.* **B19**, 4980 (1979); **B16**, 5394 (1977); *Surf. Sci.* **73**, 491 (1978).

[112]C. S. Ting, S. C. Ying, and J. J. Quinn, *Phys. Rev.* **B16**, 5394 (1977).

[113]W. Götze and J. Hajdu, *J. Phys.* **C11**, 3993 (1978); *Solid State Commun.* **29**, 89 (1979).

[114]Y. Shiwa and A. Isihara, *J. Phys.* **C16**, 4853 (1983).

The dynamic conductivity in the absence of a magnetic field was measured by Allen *et al.*[115] for Si inversion layers at 1.2 K. They used four carrier densities ranging from a high of 15.3×10^{11} cm^{-2} to a low of 1.8×10^{11} cm^{-2}. In the metallic region, the Drude expression was found to fit the data well. At low densities, its application was limited to frequencies larger than a threshold value corresponding to the activation energy of the static conductivity, which was observed only for low densities. Below the threshold frequency, the Drude form showed significant deviations at microwave frequencies. The appearance of the activated conductivity at low densities may be due to the excitation of electrons across a gap to conduction states. If so, the gap may depend on electron correlations.

Returning to the cyclotron resonance, Eq. (12.3) gives the effective mass m^* and relaxation time as follows:

$$m^* = m[1 + M_1(\omega)/\omega],$$
$$\tau = M_2^{-1}[1 + M_1(\omega)/\omega].$$

(12.5)

The real part M_1 may be determined from the imaginary part by applying the Krammers–Kronig relation. Hence, from the memory function, the two relevant parameters can in principle be evaluated. However, broadening of Landau levels must be taken into consideration in order to avoid a divergence. In 1975, Ando[116] applied his self-consistent theory of magnetoconductivity to the evaluation of the line shape of cyclotron resonance as a function of the applied magnetic field. The density dependence of the cyclotron effective mass and some other aspects of cyclotron resonance were treated by Lee *et al.*,[117] Ganguly and Ting,[118] and Isihara and Mukai.[119] More recently, Kallin and Halperin[120] studied to the lowest order in $(e^2/\kappa l)/\hbar\omega_c$ the special case in which an integral number of Landau levels is filled. They considered impurity scattering in a self-consistent but different way from Ando. On the other hand, Shiwa and Isihara[121] developed an approach that improves the RPA based on

[115]S. J. Allen, Jr., D. C. Tsui, and F. Rosa, *Phys. Rev. Lett.* **35,** 1359 (1975). See also K. C. Woo and P. J. Stiles, *Phys. Rev.* **B28,** 287 (1951).

[116]T. Ando, *J. Phys. Soc. Jpn.* **38,** 111 (1975).

[117]T. K. Lee, C. S. Ting, and J. J. Quinn, *Solid State Commun.* **16,** 1309 (1975).

[118]A. K. Ganguly and C. S. Ting, *Phys. Rev.* **B16,** 3541 (1977).

[119]A. Isihara and M. Mukai, *Phys. Rev.* **B24,** 7408 (1981); *Phys. Rev.* **B28,** 4842 (1983); *J. de Phys. Colloq.* C3, *Suppl.* **44,** C3–1449 (1983).

[120]C. Kallin and B. I. Halperin, *Phys. Rev.* **B30,** 5655 (1984); **B31,** 3635 (1985).

[121]Y. Shiwa and A. Isihara, *Solid State Commun.* **53,** 519 (1985).

Götze's truncation method.[122] In what follows, we follow the latter authors in the discussion of correlation effects.

We first note that the memory function can be expressed in terms of the density–density correlation function as follows:

$$M(\omega) = \frac{n_i}{4\pi nm} \int_0^\infty d\mathbf{q} q^3 |v(q)|^2 C(q, \omega). \qquad (12.6)$$

We introduce the dynamical variables A_i such that A_0 (or A_1) is proportional to the Fourier transform of the density (or current) and note that the correlation functions C_{ij} defined by

$$C_{ij}(\omega) = i \int_0^\infty dt e^{i\omega t}(A_i(t), A_j)$$

satisfy a matrix equation of the following form:

$$[\omega \cdot 1 - \Omega + M] \cdot C(\omega) = -1, \qquad (12.7)$$

where $\Omega_{ij} = i(\dot{A}_i, A_j)$. This equation may be solved in successive approximations. Its zeroth approximation satisfies

$$[\omega \cdot 1 - \Omega] \cdot C^{(0)} = -1.$$

By using this in Eq. (12.7), we find the next approximation in the form:

$$C_{00}^{(1)} = \frac{C_{00}^{(0)}(\omega + M(\omega))}{1 + M(\omega)C_{00}^{(0)}(\omega + M(\omega))}. \qquad (12.8)$$

The RPA may be used for the zeroth approximation:

$$C_{00}^{(0)} = \frac{1}{\omega}[\chi_0(q, \omega) - \chi_0(q, 0)], \qquad (12.9)$$

where $\chi_0(q, \omega)$ is the response function. The next approximation leads to an integral equation:

$$M(\omega) = \frac{n_i}{4\pi nm} \int_0^\infty d\mathbf{q} q^3 |v(q)|^2 \frac{C_{00}^{(0)}(q, \omega + M(\omega))}{1 + M(\omega)C_{00}^{(0)}(q, \omega + M(\omega))/\chi_0(q, 0)}. \qquad (12.10)$$

[122]W. Götze, Phil. Mag. B43, 219 (1981); J. Phys. C12, 1279 (1979); A. Gold and W. Götze, J. Phys. C14, 4049 (1981); A. Gold, S. J. Allen, C. B. A. Wilson, and D. C. Tsui, Phys. Rev. B25, 3519 (1982); A. Gold and W. Götze, Phys. Rev. B33, 2495 (1986).

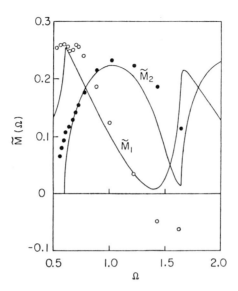

FIG. 29. Frequency dependence of the reduced memory function $\tilde{M} = M/\hbar\omega_c$ as a function of $\Omega = \omega/\omega_c$. Solid curves: Reprinted with permission from Y. Shiwa and A. Isihara, *Solid State Commun.* **53**, 519 (1985), Pergamon Journals, Ltd. Circles: Data of S. J. Allen, Jr., B. A. Wilson, and D. C. Tsui, *Phys. Rev.* **B26**, 5590 (1982).

This is a self-consistent equation to determine the memory function. Since the memory function is complex, the equation is decomposed into coupled integral equations for the real and imaginary parts. In terms of the memory function, the conductivity is given by

$$\sigma_{xx} = \frac{(ne^2/m)M_2}{(\omega - \omega_c + M_1)^2 + M_2^2},$$

$$\sigma_{xy} = \frac{(ne^2/m)(\omega - \omega_c + M_1)}{(\omega - \omega_c + M_1)^2 + M_2^2}.$$

(12.11)

For explicit results, Shiwa and Isihara adopted an impurity potential $v(q)$ given by

$$n_i |v(q)|^2 = 4\pi r_0^2 u^2 \exp[-(qr_0)^2]. \tag{12.12}$$

The frequency dependence of their theoretical memory function is compared with the data of Allen *et al.*[123] in Fig. 29. Here, the real and

[123]B. A. Wilson, S. J. Allen, Jr., and D. C. Tsui, *Phys. Rev. Lett.* **44**, 479 (1980); *Phys. Rev.* **B24**, 5887 (1981); S. J. Allen, Jr., B. A. Wilson, and D. C. Tsui, *Phys. Rev.* **B26**, 5590 (1982).

imaginary parts of the reduced memory function $\bar{M} = M/\hbar\omega_c$ are plotted against a reduced frequency $\Omega = \omega/\omega_c$. The circles represent the data. The density is 2.3×10^{11} cm^{-2}. The theoretical parameters are chosen such that $Q_c = l^2/2r_0^2 = 2.0$, $g_v = 2$, and $\bar{u} = u/\hbar\omega_c = 1.9$. The imaginary part of the memory function exhibits broad resonance near each cyclotron harmonic frequency, and the real part accordingly displays a strong dispersion. These theoretical results agree qualitatively with the experiment of Wilson et al., although in detail there are deviations.

According to the theory of Shiwa and Isihara, the memory function depends on the electron density through a parameter $\bar{r}_s = (2/g_d)r_s v^{3/2}$,

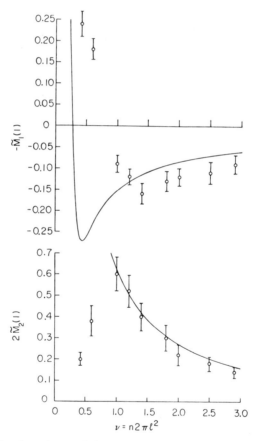

FIG. 30. Density dependence of the memory function in a Si inversion layer. Solid curves: Reprinted with permission from Y. Shiwa and A. Isihara, *Solid State Commun.* **53,** 519 (1985), Pergamon Journals, Ltd. Circles: B. A. Wilson, S. J. Allen, Jr., and D. C. Tsui, *Phys. Rev. Lett.* **44,** 479 (1980), *Phys. Rev.* **B24,** 5887 (1981).

where v is the filling factor $n/2\pi l^2$, and g_d represents the combination of the valley and spin degeneracies that are actually lifted in the present experiment. However, in Fig. 30 the memory function is plotted against the filling factor v for comparison with the data of Wilson et al.[123] The top (bottom) curve represents shifting (broadening) at $\omega = \omega_c$. The data points have been obtained by a piecewise Kramers–Kronig analysis. For the theoretical curves, the parameters are chosen such that $Q_c = 2$, $g_v = 1$, and $\tilde{u} = 2$. A new value for Q_c has been chosen because the particular data point in Fig. 29 deviates from the rest in Fig. 30. The theoretical $-\tilde{M}_1$ shows a minimum as in the data, but both the minimum value and its position are off the data. The theoretical $\tilde{M}_2$ does not show a maximum. These deviations are understandable because the theory employs a high-density approximation. Hence, the comparison must be restricted to the region $v > 1$. In this region, both $\tilde{M}_1$ and $\tilde{M}_2$ decrease due to electron correlations.

For the case $v < 1$, the system is expected to be in a highly correlated charge density or in Wigner lattice states. Therefore, an entirely new approach is necessary. Fukuyama and Lee[124] proposed defect pinning of the charge density wave for shifting and broadening. However, Cho et al.[125] did not observe such shifting.

In comparison with the data in Si inversion layers, more complicated resonance behavior has been observed in GaAs/GaAlAs. Englert et al. found that the line width of the far-infrared cyclotron resonance in these heterostructures showed oscillations between 0.070 and 0.32 T as a function of the resonance magnetic field at 5 K.[126] Figure 31 illustrates the line width of cyclotron resonance as a function of the resonance magnetic field. The ordinate, representing the full width at half maximum (FWHM), has two distinctive maxima below 5 T, one at around 3 T and the other at 1 T. From the results of SdH measurements on the same sample, it was found unambiguously that the dominant maximum at 3 T occurs when the lowest, spin degenerate Landau level is completely filled, corresponding to a filling factor $v = 2$. The one at a lower field can be attributed to $v = 4$. The inset in the graph shows the temperature dependence of the line width at 2.56 T (maximum) and at 10.74 T (minimum). An unusual temperature dependence was observed in the line width maximum: the cyclotron resonance sharpened by a factor of two when the temperature was increased from 1.5 to 40 K. The minimum

[124]H. Fukuyama and P. A. Lee, Phys. Rev. B18, 6245 (1978).

[125]A. Y. Cho, D. C. Tsui, and R. J. Wagner, Bull. Am. Phys. Soc. 25, 266 (1980).

[126]Th. Englert, J. C. Maan, Ch. Wihlein, D. C. Tsui, and A. C. Gossard, Solid State Commun. 46, 545 (1983); M. J. Chou, D. C. Tsui, and G. Weimann, Phys. Rev. B37, 848 (1988-I).

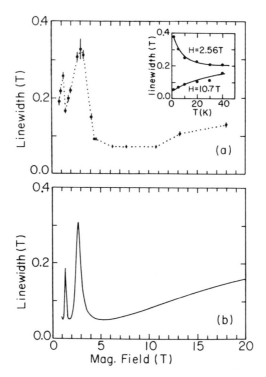

FIG. 31. Resonance magnetic field dependence of the cyclotron resonance linewidth in GaAs/GaAlAs. Reprinted with permission from Th. Englert *et al.*, *Solid State Commun.* **46,** 545 (1983), Pergamon Journals, Ltd.

showed the opposite temperature dependence: the line broadened from 0.5 to 0.17 T. In other words, the oscillations in the line width were smeared at higher tempeatures of order 40 K and approached roughly the mean value between the maxima and minima. Similar oscillations of the line width were observed in some other samples.

The resonance magnetic field plotted for several laser energies revealed that the data points for low magnetic fields lie on a straight line, corresponding to a cyclotron mass of $(0.069 \pm 0.001)m_0$. As the magnetic field increased, deviations from the straight line appeared, presumably due to nonparabolicity of the band. For magnetic fields larger than 5.5 T, the system was in the extreme quantum limit. The cyclotron line width showed a broad minimum at around $v = 1.0 \sim 0.5$ and then increased gradually at higher fields up to 18 T. Except for the data taken for wavelength $\lambda = 41 \, \mu$m, the position of the resonance was found to agree well with that expected from the nonparabolic band model. With

$\lambda = 41~\mu$m, the resonance occured at $v = 0.28$. This is slightly less than one-third, where a certain anomaly may be expected. At this frequency, the line width narrowed to approximately one-half of its average value, but the magnetic field position of the resonance was approximately 1% less than that expected from the nonparabolic band mass.

Englert *et al.* performed a model calculation of the line width under the assumption that the broadening parameter of the line width consists of oscillating and nonoscillating parts:

$$\frac{1}{\Gamma} = \frac{1}{\Gamma_0} + \frac{v_n(2 - v_n)}{\Gamma_1},$$

where Γ_0 and Γ_1 were chosen to be 0.31 T and 0.092 T so as to reproduce the data well. The filling factor was evaluated from

$$v_n = 2 \int g(\varepsilon) f(\varepsilon - \varepsilon_F)\, d\varepsilon,$$

where the density of states $g(\varepsilon)$ was assumed to be Gaussian. Their result shown in Fig. 31 agrees well with the data. A more detailed analysis, based on screening constants which depend on the filling factor v, has been given by Lassnig and Gornik.[127]

On the other hand, Schlesinger *et al.*[128] reported that the line width in cyclotron resonance at 1.3 K and $n = 1 \times 10^{11}~\mathrm{cm}^{-2}$ of GaAs/GaAsAs was essentially independent of the field from 8 to 140 kG. At a higher electron density of $3.9 \times 10^{11}~\mathrm{cm}^{-2}$ an abrupt splitting of the cyclotron absorptance was observed in the neighborhood of 45 kG. Away from this splitting, the line width remained roughly constant.

The splitting was not due to spin. With a minimum detectable level of 0.01, no absorptance was found at the second harmonic frequency. At temperatures over 20 K, the two peaks coalesced into a single absorption, and the cyclotron line width became independent of magnetic field. Contrary to Englert *et al.*, the line-width maximum (or splitting) did not necessarily occur when Landau levels were exactly filled ($v = 2, 4, \ldots$). Moreover, the critical frequency at which the splitting took place increased from $v = 2$ at $n = 1 \times 10^{11}~\mathrm{cm}^{-2}$ to -3.7 at $n = 4 \times 10^{11}~\mathrm{cm}^{-2}$, as shown in Fig. 32. The line represents $\hbar\omega_c = n^{1/2}e^2/\kappa$, where κ is the static

[127]R. Lassnig and E. Gornik, *Solid State Commun.* **47**, 959 (1983).
[128]Z. Schlesinger, S. J. Allen, J. C. M. Hwang, P. M. Platzman, and N. Tzoar, *Phys. Rev.* **B30**, 435 (1984). See also G. Lindemann, R. Lassning, W. Seidenbusch, and E. Gornik, *Phys. Rev.* **B28**, 4693 (1983).

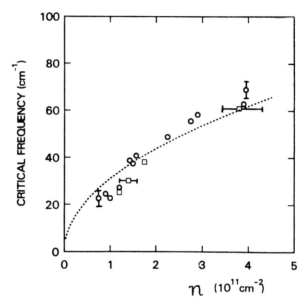

FIG. 32. Density dependence of the critical frequency. Z. Schlesinger *et al.*, *Phys. Rev.* **B30**, 435 (1984).

dielectric constant. This relation means that the average Coulomb energy and cyclotron energy are equal to each other. The data are found not to follow a constant v line. Hence, the split may be associated with magnetoplasmons.

The resonance frequency and FWHM of GaAs differ from those of Si inversion layer electrons. However, most differences are accounted for by the effective mass difference: $m = 0.069m_0$ in GaAs/GaAlAs as opposed to $0.19m_0$ in Si/SiO$_2$.

Recently, Pakulis *et al.* reported a new microwave-induced resonance in the conductance of a GaAs/GaAlAs heterostructure.[129] They observed two resonances per Landau level due to nonresonance spin transitions within each Landau level. They reported that it is the Landau-level structure itself that converts these nonresonant transitions into resonant changes in conductance. Their analysis shows that the density of states is possibly not parabolic. If it is expressed in the form

$$g(\varepsilon) \sim \exp[-|\varepsilon - \varepsilon_n|^p / 2\Gamma^p],$$

[129]E. J. Pakulis, F. F. Fang, and M. Heiblum, *In Proc. Internat. Conf. on Phys. Semicond.* (J. D. Chadi and W. A. Harrison, eds.) p. 445, Springer, 1985.

$p = 3$ gives the best account of their data. The same exponent has been suggested by Moriyama nad Kawaji[130] for the explanation of the temperature dependence of the Hall conductivity in Si MOS inversion layers.

For actual 2-D systems, transport theories based on realistic models have been given by several authors.[131] However, we mention only that in a recent article Gold[132] evaluated the frequency dependence of the scattering layer and the mobility as a function of the position of the dopant layer.

More recently, Chou et al. performed a cyclotron resonance experiment on a high-mobility GaAs/GaAlAs heterostructure at 2.3 and 4.2 K with electron densities from 1.03×10^{11} cm^{-2} to 1.55×10^{10} cm^{-2} for $\hbar\omega_c = 12.85$ meV.[126] This energy is larger than the average Coulomb energy (~ 11.9 meV) and is smaller than the (LO) phonon energy (~ 36 meV). Due to the high mobility, the line width was very narrow, making an accurate determination of the cyclotron effective mass possible. They found that this mass is constant to better than 0.025% between $v = 0.4$ and 0.14 and decreases slightly for lower v, presumably because of localization. The cyclotron lifetime is found to be 104 ps at $v = 0.4$. This lifetime is 15 times the dc scattering time. In the range, $0.08 < v < 0.4$, it follows that

$$\tau \sim n^{(1.9\pm0.1)}.$$

Chou et al. attributed this dependence to scattering due to screened residual ionized impurities in GaAs. They also shed light on the oscillatory behavior of the line width. The half width at half maximum (HWHM) of resonance lines is proportional to the CR mobility μ_{CR} at relatively low mobilities. As the mobility becomes higher, however, the mobility determined from the HWHM underestimates more and more μ_{CR}, and eventually the CR width does not reflect any change of μ_{CR}. They attributed the absence of the observed oscillation effect to this saturation effect.

13. LOCALIZATION

The conductance of actual systems depends on the dimension and on impurities. For many years its study was restricted to the case in which

[130]J. Moriyama and S. Kawaji, *Solid State Commun.* **45**, 511 (1983).

[131]S. Mori and T. Ando, *J. Phys. Soc. Jpn.* **48**, 865 (1980); J. Lee, H. Spector, and V. Arora, *Appl. Phys. Lett.* **42**, 363 (1983); *J. Appl. Phys.* **54**, 6995 (1983); H. Spector and V. Arora, *Surf. Sci.* **159**, 425 (1985).

[132]A. Gold, *Phys. Rev. B***35**, 723 (1987).

the impurity concentration is small so that the conductance is proportional to the impurity concentration. A study of the opposite case was started by Anderson,[133] who introduced the concept of localization in 1958. However, further theoretical progress did not follow immediately. In 1979, Abrahams et al.[134] reported a scaling theory that triggered remarkable progress in the study of what is now known as *Anderson localization* in disordered systems.

In this scaling theory, the conductance $\sigma_L = \sigma(L)$ of a system with length L in each direction is considered as a function of L. If the length L is large enough, the conductance should be proportional to the cross section L^{d-1} and inversely proportional to the length L, where d is the dimension. Hence, the conductance may be expressed as

$$\sigma(L) = \sigma_0 L^{d-2}. \tag{13.1}$$

Let the conductance of another system with length L' in each direction be $\sigma(L')$. Let us assume that $\sigma(L')$ depends on L' in such a way that

$$\sigma(L') = f\left(\sigma(L), \frac{L'}{L}\right), \tag{13.2}$$

where f is a universal function independent of the microscopic structure of the system. A logarithmic differentiation of this expression at $x = L'/L = 1$ leads to

$$\frac{d \ln \sigma}{d \ln L} = \beta(\sigma). \tag{13.3}$$

That is, the derivative on the left side is a function of σ only.

When the system is metallic and σ is large, Eq. (13.1) suggests that

$$\beta(\sigma) = d - 2. \tag{13.4}$$

In the opposite limit, the conductance may be considered to decrease exponentially with L. If $\sigma_L \sim \exp(-\alpha L)$ for $L \to \infty$, we obtain

$$\beta(\sigma) = \ln \sigma. \tag{13.5}$$

The variation of $\beta(\sigma)$ with σ may be assumed to be given by smoothly

[133]P. W. Anderson, *Phys. Rev.* **102**, 1008 (1958).
[134]E. Abrahams, P. W. Anderson, D. C. Licciardello, and T. V. Ramakrishnan, *Phys. Rev. Lett.* **42**, 673 (1979).

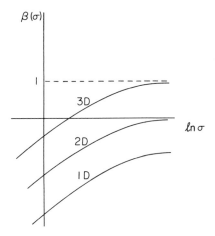

FIG. 33. Schematic σ dependence of $\beta(\sigma)$.

connecting these two limits, as shown schematically in Fig. 33. According to this graph, $\beta(\sigma)$ in 3D becomes positive above a critical conductance. Hence, the electrons are extended below a certain concentration of the impurities. On the other hand, in 1D and 2D, $\beta(\sigma)$ stays negative for all values of σ, suggesting that the electrons are localized in the limit $L \to \infty$. This seems contradictory to the experiments that show electron motion, as we discussed earlier in this chapter. However, we must remember that the above conclusion has been reached for $L \to \infty$, while all the experiments are conducted on finite systems.

Thus, let us now try to find how the conductance of a 2-D system varies with L. For large σ we may deviate from Eq. (13.5) to adopt a simple form:

$$\beta(\sigma) = -\frac{\sigma_a}{\sigma}, \qquad \sigma_a = \frac{e^2}{2\pi^2\hbar}\lambda, \tag{13.6}$$

so that $\beta(\sigma)$ vanishes in the limit $\sigma \to \infty$. Note that this form is imposed independently of material constants. Since e^2/h has the dimension of σ, λ is a number (2, as we shall discuss). From Eqs. (13.3) and (13.6), we obtain

$$\sigma(L) = \sigma_0 - \sigma_a \ln(L/L_0). \tag{13.7}$$

That is, a logarithmic decrease with L has been found.

The above consideration is for absolute zero. It may be extended to finite temperatures as follows. According to the general conductivity formula expressed by the current–current correlation function, interference of electron waves with phase coherence is crucial. At finite temperatures, electrons may be scattered elastically and diffuse about until inelastic scattering takes place and destroys phase coherence. We estimate that the elastic mean free path is given by $L_m \sim (D\tau_i)^{1/2}$, where D is the diffusion constant and τ_i is the inelastic scattering time. This time may be assumed to vary with temperature as $\tau_i \sim T^{-p}$, where p is of order 1, because $\tau_i = \infty$ at 0 K. By replacing L in Eq. (13.7) by L_m, we arrive at

$$\Delta\sigma(T) = \frac{e^2}{2\pi^2\hbar} p \ln T, \qquad (13.8)$$

where $\Delta\sigma$ represents a correction to the Drude-type conductivity due to elastic scattering. This relation states that the correction to the conductance increases logarithmically with temperature. Such a logarithmic variation has been confirmed by experiments.[135]

It must be emphasized that Eq. (13.8) refers to a correction to the Drude-type conductivity. The SdH and size oscillations illustrated in Figs. 27 and 28 have also been obtained as corrections. We shall discuss shortly the fact that effects of electron correlations can be given as corrections as well. Therefore, in order to observe the correlation effects, it is necessary to have some other similar corrections separated out.

In the region in which the logarithmic temperature (length) dependence appears in 2-D systems, the electrons are said to be *weakly localized* as a precursor to the complete localization at absolute zero ($L \to \infty$). The study of weak localization, which is favorably observed in 2D, has made rapid strides in recent years.

Weak localization corresponds to the case in which the disorder in the system is not very large and the mean free path of electrons is longer than the width of the wave function. Hence, the effect of such disorder may be treated by perturbation, although treatments beyond the Born approximation may be required.

As can be expected, interference between scattered electron waves is sensitive to an applied magnetic field. This has been dramatically demonstrated by the Aranov–Bohm effect; when a magnetic field is applied along the axis of a hollow cylinder of a 2-D film, the conductance oscillates with a flux period of $ch/2e$. If a magnetic field is perpendicular to a disordered 2-D system, spin-orbit coupling causes an increase of the magnetoconductivity, as we shall discuss later. On the other hand, in a relatively weak magnetic field, Landau orbits are spread and cause

[135]G. J. Dolan and D. D. Osheroff, *Phys. Rev. Lett.* **43,** 721 (1979); S. Kobayashi, F. Komori, Y. Ootsuka, and W. Sasaki, *J. Phys. Soc. Jpn.* **49,** 1635 (1980).

negative magnetoresistance, that is, an increase in the conductivity. This effect on weak localization is naturally dependent on the ratio of L to the magnetic length $l = (c\hbar/eH)^{1/2}$. It is significant in a weak field in contrast to the classical expectation based on the familiar form $\sigma_0/[1 + (\omega_c \tau)^2]$, where $\omega_c \tau$ is small even in a fairly strong magnetic field. A characteristic time to assess this effect is $a^{-1} = (c\hbar/(4DeH))$, and a characteristic length is of course l. Hence, if l is shorter than L_m, it is conceivable to replace L_m by l, that is τ_i by a^{-1} in Eq. (13.8) to arrive at

$$\Delta\sigma(H) = \sigma(H) - \sigma(0)$$
$$= (e^2/2\pi^2\hbar) \ln H. \tag{13.9}$$

Note that the coefficient is a universal constant originally due to the universality of $\beta(\sigma)$. A similar logarithmic dependence appears in the case of the Kondo effect, but it is dependent upon material constants.

More precisely than in Eq. (13.9), Hikami et al.[136] have shown for weak fields that the negative magnetoconductance is given by

$$\Delta\sigma(H) = \frac{e^2}{2\pi^2\hbar}\left[\psi\left(\frac{1}{2} + \frac{1}{a\tau_i}\right) - \psi\left(\frac{1}{2} + \frac{1}{a\tau}\right) + \ln\frac{\tau_i}{\tau}\right], \tag{13.10}$$

where ψ is the diagamma function and the diffusion constant $D = \varepsilon_F \tau/m$. The digamma function originates from a series of the type

$$\sum_{n=0}^{N_M} \frac{1}{(n + \frac{1}{2}) + 1/a\tau_i}.$$

The cutoff parameter N_M is of order $1/a\tau$. The digamma function yields a logarithmic function for large arguments. In particular, if $(a\tau)^{-1} > 1 > (a\tau_i)^{-1}$, Eq. (13.10) yields Eq. (13.9). Note also that for the smallest n, the denominator of the above term does not vanish but rather depends on $1/a$. This is the origin of the relaxation (negative magnetoresistance) of weak localization due to a magnetic field [cf. Eq. (13.17) below, where $q = 0$ causes divergence].

According to the experimental analysis of Kawaji and Kawaguchi[137] on silicon inversion layers at electron density $1.9 \times 10^{16}\,\text{cm}^{-2}$, the negative

[136]S. Hikami, A. I. Larkin, and Y. Nagaoka, *Prog. Theo. Phys.* **63**, 707 (1980).
[137]S. Kawaji, K. Kuboki, H. Shigeno, T. Nambu, J. Wakabayashi, J. Yoshino, and H. Sakaki, in *Proc. 17th Internat. Conf. on Semicond. Phys.* (J. D. Chadi and W. A. Harrison, eds.) p. 413. Springer-Verlag, 1985; S. Kawaji and Y. Kawaguchi, *J. Phys. Soc. Jpn.* **53**, 2868 (1984); S. Kawaji, *Progr. Theo. Phys. Suppl.* **84**, 178 (1985).

magnetoconductivity agrees well with the above formula. In the temperature range between 1.9 and 10.4 K, they fit their data by using the inelastic scattering time as an adjustable parameter. For a wide range of the magnetic field between 0 and 1.3 T, they found also that Kawabata's[138] theoretical improvements with intervalley scattering were satisfactory. In this case, the intervalley scattering time was used as another adjustable parameter. In these theories, the magnetoconductance was evaluated for finite temperatures in consideration of the competition between the magnetic length l and the mean free path L_m.

Let us now work on $\beta(\sigma)$ due to weak localization in a more quantitative way. The conductivity is given by Eq. (11.10) in terms of the velocity–velocity correlation function. The delta functions appearing in that formula can be expressed by

$$\delta(\varepsilon - \mathcal{H}) = \lim_{\delta \to 0} \frac{i}{2\pi} \left[\frac{1}{\varepsilon - \mathcal{H} + i\delta} - \frac{1}{\varepsilon - \mathcal{H} - i\delta} \right].$$

The difference on the right side is given by that of the retarded and advanced Green's functions, G_r and G_a. Therefore, using wave-number variables, we obtain

$$\sigma = -\frac{\hbar A}{2\pi} \left(\frac{e\hbar}{m} \right)^2 \int \frac{d\mathbf{k}\, d\mathbf{k}'}{(2\pi)^2} k_x k_x' [G_r(k, k') - G_a(k, k')]$$
$$\times [G_r(k', k) - G_a(k', k)]. \tag{13.11}$$

The Green functions are averaged over impurity configurations to assume the form in Eq. (11.14). Since for low temperatures, the energy of electrons stays at the Fermi energy, we neglect the energy shift in the self-energy. Hence, the self-energy beomes

$$\Sigma(\varepsilon_F) = -\frac{i\hbar}{2\tau} = -i\pi n_i V_0^2 g(\varepsilon_F), \tag{13.12}$$

where $g(\varepsilon_F)$ is the density of states at the Fermi energy, n_i is the impurity concentration and V_0 is the strength of a delta-function type impurity potential.

The integrand of Eq. (13.11) may be expressed in terms of Feynman graphs. The number of contributing Feynman graphs is reduced due to the difference in the integrand. It has been found that a significant contribution is obtained by summing over the so-called maximally crossed

[138]A. Kawabata, *J. Phys. Soc. Jpn.* **53**, 3540 (1984).

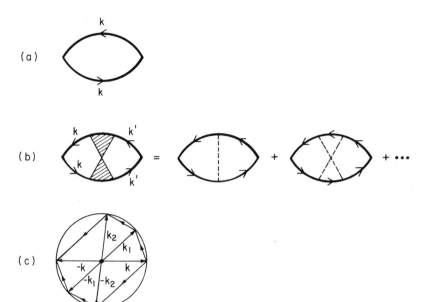

FIG. 34. Contributions to conducitivty. (a) Conribution to σ_0. Thick curves represent $G(k)$. (b) The maximally crossed diagram for $\Delta\sigma$. (c) Propagation in momentum space for (b). (Thin lines represent momentum changes starting from $\mathbf{k}$ to $-\mathbf{k}$, showing that back scattering ($q = 0$) contributes most to Eq. (13.3).) The circle represents the Fermi circle.

diagrams (Fig. 34). Their contribution is given for arbitrary dimension by

$$\sigma = \frac{\hbar}{\pi}\left(\frac{e\hbar}{mA}\right)^2 \int \frac{d\mathbf{k}\, d\mathbf{k}'}{(2\pi)^d} k_x k_x' G_r(k) G_a(k) G_r(k') G_a(k') \Lambda(\mathbf{k} + \mathbf{k}'), \quad (13.13)$$

where d is the dimension and

$$\Lambda(q) = \frac{n_i V_0^2}{1 - n_i V_0^2 \lambda(q)}, \quad (13.14)$$

and

$$\lambda(q) = \int \frac{d\mathbf{k}}{(2\pi)^d} G_r(k) G_a(\mathbf{q} - \mathbf{k}). \quad (13.15)$$

The major contribution to σ in Eq. (13.13) comes from the region in which $\mathbf{k} + \mathbf{k}'$ is small. Hence, we expand $\lambda(q)$ in q to obtain

$$\lambda(q) = \lambda(0)[1 - \tau D q^2], \quad (13.16)$$

where $D = \tau(\hbar k_F/m)^2/d$. This quadratic form leads to electron propagation that is a diffusion type. We finally obtain

$$\sigma(L) = \sigma_0 - \frac{2e^2}{\pi\hbar} \frac{1}{(2\pi)^d} \int \frac{d\mathbf{q}}{q^2}, \tag{13.17}$$

where the conductance due to ordinary scattering processes has been added. The integral is divergent if carried out in infinite space. For a finite system of length L, it is appropriate to limit the q domain such that $1/L < q < 1/L_0$ for $L_0 = \hbar k_F \tau/m$. For 2D, we arrive at

$$\sigma(L) = \sigma_0 - 2\left(\frac{e^2}{2\pi^2\hbar}\right) \ln\left(\frac{L}{L_0}\right), \tag{13.18}$$

which is of the same form as Eq. (13.7). Differentiating this formula, we find

$$\beta(\sigma) = -\frac{2}{\sigma(L)} \left(\frac{e^2}{2\pi^2\hbar}\right). \tag{13.19}$$

There are some other graphs that should be examined. However, Gor'kov et al.[139] and Hikami[139] found that to order $(1/\sigma^2)$ and $(1/\sigma^3)$, the above form is not changed. Moreover, Vollhardt and Wölfle[139] derived a form that reconciles Eq. (13.5) with Eq. (13.19).

The localization effect expressed by Eq. (13.8) may be accompanied by opposing interaction effects. For instance, spin-orbit interaction has been shown to cause positive magnetoresistance.[136,140] Poole et al.[141] and Kawaji et al.[137] reported observations of positive magnetoresistance.

Electron correlations also cause changes in the localization effect, which is expressed by Eq. (13.8). In the presence of electron–electron interaction, Finkel'shtein and others[142] have shown that Eq. (13.8) is

[139]L. P. Gor'kov, A. I. Larkin, and D. E. Khmel'nitskii, *Zh. Eksp. Teo. Fiz. Lett.* **30**, 248 (1979); D. Vollhardt and P. Wölfle, *Phys. Rev.* **B22**, 4666 (1980); S. Hikami, *Prog. Theo. Phys.* **64**, 1466 (1980).

[140]S. Maekawa and H. Fukuyama, *J. Phys. Soc. Jpn.* **50**, 2516 (1981); P. A. Lee and T. V. Ramakrishnan, *Phys. Rev.* **B26**, 4009 (1982); A. Kawabata, *J. Phys. Soc. Jpn.* **53**, 1429 (1984).

[141]D. A. Poole, M. Pepper, and A. Hughs, *J. Phys.* **C15**, L1137 (1982).

[142]A. M. Finkel'shtein, *Zh. Eksp. Teo. Fiz.* **84**, 168 (1983) [*Sov. Phys-JETP* **57**, 97 (1983)]; B. L. Altshuler and A. G. Aronov, *Solid State Commun.* **46**, 429 (1983); H. Fukuyama, Y. Isawa, and H. Yasuhara, *J. Phys. Soc. Jpn.* **52**, 16 (1983); Y. Isawa and H. Fukuyama, *J. Phys. Soc. Jpn.* **53**, 1415 (1983); H. Fukuyama, *Prog. Theo. Phys. Suppl.* **84**, 47 (1985).

replaced by

$$\Delta\sigma(T) = \frac{e^2}{2\pi^2\hbar} p_T \ln T, \qquad (13.20)$$

where

$$p_T = p + 1 - \tfrac{3}{4}F, \qquad (13.21)$$

$$F = \frac{1}{2\pi} \int_0^\pi d\theta \frac{k_s}{k_s + 2k_F \sin\dfrac{\theta}{2}}. \qquad (13.22)$$

The interaction effect is represented by $(1 - 3F/4)$. F is a function of the screening constant k_s and the Fermi wave number k_F. It decreases from 1 as k_F or electron density increases from 0. For $k_s = 1.9 \times 10^7 \text{ cm}^{-1}$ and $n = 5 \times 10^{12} \text{ cm}^{-2}$, F is approximately 0.8. Since F is a function of n, it is difficult to determine the interaction effect unless p is precisely determined.

On the experimental side, Bishop et al.[143] observed evidence for localization and the interaction effect in Si MOSFETs. Washburn et al.[143] reported a large logarithmic correction in GaSb/InAs/GaSb double heterostructures with a slope as much as 30 times the theoretical estimate. They attributed this discrepancy to an electron–hole interaction that was not included in the theory.

The electron–electron interaction also modifies the magnetoconductivity given by Eq. (13.9). This modification can be conveniently expressed in the form:

$$\Delta\sigma(H) = \frac{e^2}{2\pi^2\hbar} p_H \ln H, \qquad (13.23)$$

where

$$p_H = 1 - \frac{F}{4}. \qquad (13.24)$$

However, in the presence of spin-orbit coupling, this form is changed into $p_H = -1/2 - F/2$ so that p_H becomes negative.

Between a few gauss and approximately 1 T, the interaction effect on σ_{xx} is independent of the magnetic field and may be given by the zero-field expression. Since no effect on σ_{xy} is expected, as Fukuyama[144]

[143]D. J. Bishop, R. C. Dynes, and D. C. Tsui, Phys. Rev. **B26**, 773 (1982); S. Washburn, R. A. Webb, E. E. Mendez, L. L. Chang, and L. Esaki, Phys. Rev. **B33**, 8848 (1986).

[144]B. L. Altshuler and A. G. Aronov, Zh. Eksp. Teo. Fiz. **77**, 2028 (1979) [Sov. Phys.-JETP **50**, 968 (1979)]; B. L. Altshuler, A. G. Aronov, and P. A. Lee, Phys. Rev. Lett. **44**, 1288 (1980); B. L. Altshuler, D. Khmel'nitzkii, A. T. Larkin, and P. A. Lee, Phys. Rev. **B22**, 5142 (1980); H. Fukuyama, J. Phys. Soc. Jpn. **48**, 2169 (1980); **49**, 644 (1980); **50**, 3407 (1981). See also Ref. 64. A. Houghton, J. R. Senna, and S. C. Ying, Phys. Rev. **B25**, 2196, 6468 (1982); S. M. Girvin, M. Jonson, and P. A. Lee, Phys. Rev. **B25**, 1651 (1982).

and Houghton *et al.*[144] showed, it is easy to convert $\Delta\sigma_{xx}$ into $\Delta\rho_{xx}$. We find then that $\Delta\rho_{xx}$ is proportional to H^2. This proportionality was observed by Paalanen *et al.*[145] and also more recently by Choi *et al.*[145] in high mobility GaAs/GaAlAs. The latter authors reported that the coefficient to this quadratic dependence above 0.1 T is close to the theoretical expectation. They also observed crossover to lower dimensions. Earlier than this, Wheeler *et al.*[106] used narrow Si inversion layers to find that localization depends on sample size. Kawaji *et al.*[146] reported an observation of negative magnetoresistance in addition to positive magnetoresistance. Kawaji and Maeda[147] made an interesting observation that the magnetoresistance in Si inversion layers is independent of the field direction and is a function of $H/(T - \theta)$, where θ is negative.

IV. Quantized Hall Effect

14. INTEGRAL QUANTIZED HALL EFFECT

Concerning the relation in Eqs. (11.3) of Part III, we note that galvanomagnetic measurements may be made under the condition

$$j_y = 0. \tag{14.1}$$

The Hall conductivity σ_H is then given by

$$\sigma_H = \frac{j_x}{E_y} = -\frac{\sigma_0}{\omega_c\tau} = -\frac{nec}{H} \tag{14.2}$$
$$= 1/\rho_{yx}.$$

The condition (14.1) must be distingusihed from the requirement that

$$\sigma_{xx} = 0. \tag{14.3}$$

In this case, $j_x = \sigma_{xy}E_y$, so that

$$\sigma_{xy} = \sigma_H = \frac{j_x}{E_y} = -\frac{nec}{H}. \tag{14.4}$$

[145]M. A. Paalanen, D. C. Tsui, and J. C. M. Hwang, *Phys. Rev. Lett.* **51,** 2226 (1983); K. K. Choi, D. C. Tsui, and S. C. Palmateer, *Phys. Rev.* **B33,** 8216 (1986).
[146]S. Kawaji, H. Shigeno, J. Yoshino, and H. Sakaki, *J. Phys. Soc. Jpn.* **54,** 3880 (1985).
[147]S. Kawaji and K. Maeda, *J. Phys. Soc. Jpn.* **54,** 4712 (1985).

On the other hand, experimentally, a different condition

$$E_y = 0 \tag{14.5}$$

may be adopted. Then,

$$\sigma_{yx} = \frac{j_y}{E_x} = -\sigma_H. \tag{14.6}$$

Let us now show that Eq. (14.4) results even if we employ quantum mechanics. We remark first that in the presence of an electric field $E = E_x$ in the x direction, the Schrödinger equation for the wave function $\psi(x, y) = e^{-iky}\phi(x)$ becomes

$$\left[-\frac{\hbar^2}{2m}\frac{d^2}{dx^2} + \tfrac{1}{2}m\omega_c^2\left(x - \frac{c\hbar k}{eH}\right)^2 + eEx \right]\phi(x) = \varepsilon\phi(x). \tag{14.7}$$

The electron makes a spiral motion about the center $X = c\hbar k/eH - eE/(m\omega_c^2)$. Its ith energy level is given by

$$\varepsilon_i(E) = (i + \tfrac{1}{2})\hbar\omega_c + eEX + \tfrac{1}{2}m\left(\frac{cE}{H}\right)^2. \tag{14.8}$$

The velocity averaged by $\phi_i(x)$ is

$$\langle v_y \rangle = \frac{1}{m}\int \psi_i^*\left(\frac{\hbar}{i}\frac{\partial}{\partial y} + \frac{eHx}{c}\right)\psi_i \, d\mathbf{r}$$
$$= -cE/H. \tag{14.9}$$

Hence, the current density is $j_y = nceE/H$ so that the Hall conductivity is given by Eq. (14.6) as in the classical case.

Note in the above equation, that the result is independent of the Landau-level index i. Therefore, if i Landau levels are filled by electrons with density $n = i(eH/ch)$, then, as in Eq. (11.9) of Part III, we find

$$\sigma_H = -n\frac{ec}{H} = -i\frac{e^2}{h}. \tag{14.10}$$

At the same time, the current in the direction of the electric field E is zero. It is important to recognize that even though i is an integer, the Hall conductivity is actually proportional to n, which is a continuous variable. That is, the second equality in Eq. (14.10) holds only at a particular n.

In 1976, Kawaji and Wakabayashi[148] noticed that in strong magnetic fields up to 15 T, σ_{xx} of a Si inversion layer remained zero in finite intervals of the gate voltage about particular points that correspond to filled Landau levels. Therefore, they determined the number of "immobile" electrons from the intervals and found that it was inversely proportional to $(2i + 1)$, where i is the Landau-level index. However, it was not until 1980 that an exciting discovery was made: von Klitzing *et al.*[149] discovered that the Hall conductivity takes on quantized plateau values in the intervals where σ_{xx} vanishes. These intervals are determined by the Landau-level filling factor $v = n/(eH/ch)$, which is dimensionless. In order to vary v in Si inversion layers, n is changed through gate voltage, whereas in the case of GaAs/GaAlAs quantum wells, the magnetic field is changed. Because of the relationships between the components of the conductivity and resistivity tensors given by Eqs. (11.7) of Part III, vanishing of σ_{xx} results in $\rho_{xx} = 0$.

A typical example of experimental curves showing the quantization of the Hall conductivity is given in Fig. 35.[150] This new effect is called the *integral quantized Hall effect* (IQHE). In this figure, the upper curve illustrates the Hall resistivity ρ_{xx}, which equals $1/\sigma_{xy}$ when $\sigma_{xx} = 0$, and the lower curve represents the magnetoresistivity ρ_{xx}. The experiment was performed at a fixed carrier density and at 8 mK of a GaAs/GaAlAs quantum well. The resistivity peak at the extreme right corresponds to the first Landau level with up-spins. This is followed by a peak of down-spin $i = 1$ Landau level. The spin splitting for $i = 2$ is not quite complete. Between the resistivity peaks, the Hall resistivity takes on plateau values. Therefore, the Hall resistance shows steplike changes.

The quantization of the resistivity component, i.e., $\rho_{xy} = h/(ie^2)$ can be as precise as a few parts in 10^8.[151] Explicitly,

$$\rho_{xy} = 25812.806\,\Omega/i.$$

This precise quantization leads to a very accurate determination of the fine structure constant α, $\alpha^{-1} = 137.035\,993$, in accordance with

$$\alpha = \frac{\mu_0 c}{2} \frac{e^2}{h}, \tag{14.11}$$

[148]S. Kawaji and J. Wakabayashi, *Surf. Sci.* **58**, 238 (1976).

[149]K. von Klitzing, G. Dorda, and M. Pepper, *Phys. Rev. Lett.* **45**, 494 (1980).

[150]G. Ebert, K. von Klitzing, C. Probst, and K. Ploog, *Solid State Commun.* **44**, 95 (1982).

[151]M. E. Cage, R. F. Dzuiba, and B. F. Field, *IEEE Trans., Instrum. Meas.* **34**, 3011 (1985).

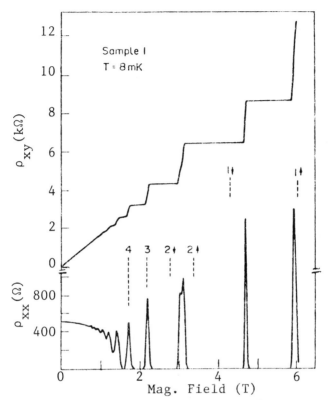

FIG. 35. Magnetic field dependencies of ρ_{xx} and ρ_{xy} of a GaAs/GaAlAs heterostructure. Reprinted with permission from G. Ebert *et al.*, *Solid State Commun.* **58**, 95 (1982), Pergamon Journals, Ltd. The numbers represent the positions of ρ_{xx} maxima and the spin polarization.

where $\mu_0 = 4\pi \times 10^{-7} H/m$ is the permeability of vacuum and c is the light velocity 299792458 m/s. Since the speed of sound c is known very precisely, the determination of the Hall plateaus provides a very accurate method of determining α. At the same time, the quantized Hall resistivity can be used as an absolute resistance standard. The resistance unit is h/e^2, which is 25812.806 Ω. Therefore, the quantized Hall effect is important from fundamental as well as practical points of view.

Each peak of ρ_{xx} occurs at the center of a Landau level, about which ρ_{xy} increases with the magnetic field. The experimental finding that ρ_{xx} or σ_{xx} vanishes between a pair of adjacent Landau levels indicates that the electrons are localized, whereas in the region in which σ_{xy} increases the electrons are mobile. The fraction of extended states within a Landau level decreases with an increasing magnetic field, but the number of these states stays approximately constant because the degeneracy of Landau levels increases with the magnetic field.

In Si inversion layers, the carrier density is controlled by the gate voltage V_g, so that the Hall plateaus appear for certain intervals of V_g. In GaAs/GaAlAs quantum wells, a magnetic field is used to change the filling factor $v = n/(eH/ch)$. Hence, the plateaus appear in plots against a magnetic field as in Fig. 35.

The Hall plateaus are horizontal at low temperatures, but above around 0.2 K their slopes, that is $d\rho_{xy}/dV_g$ in Si MOSFETs and $d\rho_{xy}/dH$ in GaAs/GaAlAs quantum wells, become measureable particularly in low magnetic fields. At the same time, the resistivity ρ_{xx}, which is smaller than $1 \, m\Omega$ at low temperatures, becomes finite. It has been found that $d\rho_{xy}/dn$ is proportional to ρ_{xx}:[152-154]

$$\rho_{xx} \sim d\rho_{xy}/dn. \tag{14.12}$$

Figure 36 illustrates the data of Chang and Tsui,[154] showing the closeness of these two quantities when plotted against the magnetic field. It has also been observed that these quantities are activated.[153-157] In the form

$$\rho_{xx} \sim \exp[-E_a/kT], \tag{14.13}$$

the activation energy E_a is approximately given by

$$E_a \sim \hbar\omega_c/2. \tag{14.14}$$

However, below 1 K, the activated resistivity changes into a hopping type, which follows approximately the theoretical expression:[156]

$$\rho_{xx} \sim \frac{1}{T}\exp[-(T_0/T)^{1/2}], \tag{14.15}$$

where T_0 is of order 7-11 meV in GaAs/GaAlAs quantum wells. Wei et

[152]K. von Klitzing, Rev. Mod. Phys. 58, 519 (1986); B. Tausendfreund and K. von Klitzing, Surf. Sci. 142, 220 (1984).

[153]V. M. Pudalov and S. G. Semenchinskii, Solid State Commun. 51, 19 (1984).

[154]A. Chang and D. C. Tsui, Solid State Commun. 56, 153 (1985).

[155]S. Kawaji and J. Wakabayashi, Solid State Commun. 22, 87 (1979); J. Moriyama and S. Kawaji, Solid State Commun. 45, 511 (1983).

[156]M. Pepper, Phil. Mag. 37, 83 (1978).

[157]H. P. Wei, A. M. Chang, D. C. Tsui, and M. Razeghi, Phys. Rev. B32, 7016 (1985); H. P. Wei, D. C. Tsui, M. A. Paalanen, and A. M. M. Pruisken, in High Mag. Fields in Semicond. Phys. (G. Landwehr, ed.) p. 11. Springer, 1987; Phys. Rev. Lett. 61, 1294 (1988); S. Kawaji and J. Wakabayashi, J. Phys. Soc. Jpn. 56, 21 (1987).

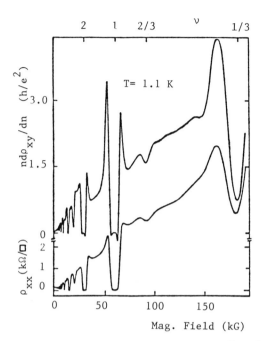

FIG. 36. Similarity between $d\rho_{xy}/dn$ and ρ_{xx}. $n = 1.5 \times 10^{11}\,cm^{-2}$ and $T = 1.1\,K$. Reprinted with permission from A. Chang and D. C. Tsui, *Solid State Commun.* **56**, 153 (1985), Pergamon Journals, Ltd.

al.[157] observed in InGaAs/InP heterostructures that both ρ_{xx} and $\Delta\rho_{xy} = |\rho_{xy} - h/2e^2|$ are activated above 10 K, but that, at low temperatures, ρ_{xx} shows hopping, and, well away from the center of $\nu = 2$, ρ_{xx}, and $\Delta\rho_{xy}$ behave differently from each other. In particular, near $\nu = 1.6$ and 2.5, ρ_{xx} no longer decreases monotonically with T, while $\Delta\rho_{xy}$ remains thermally activated. The activation energies for ρ_{xx} and $\Delta\rho_{xy}$ decrease approximately from the same peak value of 7 meV as ν moves away from 2. Concerning the different behaviors between ρ_{xx} and $\Delta\rho_{xy}$, we note that the contribution from variable-range hopping processes to ρ_{xy} has been found theoretically to be small.[158] Hence, $d\rho_{xy}/dV_g$ may be thermally activated even when the resistivity is dominated by such hoppings. Similar activation behaviors have been observed in INP heterostructures[159] and in InAs/GaSb heterostructures.[160]

Concerning the temperature variation of ρ_{xy} at the ρ_{xx} minimum, an

[158]K. I. Wysokinski and W. Brenig, *Z. Phys.* **B54**, 11 (1983).

[159]A. Briggs, Y. Guldner, J. P. Vieren, M. Voos, J. P. Hirtz, and M. Razeghi, *Phys. Rev.* **B27**, 6549 (1983).

[160]Y. Gouldner, J. P. Hirtz, A. Briggs, J. P. Vieren, and M. Voos, *Surf. Sci.* **142**, 179 (1984).

interesting relation similar to Eq. (14.12) has been obtained empirically. Yoshihiro *et al.* and Cage *et al.*[161] found that $\Delta\rho_{xy} = \rho_{xy}(T) - \rho_{xy}(0)$ satisfies a linear relation:

$$\Delta\rho_{xy} \sim -\rho_{xx}. \tag{14.16}$$

Since the QHE is characterized by plateaus, attempts have been made to determine the plateau widths. These widths may be defined by the range of the gate voltage (or magnetic field) corresponding to a specified value of the relative deviation $\delta = |\Delta\rho_{xy}/\rho_{xy}|$, for instance 0.01%. Working on the $i = 4$ plateau of a Si MOSFET, Pudalov and Semenchinskii[153] arrived at an empirical formula

$$\Delta V_g \sim 10(T_c - T)(J_c - J) + 115 \log \delta + 480 \, \text{mV},$$

where the gate voltage is given in mV, $T_c \sim 2.5$ K and $J_c \sim 15.3 A$. This relation holds for $H = 80$ kOe and for $T < T_c$ and $J < J_c$, with J being the current. There is a critical temperature or a critical current above which the Hall plateaus disappear. There is also a critical magnetic field. Störmer *et al.*[162] observed that the widths decrease with increasing magnetic field.

The plateau widths also depend on the mobility,[162-163] but in a rather complicated way: both very high and very low mobilities tend to reduce the widths. That is, there are optimum mobilities. Such mobilities are around 10,000 and 1000,000 cm^2/Vs for Si MOSFETs and GaAs/GaAlAs quantum wells, respectively. Thus, if the impurity concentration is small, the width may be expected to increase with increasing impurities. In this respect, the IQHE is quantiatively different from the fractional quantized Hall effect, which will be discussed later.

The QHE can be destroyed by increasing not only the temperature or the magnetic field but also the current,[161,164] electric field[165] etc.

[161]K. Yoshihiro, J. Kinoshita, K. Inagaki, C. Yamanouchi, J. Moriyama, and S. Kawaji, *Physica B***117,** 706 (1983); M. E. Cage, B. F. Field, R. F. Dziuba, S. M. Girvin, A. C. Gossard, and D. C. Tsui, *Phys. Rev. B***30,** 2286 (1984).

[162]H. L. Störmer, D. C. Tsui, and A. C. Gossard, *Surf. Sci.* **113,** 32 (1982).

[163]J. E. Furneaux and T. L. Reinecke, *Surf. Sci.* **142,** 186 (1984); *Phys. Rev. B***29,** 4792 (1984).

[164]V. N. Zavaritskii and V. B. Anzin, *Sov. Phys. JETP Lett,* **38,** 294 (1983); G. Ebert, K. von Klitzing, K. Ploog, and G. Weimann, *J. Phys. C***16,** 5441 (1983); F. Kuchar, G. Bauer, G. Weimann, and H. Burkhard, *Surf. Sci.* **142,** 196 (1984); H. L. Störmer, A. M. Chang, D. C. Tsui, and J. C. M. Hwang, *in Proc. Internat. Conf. on Phys. Semicond.* (D. J. Chadi and W. A. Harrison, eds.), p. 267. Springer, 1985; F. Kuchar, R. Meisels, G. Weimann, and H. Burkhard, ibid., p. 275.

[165]S. Komiyama, T. Takamatsu, S. Hiyamizu, and S. Sasa, *Solid State Commun.* **54,** 479 (1985).

Yoshihiro *et al.*[166] attributed the excess noise observed at the $i = 4$ plateau in a Si MOSFET to electron heating. A similar conclusion was reached by Komiyama *et al.*[165] concerning the breakdown of the QHE due to electric field. On the other hand, Kuchar *et al.*[167] found that the QHE can be observed at microwave frequencies as high as 30 GHz. Indeed, the effect is not restricted to the static case, as shown by Pook and Hajdu.[167]

In order to explain the QHE, edge currents have been proposed theoretically. Therefore, the question "How is the current distributed in a device at the Hall regime?" has been experimentally pursued. It has been observed by Syphers *et al.* that electrons tend to avoid the high resistant areas and by Sichel and others that the current is distributed over the entire device rather than being restricted to the edges.[168]

15. INTERPRETATION OF THE IQHE

The quantized plateau of the transverse conductivity σ_{xy} is associated with the vanishing or suddenly dipping magnetoconductivity σ_{xx}. Although the plateau width of σ_{xy} may not precisely coincide with the vanishing region of σ_{xx}, it is conceivable that the quantized plateau corresponds to a localized region between a pair of adjacent Landau levels. On the other hand, between the plateaus, σ_{xy} increases roughly linearly with electron density n, and σ_{xx} reaches a maximum. Hence, near the center of each Landau level, electron states can be extended. This view has been supported by various theoretical considerations.[169] For

[166]K. Yoshihiro, J. Kinoshita, K. Inagaki, C. Yamanouchi, T. Endo, Y. Murayama, M. Koyanagi, J. Wakabayashi, and S. Kawaji, *Surf. Sci.* **170**, 193 (1986).

[167]F. Kuchar, R. Meisels, R. Weimann, and W. Schlapp, *Phys. Rev.* **B33**, 2965 (1986); V. A. Volkov, D. V. Galchenkov, L. A. Galchenkov, I. M. Grodnenskii, O. R. Matov, S. A. Mikhailov, A. P. Senichkin, and K. V. Starostin, *Pis'ma Zh. Eksp. Teo. Fiz.* **43**, 255 (1986) [*Sov. Phys.-JETP Lett.* **43**, 326 (1986)]; W. Pook and J. Heijdu, *Phys. B***66**, 427 (1987).

[168]D. A. Syphers, F. Fang, and P. J. Stiles, *Surf. Sci.* **142**, 208 (1984); E. K. Sichel, H. H. Sample, and J. P. Salerno, *Phys. Rev.* **B32**, 6975 (1985); G. Ebert, K. von Klitzing, and G. Weimann, *J. Phys.* **C18**, L257 (1985); H. Z. Zheng, D. C. Tsui, and A. M. Chang, *Phys. Rev.* **B32**, 5506 (1985).

[169]D. C. Licciardello and D. J. Thouless, *J. Phys.* **C8**, 4157 (1975); T. Ando and H. Aoki, *J. Phys. Soc. Jpn.* **54**, 2238 (1985); T. Ando, *Prog. Theo. Phys. Suppl.* **84**, 69 (1985); *Phys. Rev. Lett.* **54**, 831 (1985); H. Aoki and T. Ando, *Surf. Sci.* **170**, 249 (1986); Y. Ono, *J. Phys. Soc. Jpn.* **51**, 237, 2055 (1982); *J. Phys. Soc. Jpn.* **52**, 2492 (1983); *J. Phys. Soc. Jpn.* **53**, 2342 (1984); *Prog. Theo. Phys. Suppl.* **84**, 138 (1985); H. Levine, S. B. Libby, and A. M. M. Pruisken, *Phys. Rev. Lett.* **51**, 1915 (1983); *Phys. Rev.* **B32**, 2636

(Footnote 169 *Continued*)

instance, Aoki and Ando[169] reported numerical simulations of the density of states for a short-range impurity potential and the inverse localization length α. This quantity appears in the expression of the Thouless number:

$$G(L) = G(0) \exp(-\alpha L). \tag{15.1}$$

The Thouless number $G(L)$ determines the magnetoconductivity of a sample of length L as follows:

$$\sigma_{xx} = \frac{e^2}{4\hbar} G(L).$$

They reported that for energies far from the center of a Landau level, the inverse localization length decreased almost linearly, and near the center, it approached zero smoothly. When it was expressed as

$$\alpha(E) \propto |E - E_i|^s, \tag{15.2}$$

the exponent s was found to be less than 2 for the lowest Landau level. However, Ono's self-consistent theory[169] supports $s = 2$. Otherwise, these two theories and Hikami's perturbation theory[169] agree with each other in that localization takes place except for the band center.

Near the center of each quantized plateau of σ_{xy}, the temperature variation of σ_{xx} may be investigated in terms of the weak localization theory outlined in Section 13. Wei et al.[157] reported that their data support the two-parameter scaling theory of Pruisken.[169] In this theory, σ_{xx} and σ_{xy} are the two independent parameters to be determined by renormalization equations similar to Eq. (13.3) of Part III. The variation of σ_{xx} is plotted against σ_{xy} in a flow diagram. In units of e^2/h, the points $(\sigma_{xx}, \sigma_{xy}) = (0, i)$ represent localization fixed points that are connected by flow lines to the delocalization fixed points at $\sigma_{xx} = i + 1/2$.

Wei et al.[157] obtained such flow diagrams working on InGaAs/InP heterostructures in the temperature range between 1.5 K and 50 K. For relatively high temperaures, they observed that σ_{xx} flows initially upward with decreasing temperature. However, this flow does not represent

(Footnote 169 Continued)

(1985); in The Quantum Hall Effect (R. E. Prange and S. M. Girvin, eds.) p. 117. Springer, 1987; S. Hikami, Prog. Theo. Phys. Suppl. **84,** 120 (1985); S. Hikami and E. Brèzin, Surf. Sci. **170,** 262 (1986); E. Brèzin, D. Gross, and C. Itzykson, Nucl. Phys. B235, 24 (1984); L. Schwitzer, B. Kramer, and A. MacKinnon, Z. Phys. B59, 379 (1985); W. Brenig, Z. Phys. B50, 305 (1983); J. Chalker, J. Phys. C16, 4297 (1983); D. E. Khmel'nitskii, JETP Lett. **38,** 553 (1983); F. Wegner, Z. Phys. B51, 279 (1983).

scaling but rather the temperature variation due to the Fermi distribution function in accordance with

$$\sigma_{\alpha\beta}(T) = -\int dE \frac{\partial f}{\partial E} \sigma_{\alpha\beta}(0).$$

At lower temperatures between 0.5 K and 4.2 K, deviations from this "classical" relation took place and flow expected from scaling was observed. More recently, they revealed that further reduction of temperature causes all flow lines to move downward towards the fixed points, as shown by the dashed lines in Fig. 37. Here, the top dotted and bottom solid lines, obtained by sweeping H, correspond to 770 mK and 50 mK, respectively. Due to inhomogeneity these curves are asymmetric and the maxima are shifted from $\sigma_{xy} = i + 1/2$. No appreciable difference was observed between the data for σ_{xy} in the range 2 and 3 and in the range 3 and 4. These results are consistent with the scaling theory.

They conducted similar experiments on GaAs/GaAlAs heterostructures, but the same scaling behavior was not observed. This is understandable because in this case the transport in the regime of the IQHE is dominated by the classical percolation rather than by quantum-mechanical localization.

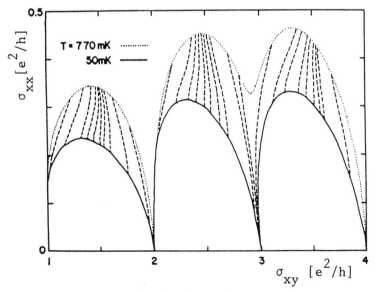

FIG. 37. Flow diagram for the IQHE in an INGaAs/InP heterostructure between 50 mK and 770 mK. H. P. Wei *et al.*, in *High Magnetic Fields in Semiconduction Physics* (G. Landwehr, ed.) (Springer, 1987).

According to Pruisken,[169] the delocalization lenght near the mobility edge E_c is expected to diverge following a power law $\xi \sim (E - E_c)^{-v}$ with a critical exponent v which is universal. The sharpness of the transition into the quantized Hall states depends on such a power law. Very recently, Wei *et al.*[157] determined the maximum of the derivative $d\rho_{xy}/dH$ in the lowest spin-down and first spin-up and -down Landau levels in the range between 0.1 K and 4.2 K to find that

$$\left(\frac{d\rho_{xy}}{dH}\right)_{max} \sim T^{-\kappa} \qquad \kappa = 0.42 \pm 0.04 \tag{15.3}$$

for all these levels. Moreover, a half-width ΔH of ρ_{xx}, defined as the distance between the two extrema in $d\rho_{xx}/dH$, vanishes as T^κ with the same exponent κ. This exponent is related to v through $\kappa = p/2v$, where p is the exponent of the inelastic scattering rate $\tau_i \sim T^{-p}$ as in Section 13.

On the other hand, Kawaji and Wakabayashi[157] claimed that their data on the spin/valley split lowest Landau level in a Si MOSFET are in accord with the theory of Aoki and Ando.[169] In this theory, σ_{xx} and σ_{xy} are not independent parameters but are functionally related to each other as indicated by their approximate relation:

$$\left(\frac{\sigma_{xx}}{\sigma_{max}}\right)^2 + 4\left[\frac{\sigma_{xy}}{(-e^2/h)} - (j + \tfrac{1}{2})\right]^2 = 1,$$

where [cf. Eq. (15.1), and also Eq. (11.24) of Part III]

$$\sigma_{xx}(\varepsilon, L) = \sigma_{max} \exp(-\alpha(\varepsilon)L)$$

$$\sigma_{max} = (j + \tfrac{1}{2})\frac{e^2}{\pi^2\hbar}.$$

However, it seems necessary to reexamine the agreement. In any case, note that Hikami[169] obtained $\sigma_{max} = 1.4e^2/(2\pi^2\hbar)$ for $j = 0$.

In 1981, Laughlin[170] forwarded an elegant theory of the IQHE based on gauge invariance and taking into consideration the extended/localized structure of the Landau levels. Subsequently, further considerations[170–172]

[170]R. B. Laughlin, *Phys. Rev.* B23, 5632 (1981); B. I. Halperin, *Phys. Rev.* B25, 2185 (1982); N. Byers, and C. N. Yang, *Phys. Rev. Lett.* 7, 46 (1961).
[171]R. Joynt and R. F. Prange, *Phys. Rev.* B29, 3303 (1984).
[172]O. Heinonen and P. L. Taylor, *Phys. Rev.* B28, 6119 (1983); R. F. Kazarinov and S. Luryi, *Phys. Rev.* B25, 7626 (1982); G. F. Giuliani, J. J. Quinn, and S. C. Ying, *Phys. Rev.* B28, 2969 (1983); A. Isihara, *Surf. Sci.* 170, 267 (1986).

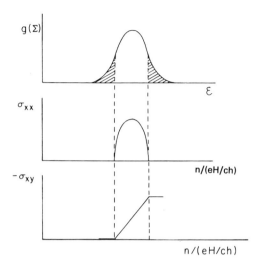

FIG. 38. Structure of broadened Landau levels and the density of states.

were reported. In view of these theories and also of the superconducitivty theory of Byers and Yang,[170] let us try to interpret the IQHE.

We assume that the Landau levels are broadened by impurities and that near the center of each broadened level there is an extended region, as schematically illustrated in Fig. 38. The shaded areas represent localized regions. Let us consider the case in which the Fermi energy lies in the ith localized region below which all i Landau levels are completely filled. In this situation, no current will flow in the (x) direction of the electric field. In the perpendicular (y) direction, let us determine the Hall current by using a Corbino disc, as shown in Fig. 39.

Because of the geometry, we use the condition $E_y = 0$ for the Hall current. The circumference of the disc is assumed to be long enough so that the coordinates x and y constitute a rectangular coordinate system.

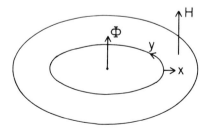

FIG. 39. Electron in a Corbino disk.

The metallic disc is equivalent to a rectangle when it is cut radially, and the dependence of the circumference on position x is negligible. Let this circumference be $2\pi R$. Of course, the electrons are subject to a perpendicular uniform magnetic field H.

If the wave function of an electron is extended, and if an electron circulates the disc once, the magnetic flux passing through its closed orbit at R must be quantized:

$$\pi R^2 H = \text{integer} \times \phi_0,$$

where $\phi_0 = ch/e$ is the flux quantum.

In order to generate the Hall current, we send in a magnetic flux Φ through a thin solenoid at the center. This flux is confined within the solenoid and does not produce a magnetic field on electrons. Therefore, we can require rot $\mathbf{A} = 0$, and hence $\mathbf{A} = \text{grad } \chi$, where $\mathbf{A}$ is the vector potential and χ is single valued. We find that $\Delta \chi = \Phi$. The Schrödinger equation will include the vector potential. However, by transforming the wave function such that

$$\psi_1 = \psi \exp\left[i\frac{e}{ch}\chi(r)\right],$$

the vector potential can be eliminated. However, when one electron moves around the disc once with the positions of all other electrons fixed, ψ_1 will change such that

$$\psi \rightarrow \psi_1 \exp\left[i\frac{e}{ch}\Phi\right],$$

because χ is single valued. Hence we find

$$\Phi = \text{integer} \times \phi_0,$$

Thus, altogether the total flux is quantized:

$$\pi R^2 H + \Phi = j\phi_0, \tag{15.4}$$

where j is an integer.

Now if Φ is increased for a given j, R must decrease. That is, the electron moves inward. In particular, if Φ is increased by ϕ_0, the electron state will be shifted to the $(j-1)$ state. Other than this, there will be no remaining effect. However, if there is an electric field E in the x direction, the dependence of one-electron energy on the center of the harmonic motion suggests that there will be a change of energy $\Delta \varepsilon = eEL$

when the electron moves a distance L from the outer edge to the inner edge of the system. The current density j_y due to the electron is obtained from its energy change:

$$j_y = \frac{c}{L} \frac{\partial \varepsilon}{\partial \Phi}. \tag{15.5}$$

Thus, by replacing $\partial \varepsilon / \partial \Phi$ by $\Delta \varepsilon / \phi_0$, we get

$$j_y = \frac{e^2}{h} E. \tag{15.6}$$

Since the same electron transfer takes place in all i Landau levels, the total current is i times the above value. In accordance with Eq. (14.6), we find

$$\sigma_H = -\frac{e^2}{h} i. \tag{15.7}$$

The above argument does not explicitly consider edge and impurity effects. In fact, it is based on the assumption that the distances between the electron position R and the edges are larger than the cyclotron radius and that the primary effect of impurities is in causing broadening of Landau levels. These effects have been discussed in detail by Halperin[170] and by Joynt and Prange.[171] Halperin did not explicitly use Gauge invariance but rather evaluated the current directly to arrive at Eq. (15.7). A similar direct evaluation of the current was made by Heinonen and Taylor.[172]

In interpreting the IQHE, the existence of extended and localized states in Landau levels is generally assumed. Following Kazarinov and Luryi,[172] the extended states may be pictured as those electron orbits that encircle the entire Corbino disk and the localized states as those surrounding impurities. Among other theories, Giuliani et al.[172] discussed how an adiabatic change of the magnetic flux modifies the energy spectrum, and Isihara[172] extended the memory functional approach discussed in Section 12 to find a solution of a nonlinear integral equation for the real part of the memory function that corresponds to the IQHE.

On the other hand, Thouless and others[173] investigated magnetic Bloch states of electrons in a periodic potential under the assumption that the magnetic flux is commensurate with the unit cell of the potential such that p unit cells contain q flux quanta. Omitting the particle indices for

[173]Q. Niu, D. J. Thouless, and Y. Wu, *Phys. Rev.* **B31**, 3372 (1985); D. J. Thouless, *Phys. Rev.* **B31**, 8305 (1985); M. Kohmoto, *Ann. Phys.* (N.Y.) **160**, 343 (1985).

simplcity, let us start with the Hamiltonian:

$$\mathcal{H} = \frac{1}{2m}\left[\left(\frac{\hbar}{i}\frac{\partial}{\partial x}\right)^2 + \left(\frac{\hbar}{i}\frac{\partial}{\partial y} + \frac{eHx}{c}\right)^2\right] + V(x, y),$$

where the potential $V(x, y)$ is periodic. The transverse conducitivty is determined by the velocity–velocity correlation function such that

$$\sigma_{xy} = \frac{ie^2\hbar}{L^2}\sum_n \frac{(v_x)_{0n}(v_y)_{n0} - (v_y)_{0n}(v_x)_{n0}}{(E_0 - E_n)^2}, \tag{15.8}$$

where i is the imaginary unit and the velocities v_x and v_y are

$$v_x = \frac{1}{m}\left(\frac{\hbar}{i}\right)\frac{\partial}{\partial x}, \qquad v_y = \frac{1}{m}\left(\frac{\hbar}{i}\frac{\partial}{\partial y} + \frac{eHx}{c}\right).$$

We assume that the magnetic flux per unit cell is a rational number and that the wave function satisfies the generalized Bloch conditions:

$$\psi(x + L, y) = \exp(ik_x L)\exp(iyL/l^2)\psi(x, y),$$
$$\psi(x, y + L) = \exp(ik_y L)\psi(x, y).$$

We make a unitary transformation

$$\tilde{\psi} = \exp(-ik_x x - ik_y y)\psi,$$

so that the Hamiltonian becomes

$$\mathcal{H} = \frac{1}{2m}\left(\frac{\hbar}{i}\frac{\partial}{\partial x} + \hbar k_x\right)^2 + \frac{1}{2m}\left(\frac{\hbar}{i}\frac{\partial}{\partial y} + \frac{eH}{c}x\right)^2 + V(x, y).$$

The velocities in Eq. (15.8) can be expressed as

$$v_x = \frac{1}{\hbar}\frac{\partial\mathcal{H}}{\partial k_x}, \qquad v_y = \frac{1}{\hbar}\frac{\partial\mathcal{H}}{\partial k_y}.$$

According to first order perturbation theory,

$$\delta\tilde{\psi}_0 = \sum_n{}' \frac{1}{\varepsilon_0 - \varepsilon_n}\left(\frac{\partial\mathcal{H}}{\partial k_j}\right)_{n0}\delta k_j\tilde{\psi}_n.$$

Thus, Eq. (15.8) can be expressed as

$$\sigma_{xy} = \frac{ie^2}{\hbar(2\pi)^2} \int d\mathbf{k} \int d\mathbf{r} \left[\frac{\partial \bar{\psi}_0^*}{\partial k_x} \frac{\partial \bar{\psi}_0}{\partial k_y} - \frac{\partial \bar{\psi}_0^*}{\partial k_y} \frac{\bar{\psi}_0}{\partial k_x} \right],$$

where the integration is carried out in the unit cells in r and k space. This integral is equal to the loop integral around the unit cell:

$$\sigma_{xy} = \frac{ie^2}{4\pi\hbar} \oint dk_j \int d\mathbf{r} \left[\bar{\psi}_0^* \frac{\partial \bar{\psi}_0}{\partial k_j} - \frac{\partial \bar{\psi}_0^*}{\partial k_j} \bar{\psi}_0 \right]. \tag{15.9}$$

For nonoverlapping subbands, ψ is a single-valued analytic function in the unit cell and can change only by an r-independent phase factor θ. Hence, the integrand reduces to $\partial\theta/\partial k_j$, and the integral is $4\pi i$, where i is an integer. Avron et al.[174] recognized that the integral of this type belongs to the first Chern class of the mapping of the Brillouin zone on a torus geometry onto a complex space. Thouless et al.[173] then showed how the integer i can be determined for a weak periodic potential. They also discussed how the above argument, known as topological theory, can be modified when the periodic potential and the flux are not commensurate or when Coulomb interaction is present. The argument is insensitive to small perturbations such as irregularities and variation in the magnetic field. On the other hand, it depends on the existence of energy gaps.

Since around 1955, numerous studies have been made on magnetic Bloch states. It has been found that if the number of flux quanta passing through a unit cell is a rational number p/q, each Landau level is split into q subbands. If it is not a rational number, infinitely many subbands may appear but with exponentially small gaps. Generally, energy bands are strongly discontinuous functions of the flux per unit cell, but the gaps seems to vary rather continuously. The structure of magnetic Bloch states is important to our next subject, and we shall comment on subband splitting again towards the end of this chapter.

16. FRACTIONAL QUANTIZED HALL EFFECT

In 1982, Tsui et al.[175] discovered that the diagonal part of the resistivity tensor of high-mobility GaAs/GaAlAs heterostructures dips down to

[174]R. Avron, R. Seiler, and B. Simon, *Phys. Rev. Lett.* **51**, 51 (1983).

[175]D. C. Tsui, H. L. Störmer, and A. C. Gossard, *Phys. Rev. Lett.* **48**, 1559 (1982); R. Willett, J. P. Eisenstein, H. L. Störmer, D. C. Tsui, A. C. Gossard, and J. H. English, *Phys. Rev. Lett.* **59**, 1776 (1987). For even q states, see also Ebert et al. and Clark et al. in Ref. 176.

zero, while the Hall resistivity takes on a plateau value $\rho_{xy} = 3h/e^2$ at $v = 1/3$, where v is a filling factor of the lowest Landau level. Subsequently, the same phenomenon, now called the *fractional quantized Hall effect* (FQHE), has been observed at $v = 4/3$, $5/3$, $7/3$, $1/5$, $2/5$, $3/5$, $4/5$, $7/5$, $8/5$, etc.[176,177]

This phenomenon resembles the IQHE, but a small increase in impurity concentration reduces the plateaus in the FQHE, whereas it widens in the IQHE. Besides, the FQHE is observed in only high-mobility samples at very low temperatures and at rather low electron densities. Therefore, it has been associated with strong electron correlations. At finite temperatures, the magnetoresistivity is activiated in a plateau region of the Hall resistivity. Hence, in the FQHE, the electrons may be considered to be in states that are separated by finite energy gaps from the single-particle energy states. We might say that the electrons have condensed into highly correlated new states.

For the FQHE at $v = p/q$, the term "$1/q$ series" is used for the states of the same q. The strength of the minima decreases with increasing q among different q series, and within a given q series, it decreases with increasing p. However, at $q = 3$, the $1/3$ and $2/3$ minima have the same strength when they are shifted to same magnetic field by varying the electron density. The $2/3$ state is viewed as the electron–hole conjugate of the $1/3$ state, as we discuss later. However, we remark that a study of a 2-D hole system by Mendez *et al.*[177] showed that the $4/5$ state is less stable than the $1/5$ state, indicating possible asymmetry between the two states.

The FQHE is not restricted to the lowest Landau level.[177,178] For instance, Mendez *et al.* in a later article, reported quantization of ρ_{xy} at $v = 4/3$ and $5/3$ and well-defined structures of ρ_{xx} at $v = 7/3$ and $9/3$. This feature can also be seen in Fig. 40, which represents the recent data of Willett *et al.*[175] on GaAs/GaAlAs heterostructures at around 150 mK and carrier density $n = 3.0 \times 10^{11}$ cm^{-2}. They used a hybrid magnet with fixed base field so that the figure is composed of four different traces with breaks at around 12 T. The Landau-level index i and the filling factor v

[176]H. L. Störmer, A. Chang, D. C. Tsui, J. C. M. Hwang, A. C. Gossard, and W. Wiegmann, *Phys. Rev. Lett.* **50**, 1953 (1983); E. E. Mendez, W. I. Wang, L. L. Chang, and L. Esaki, *Phys. Rev.* **B28**, 4886 (1983); G. Ebert, K. von Klitzing, J. C. Maan, G. Remenyi, C. Probst, G. Weimann, and W. Schlapp, *J. Phys.* **C17**, L775 (1984); R. C. Clark, R. J. Nicholas, A. Usher, C. T. Foxon, and J. J. Harris, *Surf. Sci.* **170**, 141 (1986).

[177]E. E. Mendez, W. I. Wang, L. L. Chang, and L. Esaki, *Phys. Rev.* **B30**, 1087 (1984); E. E. Mendez, L. L. Chang, M. Heiblum, L. Esaki, M. Naughton, K. Martin, and J. Brooks, *Phys. Rev.* **B30**, 7310 (1984).

[178]A. H. MacDonald, *Phys. Rev.* **B30**, 3550 (1984).

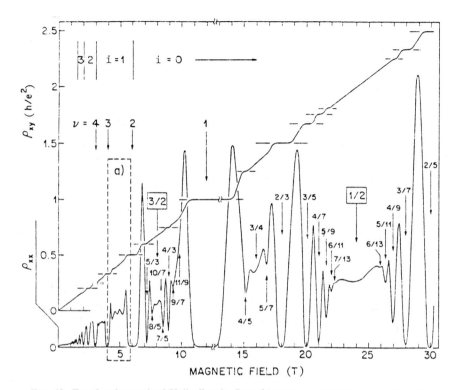

FIG. 40. Fractional quantized Hall effect in GaAs/GaAlAs. R. Willett *et al.*, *Phys. Rev. Lett.* **59**, 1776 (1987). ν: filling factor, i: Landau index.

are indicated above the curves. As we see, there are many well-defined fractional states which occur at odd q values. However, they observed that at $\nu = 5/2$ and at temperatures below 100 mK, a fractional Hall plateau develops at $\rho_{xy} = (h/e^2)/(5/2)$ in area (*a*), which is enlarged in Fig. 41. No equivalent even-denominator quantization was observed in the lowest Landau level, but the existence of other even-denominator states and their relation with the odd-denominator states remain to be seen. The appearance of such an even-denominator state is somewhat exceptional, and our description of the FQHE will follow the theoretical view that q is odd.

The quantization at $\nu = 1/3$ is better than one part in 10^4.[179] Near the minima, the resistivity is activated. Since the magnetoresistivity ρ_{xx} is much smaller than the Hall resistivity ρ_{xy} in these regions, the conduc-

[179]D. C. Tsui, H. L. Störmer, J. C. Hwang, J. S. Brooks, and M. J. Naughton, *Phys. Rev.* **B28**, 2274 (1983).

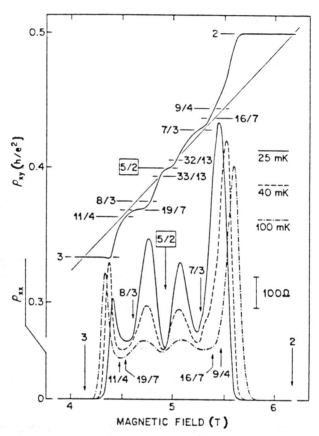

FIG. 41. Magnetoresistivity ρ_{xx} and Hall resistivity ρ_{xy} in Section (a) of Fig. 40. R. Willett *et al.*, *Phys. Rev. Lett.* **59**, 1776 (1987).

tivity and resistivity differ only by a constant factor in accordance with $\sigma_{xx} = \rho_{xx}/(\rho_{xx}^2 + \rho_{xy}^2) \sim \rho_{xx}/\rho_{xy}^2$. Thus, these two quantities are expected to show the same temperature dependence that has been used to determine the activation energies in the FQHE. Chang *et al.*[180] reported that the temperature dependence of ρ_{xx} near $v = 2/3$ is an activation type and that the activation energy is a function of v with a maximum at $v = 2/3$, with its value being $\Delta_{max} = 0.830 \pm 0.03$ K. Using a smoothly varying curve fit without a break in the plot of ρ_{xx} against $1/T$, they suggested a single (rather than two) activation energy Δ for the $v = 1/3$

[180]A. M. Chang, M. A. Paalanen, D. C. Tsui, H. L. Störmer, and J. C. M. Hwang, *Phys. Rev. B***28**, 6133 (1983); I. V. Kukushkin and V. B. Timofeev, in *High Magnetic Fields in Semicond. Phys.* (G. Landwehr, ed.), p. 136. Springer, 1987.

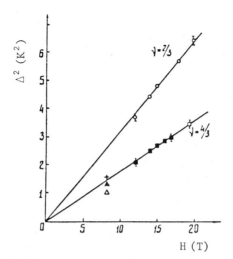

Fig. 42. Magnetic field dependence of the activation energy Δ of $\rho_{xx} \sim \exp(-\Delta/kT)$ at $v = 2/3$ and $4/3$ in Si MOSFETs for given mobility μ. I. V. Kukushkin and V. B. Timofeev, *High Magnetic Fields in Semiconductor Physics* (G. Landwehr, ed.) (Springer, 1987). Circles: $\mu = 3.5 \times 10^4 \, \text{cm}^2/\text{s}$. Squares: $\mu = 2.7 \times 10^4 \, \text{cm}^2/\text{s}$. Triangles: determined from conductivity. Cross: Spectroscopic data point.

series, which is a function of the magnetic field. However, Kawaji *et al.*[181] were led from their data to two activation energies. According to these authors, it is not possible to accurately measure σ_{xx} by using quasi-Corbino geometry specimens as Boebinger *et al.*[181] did. Their reason is that the source-drain current could include a contribution from the Hall current since the electrode is not closed. From their data for $v = 2/3$, they concluded that the activation energy cannot be scaled by the magnetic field alone. In any case, the prefactor for their second activation term, which becomes important at low temperatures, was found to be very small in some of their very highly mobile samples and even sample dependent. Therefore, they used the activation energy determined at relatively high temperatures or by the slope of $\log \rho_{xx}$ ploted against $1/T$ for further analyses to arrive at the conclusion that the activation energy shows magnetic field and v dependencies similar to those of Boebinger *et al.*[181] and Chang *et al.*[180] Note that Boebinger *et al.* used $\rho_{xx} \sim \exp(-\Delta/2kT)$, which is somewhat more theoretical than the conventional form $\rho_{xx} \sim \exp(-\Delta/kT)$ used by Chang *et al.* Theoretically, Δ

[181]S. Kawaji, J. Wakabayashi, J. Yoshino, and H. Sakaki, *J. Phys. Soc. Jpn.* **53,** 1915 (1984); G. S. Boebinger, A. M. Chang, H. L. Störmer, and D. C. Tsui, *Phys. Rev. Lett.* **55,** 1606 (1985); J. Wakabayashi, S. Kawaji, J. Yoshino, and H. Sakaki, *J. Phys. Soc. Jpn.* **55,** 1319 (1986).

consists of the quasiparticle and quasihole excitation energies, as we shall discuss later. For Si MOSFETs, Kukushkin and Timofeev[180] observed that Δ in the conventional form of ρ_{xx} shows an $H^{1/2}$ proportionality as in Fig. 42; Δ depends also on the mobility μ in a systematic way. They found that

$$\Delta = \tilde{\Delta}_v(1 - \mu_0/\mu)(e^2/\kappa l),$$

where μ_0 is constant. The dimensionless activation energy $\tilde{\Delta}_v = 0.018$ ($v = 1/3$), 0.013 ($v = 2/3, 4/3$), and 0.006 ($v = 4/5$).

17. $v = 1/m$ FRACTIONAL STATE

In 1983, Laughlin[182] associated the $v = 1/m$ quantized Hall state, where m is an odd integer, with a new liquidlike electron state based on a Jastrow-type trial wave function because it leads to a lower ground-state energy than the Wigner lattice and to a constant density. His wave function is

$$\psi_m = \prod_{i<j}(z_i - z_j)^m \exp\left(-\frac{1}{4}\sum_{i=1}^{N}|z_i|^2\right), \qquad (17.1)$$

where $z_i = x_i + iy_i$ is the coordinate of the ith electron in units of magnetic length l, $m = 1/v$ is an odd integer due to Fermi statistics, and $v = n(2\pi l^2)$ is the filling factor of the lowest Landau level. This wave function consists only of states in the lowest Landau level with angular momentum specified by m. The total angular momentum for N electrons is given by

$$L = \tfrac{1}{2}mN(N - 1). \qquad (17.2)$$

The polynomial part of the wave function can be expressed by the mth power of the van der Mond determinant given by

$$\prod_{i<j}(z_i - z_j) = \begin{vmatrix} z_1^0 & z_1^1 & z_1^2 & \cdots & z_1^{N-1} \\ z_2^0 & z_2^1 & z_2^2 & \cdots & z_2^{N-1} \\ \vdots & & & & \vdots \\ z_N^0 & z_N^1 & z_N^2 & \cdots & z_N^{N-1} \end{vmatrix}$$

$$= D_{0,1,2,\ldots,N-1}(z_1, z_2, \ldots, z_N). \qquad (17.3)$$

[182]R. B. Laughlin, *Phys. Rev. Lett.* **50**, 1395 (1983). See also Ref. 194 and Z. Gan and Z. Su, *Phys. Rev.* **B35**, 5702 (1987).

Note that the Slater determinant consisting of angular momentum states $l_1, l_2, \ldots, l_N$ of electrons $1, 2, \ldots, N$, respectively, is represented by $D_{l_1, l_2, \ldots, l_N}$ apart from the exponential factor appearing in Eq. (17.1).

Laughlin's wave function has an important connection with a classical one-component plasma (OCP). If we write

$$|\psi_m|^2 = \exp[-\beta\Phi], \tag{17.4}$$

with $\beta = 1/m$ playing the role of $\beta = 1/kT$, we find

$$\Phi = -\sum_{i<j} 2m^2 \ln |z_i - z_j| + \tfrac{1}{2}m \sum_i^N |z_i|^2, \tag{17.5}$$

with Φ defined by this expression being nothing but the potential energy of the 2-D OCP consisting of N identical particles with charge $\sqrt{2}\, m$. For $l = 1$, the first term is the mutual repulsion and the second term represents the interaction with a cloud of neutralizing charges that are uniformly distributed in the system with charge density $1/(\sqrt{2\pi})$. For certain densities, such a plasma is expected to have a uniform charge distribution. In fact, Monte Carlo simulations[183] have shown that the OCP is a fluid if $\Gamma = (\pi n)^{1/2} e^2 \beta = (2m)^{1/2}$ is less than 140. It forms a hexagonal crystal above this Γ. Hence, Laughlin's state is also expected to have a constant charge density and may be associated with a liquidlike electron state. $|\psi_m|^2$ corresponds to a system of a uniform density of $1/(2\pi m)$.

Laughlin found that the projection of ψ_m to the true lowest-energy eigenstate of angular momentum $3m$ is 0.99946 and 0.99468, respectively, for $m = 3$ and 5. Therefore, even though it is a trial wave function, it is very close to the true ground state. Based on the OCP analogue, he also found that ψ_m has a lower energy than the corresponding CDW state. If the radial distribution function of the OCP is $g(r)$, the energy per electron is given by

$$\varepsilon_g = \int_0^\infty n \frac{e^2}{r} [g(r) - 1] \pi r \, dr. \tag{17.6}$$

For large Γ, this energy may be approximated by the ion disk energy

$$\varepsilon_g = -n \int \frac{e^2}{r} d^2 r + \frac{n^2}{2} \int \int \frac{e^2}{r_{12}} d^2 r_1 \, d^2 r_2$$
$$= (4/3\pi - 1) 2e^2 / R,$$

[183]J. M. Caillol, D. Levesque, J. J. Weis, and J. P. Hansen, *Stat. Phys.* **28**, 25 (1982).

where the integration is carried over a disk of radius $R = (\pi n)^{-1/2}$. At $\Gamma = 2$, which corresponds to a full Landau level, the exact result[184]

$$g(r) = 1 - \exp[-(r/R)^2] \qquad (17.7)$$

leads to

$$\varepsilon_g = -\tfrac{1}{2}\pi^{1/2}\frac{e^2}{R}. \qquad (17.8)$$

For $m = 3$ and 5, he adopted the modified hypernetted chain results[183] to obtain $\varepsilon_g = -0.4156 \pm 0.0012$ and -0.3340 ± 0.0028, respectively, in units of $e^2/\kappa l$. More recently, Levesque et al.[185] used an improved OCP radial distribution function and reported that for the $v = 1/m = 1/3$ state,

$$\varepsilon_g = -0.4100 \pm 0.0001 \qquad (v = 1/3). \qquad (17.9)$$

By using Monte Carlo calculations, Morf and Halperin[186] arrived at the same ground-state energy. These energies are lower than the corresponding values for the CDW state, which has -0.389 and -0.322, respectively, for $m = 3$ and 5.[187]

For the $v = 1/m$ state, Laughlin's wave function approaches zero as r^m, and, therefore, the radial distribution function vanishes as r^{2m} when two electrons close in on each other. Such a strong exclusion of overlappings results in the above low energies.

It is instructive to represent Laughlin's state in second quantization. For the basis states, we can use the Slater determinants of elements

$$\phi_l(z) = (2^{l+1}\pi l!)^{-1/2}z^l \exp[-|z|^2/4]. \qquad (17.10)$$

A set of angular-momentum quantum numbers can be used to represent ψ_m in terms of basis states $|l_1, l_2, l, \ldots, l_N\rangle$ for a system of N particles such that

$$\psi_m = \sum_{l_1 < l_2 < \ldots < l_N} A(l_1, l_2, \ldots, l_N)\,|l_1, l_2, \ldots, l_N\rangle, \qquad (17.11)$$

where the l_i, representing angular-momentum eigenvalues, satisfy the conditions

$$0 \le l_1 < l_2 \ldots < l_N \le m(N-1) \qquad (17.12)$$

$$\sum_j l_j = \tfrac{1}{2}m(N-1). \qquad (17.13)$$

[184]B. Jancovici, Phys. Rev. Lett. **46**, 386 (1981).
[185]D. Levesque, J. J. Weis, and A. H. MacDonald, Phys. Rev. B**30**, 1056 (1984).
[186]R. Morf and B. I. Halperin, Phys. Rev. B**33**, 2221 (1986).
[187]D. Yoshioka and H. Fukuyama, J. Phys. Soc. Jpn. **47**, 394 (1979).

The first condition specifies the range of the angular momentum variables, and the second condition represents the conservation of the total angular momentum. We find then[188]

$$A(l_1, l_2, \ldots, l_N) = (l_1! \, l_2! \ldots l_N!)^{1/2} C(l_1, l_2, \ldots, l_N), \quad (17.14)$$

$$C(l_1, l_2, \ldots, l_N) = \sum_{n_{jk}=0}^{N-1} \prod_{j=1}^{N} \delta\left(\sum_{k=1}^{m} n_{jk}, l_j\right) \prod_{k=1}^{m} \varepsilon(n_{1k}, n_{2k}, \ldots, n_{Nk}),$$

$$(17.15)$$

where j and k in the first sum run respectively from 1 to N and from 1 to m. Hence, there are actually Nm sums in which each n_{jk} varies from 0 to $N-1$. Kronecker's delta is designated by $\delta(\ldots)$, and $\varepsilon(\ldots)$ takes on the same sign as the anti-symmetric unit tensor of elements $n_{1k}, \ldots, n_{Nk}$.

The electrons are distributed in space following the distribution function:

$$\rho(z) = \frac{1}{2\pi l^2} \sum_{s=0}^{m(N-1)} \langle c_s^* c_s \rangle \frac{1}{s!} \left(\frac{|z|^2}{2l^2}\right)^s \exp(-|z|^2/2l^2), \quad (17.16)$$

where the magnetic length l has been restored to show the dimension of $\rho(z)$, and

$$\langle c_s^* c_s \rangle = \frac{1}{M(N-1)!} \sum_{l_2, l_3, \ldots, l_N} [A(s, l_2, l_3, \ldots, l_N)]^2. \quad (17.17)$$

This function determines the distribution in the angular momentum space. The annihilation (creation) operator of an electron in the orbital s is $c_s(c_s^*)$, and M is the normalization factor given by

$$M = \frac{1}{N!} \sum_{l_1, l_2, \ldots l_N} [A(l_1, l_2, \ldots, l_N)]^2, \quad (17.18)$$

where the probability amplitude $A(l_1, l_2, \ldots, l_N)$ has been redefined such that it is antisymmetric and vanishes unless the total angular momentum is conserved. As can be seen from Eq. (17.11), it was originally defined for $l_1 < l_3 < \ldots < l_N$. Since the Slater determinant $|l_1, l_2, \ldots, l_N\rangle$ is antisymmetric, Laughlin's state is invariant under exchanges of the l's.

The ground-state energy consists of the direct-type and exchange-type

[188]K. Takano and A. Isihara, *Phys. Rev.* B**34**, 1399 (1986).

interaction energies, ε_{ee}^{d} and ε_{ee}^{x}, respectively, plus the electron–background and background–background energies, ε_{eb} and ε_{bb} respectively:

$$\varepsilon_{g} = \varepsilon_{ee}^{d} + \varepsilon_{ee}^{x} + \varepsilon_{eb} + \varepsilon_{bb}. \qquad (17.19)$$

By using the second quantization representation discussed above, Takano and Isihara[188] obtained exact and explicit formulae for the right-side energies. The ground-state energy for the disc geometry is found to be -0.386505 ($N = 3$), -0.388855 ($N = 4$), -0.390255 ($N = 5$), and -0.391517 ($N = 6$) in units of $e^{2}/\kappa l$. The background charges are assumed in these results to be distributed uniformly in the disc at origin with radius $R = (2mN)^{1/2}l$. Hence, the electron density is $n = 1/(2\pi ml^{2})$. Note that these energies are higher than the limiting value -0.4100 because of the small number of electrons. The breakdown of the above energies is listed below.

TABLE I.   GROUND-STATE ENERGY OF LAUGHLIN'S 1/3 STATE

N	3	4	5	6
ε_{ee}^{d}	0.288996	0.390116	0.478387	0.556933
ε_{ee}^{x}	-0.049224	-0.062467	-0.072243	-0.079534
ε_{ee}	0.239772	0.327649	0.406143	0.477399
ε_{eb}	-1.226488	-0.409568	-1.571267	-1.717742
ε_{bb}	0.600211	0.693064	0.774869	0.848826
ε_{g}	-0.386505	-0.388855	-0.390255	-0.391517
		$(\varepsilon_{ee} = \varepsilon_{ee}^{d} + \varepsilon_{ee}^{x})$		

It is interesting to compare these energies with those of some other representations. First, for small systems, the energy can be obtained numerically by direct diagonalization of the Hamiltonian. For instance, Yoshioka et al.[189] obtained -0.4152, -0.4127, and -0.4128 for $N = 4$, 5, and 6, respectively, by using a periodic boundary condition for the rectangular geometry. Second, Haldane and Rezayi[190] introduced the spherical representation:

$$\psi_{m}^{HR} = (u_{i}v_{j} - v_{i}u_{j})^{m}, \qquad (17.20)$$

[189]D. Yoshioka, B. I. Halperin, and P. A. Lee, Phys. Rev. Lett. 50, 1219 (1983); D. Yoshioka, Phys. Rev. B29, 6833 (1984); W. P. Su, Phys. Rev. B30, 1069 (1984).
[190]F. D. M. Haldane, Phys. Rev. Lett. 50, 1395 (1983); F. D. M. Haldane and E. H. Rezayi, Phys. Rev. Lett. 54, 237 (1985).

where

$$(u, v) = (\cos \tfrac{1}{2}\theta e^{i\phi/2}, \sin \tfrac{1}{2}\theta e^{-i\phi/2}) \tag{17.21}$$

are spinor variables describing particle coordinates. ψ_m^{HR} is a polynomial of degree $2L = m(N - 1)N \equiv 2NS$. The electrons are distributed uniformly at density $(2\pi ml^2)^{-1}$ in the limit $N \to \infty$ on the sphere of radius $R = l(L/N)^{1/2}$. The surface area of this sphere is $4\pi l^2 L/N = 2\pi l^2 mN$ in this limit. The total flux is $\phi_0 m(N - 1)$ with $\phi_0 = ch/e$ for the flux quantum. Hence, $2S$ is the flux per electron in units of ϕ_0. The total angular momentum is zero in the ground state, as can be seen from Eq. (17.21).

Haldane and Rezayi obtained -0.449954 for the ground-state energy at $N = 6$ and $v = 1/3$ compared with the exact value -0.450172. They found also that the ground-state energy increases with N. However, we must note that the radius of the sphere varies with N. Therefore, even though there is no boundary effect, it becomes necessary to compare the energies of the finite systems under the condition that the areal density $N/(4\pi R^2) = N/(4\pi l^2 S)$ is always n. Hence, a new energy unit, $e^2/\kappa l'$, $l' = [(N - 1)/N]^{1/2}l$, may be adopted.

Note that the results of Morf and Halperin based on Monte Carlo simulations[186] indicate a gradual decrease of the ground-state energy when N is increased. For $v = 1/3$, they evaluated the energy for $N = 20$, 30, 42, 72, and 144, and by using polynomial fits, they arrived at an empirical formula:

$$\varepsilon_g = -0.4101 + 0.06006N^{-1/2} - 0.0423N^{-1}. \tag{17.22}$$

By a separate Monte Carlo calculation, they also obtained -0.410 ± 0.001 as the ground-state energy in the thermodynamic limit, in close agreement with the OCP result of Levesque et al.[185] These authors evaluated the interaction energy on the basis of the OCP radial distribution function. By using as many as 256 electrons, they fit their results into the range $1 \le m \le 20$ empirically to the following expression:

$$\varepsilon_g = -\frac{0.782133}{m^{1/2}} \left[1 - \frac{0.211}{m^{0.74}} + \frac{0.012}{m^{1.7}} \right]. \tag{17.23}$$

This empirical formula shows that the ground-state energy decreases also with m. The first term here is the limiting value representing the energy of a triangular Wigner lattice. The Wigner lattice is preferred in the low-density limit, but for $v > 0.1$ or $m < 10$, liquidlike states have lower energies. Whether or not the system is in liquidlike states is seen from the behavior of the radial distribution function. When plotted against m, the energy given by Eq. (17.23) crosses the curve of the first term at around m = 10, indicating a phase change. The formula produces the following ground-state energy for $m = 5$:

$$\varepsilon_g = -0.3277 \pm 0.0002, \qquad (\nu = 1/5). \qquad (17.24)$$

For $m = 1$, Laughlin's state is represented by the Slater determinant of Eq. (17.3) and corresponds to the case in which all the angular momentum states from 0 to $N - 1$ are occupied. Since in this case $n = eH/(ch) = H/\phi_0$, each electron carries the flux quantum ϕ_0. If $\nu = 1/m$, the electrons are spread more widely from 0 to $m(N - 1)$ in the angular momentum space, as indicated in Eq. (17.12). If the electrons are distributed uniformly in this space, their spacing is given by m. That is, every mth state in the angular momentum space is occupied. For the above angular momentum range, there is no other uniform distribution of the electrons if N is finite. It has been found that ε_g shows cusp-type minima at odd m.

On the basis of numerical analyses of small systems, Laughlin,[182] and Yoshioka et al.[189] have shown that Laughlin's state is very close to the exact ground state. Indeed, the overlapping integrals of Laughlin's state with the exact ground state are 0.979 and 0.974 for $m = 3$ and 5, respectively. It is a Jastrow-type trial function and yet does not include any variational parameter. Therefore, it does not allow a variational approach to the true ground-state. We note also that Laughlin's state represents the strongest mutual exclusion for $\nu = 1/m$ but is independent of Coulomb interaction. On the other hand, Yoshioka et al.[189] noticed that the true ground state may deviate from the m-fold zero property of Laughlin's state.

For these reasons, and in order to approach the true ground-state, Takano and Isihara[191] have proposed a new variational wave function based on the following criteria:

(*i*) It includes many-body correlations and yet is close to Laughlin's state.

(*ii*) It is constructed on states of the lowest Landau level such that the total angular momentum is conserved and is given by $L = mN(N - 1)/2$.

[191]K. Takano and A. Isihara, in *Anderson Localization* (T. Ando and H. Fukuyama, eds.), p. 268, Springer, 1988.

We note that the power m in Laughlin's state ψ_m must be odd in order to satisfy the Pauli principle. If we write $m = 2p + 1$, the integer p depends on $v = 1/m$, but the integer 1 is common to all the fractions v of the FQHE. Therefore, in modification of Laughlin's state, a new trial wave function is introduced such that

$$\Psi = \prod_{i<j} (z_i - z_j)(f_{ij})^p (f_{ji})^p \exp\left[-\sum_k |z_k|^2/4 \right], \qquad (17.25)$$

where f_{ij} is a function of the z_i coordinates. Condition (ii) suggests that it is a linear function of the z_i's and may be chosen as

$$f_{ij} = z_i - z_j + c \sum_{k \neq i,j} z_k, \qquad (17.26)$$

where the constant c should be small because Laughlin's state is close to the true ground state. Note that the product $f_{ij}f_{ji}$ is symmetric with respect to particle exchanges. The Pauli exclusion is taken care of by the first factor in the wave function.

The above trial wave function is not a Jastrow-type function consisting only of electron pairs but includes a variational parameter c. Yet, it still conserves the total angular momentum and does not mix Landau levels. The c term in f_{ij} or f_{ji} includes the coordinates of all the electrons but i and j. The magnitude of c may be dependent on Coulomb interaction. As such, the c term represents the influence of the distribution of charges other than the chosen pairs on the Coulombic exclusion. However, this influence vanishes when the center of gravity of these other electrons is at the origin, because their combined effect will be compensated by neutralizing charges that are distributed symmetrically about the origin.

Although it is difficult to speculate on the role played by the c term, the use of the OCP potential energy is somewhat helpful. The potential energy is given to first order in c by

$$\Phi = -m^2 \sum_{i<j} \ln |z_i - z_j| + \frac{m}{4} \sum_i |z_i|^2 + \frac{m-1}{2m} \sum_{i<j} \frac{|c|^2 |\sum_k' z_k|^2}{|z_i - z_j|^2}. \qquad (17.27)$$

This expression indicates that the c term plays the role of adding a repulsive potential to the system. This potential depends on the center of gravity of charges other than chosen pairs and decreases faster than the Coulomb potential. It depends also on the density $(1 - v)$ of the background charges.

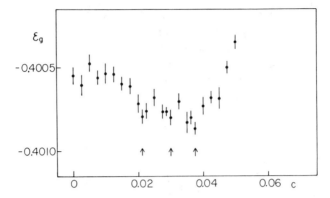

FIG. 43. Ground-state energy of the variational wave function in Eq. (17.25). K. Takano and A. Isihara, in *Anderson Localization* (T. Ando and H. Fukuyama, eds.), p. 268, Springer, 1988. For $N = 30$. The point $c = 0$ corresponds to Laughlin's wave function.

Figure 43 illustrates the Monte Carlo results of Takano and Isihara[191] for $N = 30$. Note that for small values of the variational parameter c, the energy is reduced as c increases from 0, at which its value -0.4005 corresponds to Laughlin's state for $N = 30$. Especially interesting is the appearance of three energy minima -0.4008, -0.4008, and -0.4009 at $c = 0.0213$, 0.030, and 0.038, respectively. These minima might coincide with each other in the bulk limit, leading to a threefold degenerate ground state for $v = 1/3$. On the other hand, extending the topological theory to the FQHE, Niu *et al.*[173] discussed the possibility that the $v = 1/m$ state is m-fold degenerate in agreement with Yoshioka *et al.* and with Su.[189] The same conclusion has been reached by Tao and Haldane[192] and is also based on the periodic boundary condition. These authors suggested the same m-fold degeneracy for the $v = p/m$ states.

Laughlin's wave function is designed for fermions. However, by replacing the Slater determinants by permanents and with even m, it can be used for bosons. On the basis of their exact formulae, Takano and Isihara[191] obtained the ground-state energy for even m Laughlin states: At $v = 1/2$, $\varepsilon_g = -0.434980$ ($N = 3$), -0.440338 ($N = 4$), and at $v = 1/4$, $\varepsilon_g = -0.349422$ ($N = 3$), -0.350729 ($N = 4$). These energies appear relatively low, indicating that if somehow electron pairs are formed, they might be in new Laughlin state. However, we must remember that these energies are expressed in units of $e^2/\kappa l$. Hence, if the charge is doubled, the energy is increased by a factor of $2^{5/2}$.

[192]R. Tao and F. D. M. Haldane, *Phys. Rev. B***33**, 3844 (1986).

18. HIERARCHY OF FRACTIONAL STATES

Let us introduce a small perturbation to the $1/m$ quantized Hall state by expanding the system slightly at a fixed magnetic field. Even if v deviates from the exactly quantized value of $1/m$, the electrons will try to stay in the same state ψ_m and a charge deficiency will be created. In the incompressible fluid, this creation would cost energy because the original $1/m$ state is stable.

A quasihole may be created by a gedanken experiment in which the system is pierced at z_0 by an infinitely thin solenoid for passing through a flux quantum ϕ_0 adiabatically. Without including the normalization factor, the state with one quasihole at z_0 is represented by

$$\Psi_m^{+1} = \prod_{i=1}^{N} (z_i - z_0) \psi_m(z_1, z_2, \ldots, z_N). \tag{18.1}$$

This can be seen by applying such an OCP analogue as expressed by Eq. (17.4) to $|\Psi_m^{+1}|^2$. This can be expressed as

$$|\Psi_m^{+1}|^2 = \exp[-\Phi^{+1}/m],$$

$$\Phi^{+1} = -2m^2 \sum \ln |z_i - z_k| + \frac{m}{2} \sum |z_i|^2 - 2m \sum \ln |z_i - z_0|. \tag{18.2}$$

It represents the same plasma as before except for the quasihole at z_0, with its charge being $|e|/m$ (and thus "quasi").

The above wave function can formally be introduced by applying an operator $S^+(z_0)$ to ψ_m in such a way that

$$\Psi_m^{+1} = S^+(z_0) \psi_m, \tag{18.3}$$

where the operator $S^+(z_0)$ for creating a quasihole at z_0 can be expressed as

$$S^+(z_0) = \prod_{j}^{N} (a_j^\dagger - z_0)$$

$$a_j = \frac{x_j + iy_j}{2} + \left(\frac{\partial}{\partial x_j} + i \frac{\partial}{\partial y_j} \right) \tag{18.4}$$

$$a_j^\dagger = \frac{x_j - iy_j}{2} + \left(\frac{\partial}{\partial x_j} - i \frac{\partial}{\partial y_j} \right).$$

The conjugate operators $S^-(z_0)$, which creates a quasiparticle, is given by

$$S_m^- = \prod_j^N (a_j - z_0^*).$$ (18.5)

Hence, the state with one quasiparticle at z_0 in the $1/m$ fractional state is represented by the wave function:

$$\Psi_m^{-1} = \prod_j^N \left(2\frac{\partial}{\partial z_j} - z_0^*\right)\psi_m(z_1, z_2, \ldots, z_N).$$ (18.6)

The differentiation is performed only on the polynomial part of ψ_m. Note that the adjoint of z is $2(\partial/\partial z)$. A formal aspect of defining quasihole and quasiparticle wave functions has been given by Girvin[193] in the framework of Bargmann's representation[193] of the rotation group.

These quasiholes or quasiparticles carry charges. However, like the electrons in the original $1/m$ state, they can be localized and carry no current in the direction of the electric field. It is then conceivable that they form new fractional states when their densities reach certain values. Let us imagine that M such quasiholes have been created in the state ψ_m. The corresponding wave function may be written as

$$\Psi_m^{+M} = \prod_{i<k}^M (Z_i - Z_k)^\theta \exp\left(-\sum |Z_k|^2/4l_h^2\right) \prod_j^N (z_j - Z_k)\psi_m(z_1, z_2, \ldots, z_N),$$ (18.7)

where the Z_i represent quasihole positions and l_h is their cyclotron radius. The parameter θ is either 0 or 1, depending on whether the quasiholes obey Bose statistics or Fermi statistics, respectively. We shall not specify their statistics here because it is not required in the following discussion. However, we comment that in the disc geometry they can be considered as fermions[194] or anions (particles obeying fractional statistics),[195,196] and in the spherical geometry as bosons.[197] The statistics and charges of

[193]V. B. Bargmann, *Rev. Mod. Phys.* **34**, 829 (1962); S. M. Girvin, *Phys. Rev.* **B29**, 6012 (1984); S. M. Girvin and T. Jach, *Phys. Rev.* **29**, 5617 (1984).

[194]R. B. Laughlin, *Surf. Sci.* **142**, 163 (1984).

[195]B. I. Halperin, *Phys. Rev. Lett.* **52**, 1583 (1984).

[196]F. Wilczek, *Phys. Rev. Lett.* **49**, 957 (1982); D. Arovas, J. R. Schrieffer, and F. Wilczek, *Phys. Rev. Lett.* **53**, 722 (1984); R. Tao and Y. Wu, *Phys. Rev.* **B31**, 6859 (1985); R. Tao, *Phys. Rev.* **B33**, 2937 (1986).

[197]F. D. M. Haldane, *Phys. Rev. Lett.* **51**, 605 (1983).

quasiparticles entering the fractional states have been discussed in detail by Wilczek and others.[196]

The zeros of the wave function may conveniently be called vortices. These vortices are attached to either electrons or holes. The magnetic field requires $1/(2\pi l^2)$ vortices per unit area with respect to the electrons. When the magnetic field is kept constant in the process of creating quasiholes, the number of electron vortices of the above wave function is given by

$$mn + n_h = g_e,$$ (18.8)

where n_h is the quasihole density and $g_e = eH/ch$. Note here that the electron density has been decreased from g_e/m.

The quasiholes produce zeros of the wave function also. The number of quasihole vortices is determined by

$$n + \theta n_h = g_h$$
$$= 1/(2\pi l_h^2).$$ (18.9)

We now postulate that when the concentration of these quasiholes reaches a particular value, they form a new Laughlin state, as originally the electrons did at $v = 1/m$. Their wave function then should have the same form as ψ_m and is given by

$$\psi_{p_1+\theta} = \prod_{i<k} (Z_i - Z_k)^{p_1+\theta} \exp\left(-\sum_k \frac{|Z_k|^2}{4l_h^2}\right),$$ (18.10)

where p_1 has to be an even integer in order to be consistent with the original statistics, which could have been chosen in Eq. (18.7). The filling factor of these quasiholes should then be given by

$$v_h = \frac{n_h}{g_h} = \frac{1}{p_1 + \theta}.$$ (18.11)

Let us now try to find out how this fraction translates itself into a new filling factor of electrons. We can introduce n_h and g_h, as determined from Eqs. (18.8) and (18.9), into the above equation to arrive at the electron's filling factor at which the quasiholes condense into a liquidlike Laughlin state. It is given by

$$v = \frac{1}{m + \dfrac{1}{p_1}}.$$ (18.12)

In particular, if $m = 1$ and $p = 2$, then $v = 2/3$. Since ψ_1 represents the state in which the lowest Landau level is completely filled, the $v = 2/3$ fractional state is the electron-hole conjugate of the 1/3 state. If $m = 3$ and $p_1 = 2$, we find a fractional state at $v = 2/7$.

Similarly, in the case of quasiparticles, the condensation takes place at

$$v = \cfrac{1}{m - \cfrac{1}{p_1}}. \tag{18.13}$$

In particular, if $p_1 = 2$ and $m = 3$, we arrive at the 2/5 fractional state, More generally, Haldane[197] has expressed the hierarchy of the fractional states by the continued fraction:

$$v = \cfrac{1}{m + \cfrac{\alpha_1}{p_1 + \cfrac{\alpha_2}{p_2 + \cdots}}}$$

where α's are either $+1$ or -1 and p's are given integers.

As in the case of the $v = 1/m$ state, it is very important to evaluate the ground-state energy for these hierarchical states in comparison with arbitrary occupations in order to confirm that they are indeed stable fractional states. Among a few possibilities, Halperin and his colleagues have constructed trial wave functions for the 2/5 state based on the model in which the electrons somehow form pairs. For example, for the spherical geometry, Morf and Halperin[198] recently arrived at a bulk limit of the ground-state energy per electron of the 2/5 state:

$$\varepsilon_g = -0.430 \pm 0.003 \qquad (v = 2/5). \tag{18.15}$$

This value improves their previous result of -0.414 on the disc geometry.[186] According to these and other analyses, the hierarchical states are indeed stable.

The creation of a quasihole or a quasiprticle requires energy. Let us denote their energies by $\bar{\varepsilon}_+$ and $\bar{\varepsilon}_-$, respectively. The evaluation of these energies is important because it leads to the excitation energy of neutral excitation, which can be compared with the experimental activation energy of ρ_{xx} that is observed at finite temperatures in accordance with

$$\rho_{xx} \sim \exp(-\Delta/2kT). \tag{18.16}$$

[198]R. Morf and B. I. Halperin, Z. Phys. B68, 391 (1987); T. Chakraborty, Phys. Rev. B31, 4026 (1985); G. Fano, F. Ortolani, and E. Colombo, Phys. Rev. B34, 2670 (1986).

Theoretically, Δ is given by

$$\Delta = \bar{\varepsilon}_+ + \bar{\varepsilon}_- . \qquad (18.17)$$

Unfortunately, the calculation is difficult and the theoretical estimates of Δ are widely spread. For instance, at $\nu = 1/3$ and in units of $e^2/\kappa l$, the HNC results of Laughlin[194] and of Chakraborty[198] are in the range $\bar{\varepsilon}_+ = 0.025 - 0.03$; the result of Fano et al.[198] based on extrapolation from exact treatment of small systems is 0.0772, whereas the Monte Carlo result of Morf and Halperin[186] is 0.0698 ± 0.0033. More recently, Morf and Halperin[198] used an improved pair wave function on the spherical geometry to obtain $\bar{\varepsilon}_+ = 0.070 \pm 0.003$ and $\bar{\varepsilon}_- = 0.0224 \pm 0.0016$. Their excitation energy is 0.092 ± 0.004. This value is close to their previous result of 0.099 ± 0.009. However, it is smaller than the 0.106 of Girvin et al.[199] obtained by the single-mode approximation, the 0.105 of Haldane and Rezayi[190] estimated from small systems, and the 0.114 of McDonald and Girvin[199] based on correctly normalized Laughlin-type wave functions. On the experimental side, the activation energy at the center of the Hall plateau of $\nu = 1/3$ is

$$\Delta = 0.02 \sim 0.03. \qquad (18.18)$$

In terms of temperature, Δ is of the order of 6 K. This value is much smaller than the activation energy E_a in the case of the IQHE, where the excitations are associated with the mobility gap between Landau levels. In the present case of the FQHE, excitations occur within the lowest Landau level.

Considerable efforts have been made to explain the discrepancy between the theoretical and experimental excitation energies. Among the mechanisms that have been proposed are the effects due to finite thickness, reversed spins, impurities, admixture of higher Landau levels, etc. These effects tend to reduce the theoretical excitation energy.[198–200]

19. Collective Excitations and Closing Remarks

We have discussed the fact that in the fractionally quantized Hall regime the electrons are in liquidlike states. Therefore, it is conceivable

[199]S. M. Girvin, A. H. MacDonald, and P. M. Platzman, Phys. Rev. B33, 2481 (1986); A. H. MacDonald and S. M. Girvin, Phys. Rev. B34, 5639 (1986); S. M. Girvin and A. H. MacDonald, Phys. Rev. Lett. 58, 1252 (1987); A. H. MacDonald and S. M. Girvin, Phys. Rev. B33, 4009 (1986); M. Rasolt and A. H. MacDonald, Phys. Rev. B34, 5530 (1986).
[200]D. Yoshioka, J. Phys. Soc. Jpn. 55, 885 (1986); T. Chakraborty, P. Pietilainen, and F. C. Zhang, Phys. Rev. Lett. 57, 130 (1986).

that elementary excitations in this liquid may be described in a way similar to those in liquid helium. Recently, Girvin et al.[199] have shown that such a description is indeed possible and elucidates the nature of the FQHE. Their theory is based on the single-mode approximation, which is similar to Feyman's theory of liquid helium, and has the advantage of being able to provide energy-dispersion relations of elemetnary excitations.

For our purpose, it is convenient to describe Feyman's theory in terms of the f-sum rule with the single-mode approximation. Let us adopt the natural unit in which $\hbar = 1$ and $2m = 1$ and describe this theory briefly. In the single-mode approximation, the dynamical structure factor

$$S(k, \omega) = \sum_j (\rho_k^*)_{j0}^2 \delta(\omega - \omega_{j0}) \tag{19.1}$$

is expressed by

$$S(k, \omega) = NS(k)\delta(\omega - \Delta), \tag{19.2}$$

where ρ_k is the Fourier transform of the local density

$$\rho_k = \sum_i \exp(-i\mathbf{k} \cdot \mathbf{r}_i). \tag{19.3}$$

If the interaction in the Hamiltonian $\mathcal{H}$ of the system is pairwise, the f-sum rule states that

$$\sum_j \frac{1}{k^2} [(\rho_k^*)_{j0}]^2 \omega_{j0} = N. \tag{19.4}$$

This relation can be shown by taking the matrix element of the double commutation

$$[\rho_k^*, [\rho_k, \mathcal{H}]] = 2Nk^2. \tag{19.5}$$

The static structure factor $S(k)$ is given by integrating $S(k, \omega)$ over ω. It is related to the radial distribution function $g(r)$ through the equation:

$$S(k) = 1 + n \int d\mathbf{r} \exp(-i\mathbf{k} \cdot \mathbf{r})[g(r) - 1] + (2\pi)^2 n\delta(k). \tag{19.6}$$

The excitation energy Δ is then given by

$$\Delta(k) = \frac{f(k)}{S(k)}, \tag{19.7}$$

where $f(k)$ is

$$f(k) = \frac{1}{2N} \langle 0| \rho_k^*, [\rho_k, \mathcal{H}] |0 \rangle = k^2. \tag{19.8}$$

Hence, the excitation energy is determined by the first moment of the dynamical structure factor. It represents the average energy of the excitations that are related to the density fluctuations described by ρ_k.

The case of the FQHE differs from the case of liquid helium in that the electrons obey Fermi statistics, interact with each other through long-range Coulomb forces, are subject to a strong magnetic field, and are in Landau levels. The effect is clearly not related to the excitations to high Landau levels, but is primarily a phenomenon within the lowest Landau level. Therefore, it is appropriate to use a Hilbert space constructed within the lowest Landau level. Fortunately, a formal theory relevant to such a space has been developed.[193] Let us now look at this theory and then present the theory of Girvin et al.[199] For convenience, we adopt the symmetric gauge and set the magnetic length to 1.

We note that the eigenfunctions of the lowest Landau level are

$$\phi_l = \frac{z^l}{[2^{l+1}\pi l!]^{1/2}} \exp(-|z|^2/4). \tag{19.9}$$

We consider a Hilbert space of entire analytic functions. Such functions may be expanded in power series such that

$$f(z) = \sum_n a_n z^n. \tag{19.10}$$

Note that powers z^n appear in the polynomial part of the eigenfunctions.

Next, we define a scalar product of two analytic functions by

$$(f, g) = \int d\mu(z) f^*(z) g(z), \tag{19.11}$$

where

$$d\mu(z) = \frac{1}{2\pi} dx \, dy \, \exp(-|z|^2/2). \tag{19.12}$$

This measure is used in view of the form of the above eigenfunctions. All the functions belonging to the Hilbert space must satisfy

$$(f, f) \geq 0. \tag{19.13}$$

In particular, we find

$$(z^m, z^l) = \begin{cases} 0, & (m \neq l) \\ m!, & (m = l). \end{cases} \tag{19.14}$$

All the operators must map the functional space onto itself. The differential operator d/dz may not necessarily satisfy this requirement, but if the analytic functions are restricted to polynomials, there will be no problem. Therefore, we set a rule that the operators in this space operate only on the polynomial part of the eigenfunctions. Hence, for

$$f_l = \frac{z^l}{(2^l l!)^{1/2}}, \tag{19.15}$$

we obtain

$$z f_l = [2(l + 1)]^{1/2} f_{l+1}$$
$$\frac{d}{dz} f_l = \left(\frac{2}{l}\right)^{1/2} f_{l-1}. \tag{19.16}$$

Thus

$$a^\dagger = z/\sqrt{2} \qquad a = \sqrt{2} \frac{d}{dz} \tag{19.17}$$

are creation and annihilation operators, and the conjugate of z is $z^\dagger$:

$$z^\dagger = 2 \frac{d}{dz}. \tag{19.18}$$

Hence, we introduce the projection of z^* onto the lowest Landau level by

$$\overline{z^*} = z^\dagger = 2 \frac{d}{dz}. \tag{19.19}$$

Clearly, these relations can be extended to many variables. The operators z_k and $D_k = 2 \partial/\partial z$ satisfy the commutation relations

$$[z_k, z_j] = 0, \qquad [D_k, D_j] = 0, \qquad [D_k, z_j] = \delta_{kj}. \tag{19.20}$$

Furthermore, z_k and D_k are adjoint in the sense that

$$(z_k f, g) = (f, D_k g). \tag{19.21}$$

They can be considered as ladder operators for angular momentum because they change the power of z if operated only on the polynomial part of the eigenfunctions. Since z^* and z do not commute, the former is placed to the left in the normal order.

The density operator is appropriately expressed as

$$\rho_k = \sum_{j}^{N} \exp\left[-\frac{ik}{2}z_j^* - \frac{ik^*}{2}z_j\right]. \tag{19.22}$$

Its projection onto the lowest Landau level is given by

$$\bar{\rho}_k = \sum_{j}^{N} \exp\left(-ik\frac{\partial}{\partial z_j}\right)\exp\left(-\frac{ik^*}{2}z_j\right). \tag{19.23}$$

Note the normal order on the right side. The Hamiltonian can also be projected. Since the kinetic energy can be ignored under a strong magnetic field, we are concerned only with the Coulomb potential:

$$\Phi = \frac{1}{2}\int \frac{d\mathbf{q}}{(2\pi)^2}u(q)\sum_{i<j}^{N}\exp[i\mathbf{q}\cdot(\mathbf{r}_i - \mathbf{r}_j)].$$

Its projection may be written as

$$\bar{\Phi} = \frac{1}{2}\int \frac{d\mathbf{q}}{(2\pi)^2}u(q)[\bar{\rho}_q\bar{\rho}_q - e^{-q^2/2}]. \tag{19.24}$$

The projected oscillator strength $\bar{f}(k)$ is obtained from the double commutator $[\bar{\rho}_k^{\dagger}, [\bar{\rho}_k, \Phi]]$ as

$$\bar{f}(k) = \tfrac{1}{2}\sum_{q}u(q)[e^{q^*k/2} - e^{qk^*/2}]\times[\bar{S}(q)e^{-k^2/2}(e^{-k^*q/2} - e^{-kq^*/2})$$
$$+ \bar{S}(q + k)(e^{k^*q/2} - e^{kq^*/2})], \tag{19.25}$$

where the projected static structure factor $\bar{S}(k)$ is given by

$$\bar{S}(K) = \frac{1}{N}\langle 0|\bar{\rho}_k^{\dagger}\bar{\rho}_k|0\rangle, \tag{19.26}$$

and $|0\rangle$ is the ground state represented by the Hilbert space of analytic functions. Note that Eq. (19.26) can be expressed as

$$\bar{S}(k) = S(k) - (1 - e^{-|k|^2/2}), \tag{19.27}$$

because

$$\overline{\rho_k^\dagger \rho_k} = \bar{\rho}_k^\dagger \bar{\rho}_k + (1 - e^{-|k|^2/2}).$$

Finally, the projected excitation energy in the single-mode approximation is given by

$$\Delta(k) = \frac{\bar{f}(k)}{\bar{S}(k)}. \tag{19.28}$$

By expanding Eq. (19.25) in powers of k, we find that the oscillator strength $\bar{f}(k)$ varies as $|k|^4$ at long wavelengths. Hence, a finite gap can exist at $k = 0$ if $\bar{S}(k)$ varies as $|k|^4$. In fact, one can show that its limiting form is

$$\bar{S}(k) = \frac{1-v}{8v}|k|^4 + \ldots . \tag{19.29}$$

In order to see the long wavelength behavior of $\bar{S}(k)$, we evaluate the moments of the correlation function defined by

$$M_n = \rho \int d\mathbf{r} \left(\frac{r}{2}\right)^n [g(r) - 1]. \tag{19.30}$$

By expanding the correlation function in terms of the single-particle eigenfunctions, the first two moments are found to be

$$M_0 = v^{-1}[\langle Nn_0 \rangle - \langle N \rangle \langle n_0 \rangle] - 1$$
$$M_1 = v^{-1}[\langle (L+N)n_0 \rangle - \langle L+N \rangle \langle n_0 \rangle] - 1, \tag{19.31}$$

where $N = \sum_{m=0}^{\infty} n_m$ is the total number of particles and $L = \sum_{m=0}^{\infty} mn_m$ is the total angular momentum. Since these are conserved,

$$M_0 = M_1 = -1. \tag{19.32}$$

These moments lead to the proportionality $\bar{f}(k) \sim |k|^4$.

In order to evaluate the excitation energy explicitly, Girvin et al.[199] adopted the following analytic form for the radial distribution function:

$$g(r) = 1 - e^{-r^2/2} + \sum_{n=0}^{\infty}{}' \frac{2}{n!} \left(\frac{r^2}{4}\right)^n c_n e^{-r^2/4} \tag{19.33}$$

where the coefficients c_n are unknown and the prime indicates that the sum goes only over odd integers in reflection of Fermi statistics. For

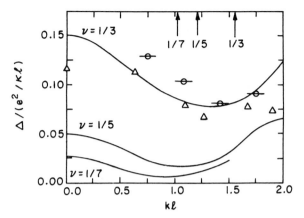

FIG. 44. Dispersion curves of elementary excitations. S. M. Girvin *et al.*, *Phys. Rev.* *B33*, 2481 (1986).

$v = 1/m$ the coefficients c_n are constrained by the charge-neutrality, perfect screening, and compressibility sum rules. These lead to

$$\sum_{n=1}^{\infty}{}' c_n = \frac{1-m}{4}$$

$$\sum_{n=1}^{\infty}{}' (n+1)c_n = \frac{(1-m)}{8} \qquad (19.34)$$

$$\sum_{n=1}^{\infty}{}' (n+2)(n+1)c_n = \frac{(1-m)^2}{8}.$$

Under these constraints. Girvin *et al.* fit a finite number of the coefficients to the Monte Carlo results of Caillol *et al.*[183] and Levesque *et al.*[185]

Figure 44 illustrates their results for $v = 1/3$, $1/5$, and $1/7$. Circles represent the numerical results of Haldene and Rezayi[190] based on the spherical geometry for 7 particles at $v = 1/3$, and traingles are from a 6-particle system with a hexagonal unit cell. Horizontal bars represent errors due to the uncertainty in converting angular momentum on the sphere to linear momentum. Arrows indicate the magnitude of the primitive reciprocal-lattice vector of the corresponding Wigner lattice. As we see, there is a finite energy gap at $k = 0$, and a roton-type minimum at a larger k. As v decreases, these energies become less and less as a precursor to the collapse of the gap that occurs at the critical density where the Wigner crystal becomes more stable. This density has been estimated to be $1/6.5$.[201] Their excitation energy $\Delta(0)$ at $k = 0$ and

[201]P. K. Lam and S. M. Girvin, *Phys. Rev. B30*, 473 (1984); *Phys. Rev.* **31**, 613(E) (1984).

$v = 1/3$ was compared with some other results in the previous section. Note that $\Delta(0)$ is approximately twice the minimum roton energy and that the roton minimum is broad. Hence, two rotons of opposite momenta might form a bound state below the one-phonon state.

As in the case of Feynman's dispersion relation for liquid helium, the above single-mode theory breaks down at large k. Nevertheless, Girvin et al. note that

$$\bar{f}(k) = -2e^{-|k|^2/2}[\varepsilon_g(v) - v\varepsilon_g(1)] \tag{19.35}$$

$$= 2e^{-|k|^2/2}\varepsilon_{co}(v), \tag{19.36}$$

where $\varepsilon_{co}(v)$ is the cohesive energy per particle, to arrive at the exact limiting result

$$\bar{\Delta} = \lim_{k\to\infty} \frac{\bar{f}(k)}{\bar{S}(k)} = \frac{2\varepsilon_{co}}{1-v}. \tag{19.37}$$

For $v = 1/3$, $\bar{\Delta} = 0.603$. For the large k limit of the excitation energy, an improvement on the above result may be obtained by producing a single quasielectron–quasihole pair[202] in the state. Since such an exciton would carry the momentum kl if the separation is mkl, the energy at finite k would be reduced from $\Delta(\infty)$ by the Coulomb attraction between the two particles. Hence, recovering the magnetic length l, we find

$$\Delta(k) = \Delta(\infty) - \frac{1}{m^3kl}. \tag{19.38}$$

If the roton minimum is used for $\Delta(k)$, this leads to

$$\Delta(\infty) = \Delta(k_{min}) + \frac{1}{m^3k_{min}l}. \tag{19.39}$$

They arrived at $\Delta(\infty) = 0.106$ for $v = 1/3$, as quoted at the end of the previous section. As remarked there, this and other theoretical excitation energies are significantly larger than the experimental energy. Girvin et al. found that the excitation energy based on the single mode approximation is reduced considerably by finite thickness and is expected to be proportional to the square root of the magnetic field. There are still some other important factors that are expected to affect the excitation energy. For example, finite temperature and disorder would reduce the excitation

[202]C. Kallin and B. I. Halperin, Phys. Rev. B30, 5655 (1984); R. B. Laughlin, Physica B126, 254 (1984).

energy.[198–202] Therefore, further studies of the excitation energies are needed.

On the other hand, Platzman et al.[203] studied σ_{xx} based on the single-mode approximation. In the conductivity formula given by Eq. (12.3) of Part III together with Eq. (12.4), the imaginary part of the inverse dielectric function is given by

$$-\frac{1}{\pi u(q)} \operatorname{Im} \frac{1}{\varepsilon(q, \omega)} = S(q\, \omega). \qquad (19.40)$$

Therefore, the memory function M can be obtained. The conductivity in the static limit is given by

$$\sigma_{xx} \sim -\sigma_{xy} \lim_{\omega \to 0} \frac{M(\omega)}{\omega_c}. \qquad (19.41)$$

They observed that the imaginary part of the memory function is activated. However, since the single-mode approximation leads to the excitation energy, which is almost five times larger than the experimental value, improvments on the approximation such as the renormalization of the roton energy by impurities must be made before comparing the conductivity with the experiment.

Our discussions in this and previous sections lead to the conclusion that the FQHE represents a macroscopic manifestation of quantum states of interacting electrons in the lowest Landau level. We have discussed that this effect is associated with the downward cusps in the ground-state energy at $v = 1/m$, where the electrons fall into liquidlike states. We have interpreted the plateaus at $v = p/q$ by showing how these states could be generated. However, there are theories that do not belong to this line of work.

For example, Chui[204] proposed a solidlike wave function in combination with a Wigner lattice function and Laughlin's function. In the harmonic approximation, they obtained some low energies. However, the OCP relation shows that at $v = 1/m$, the electrons prefer not to be on lattice sites. On the other hand, their wave function allows a smooth interpolation to a Wigner lattice.

Mention must be made of theories on magnetic Bloch states. The

[203]P. M. Platzman, S. M. Girvin, and A. H. MacDonald, *Phys. Rev.* **B32**, 8458 (1985). See also A. Gold, *Phys. Rev.* **B33**, 5959 (1986).

[204]S. T. Chui, T. M. Hakim, and K. B. Ma, *Phys. Rev.* **B33**, 7110 (1986); S. T. Chui, *Phys. Rev.* **35**, 7787 (1987).

quantum mechanics of electrons in a periodic potential and simultaneously in a magnetic field has been a subject of extensive study since around 1955, i.e., long before the discovery of the QHE. It has been revealed that if the flux passing through a unit cell of such a potential is a rational number p/q (in units of the flux quantum ϕ_0), where p and q are mutual prime numbers, the potential splits each Landau level into p subbands. Equivalently, the magnetic field splits each Bloch band into q subbands.

Thouless et al.[205] used these and other results to interpret the quantized Hall effect. They discussed the fact that Eq. (15.9) could be modified into an integral over phases $\theta = \alpha L$ and $\phi = \beta L$, thus accomodating various boundary conditions. If the ground state is nondegenerate, the integral is the same invariant as before, leading $\sigma_{xy}/(e^2/h)$ to an integer. If the ground state is m-fold degenerate, the average over similar integrals contributed from all the subbands must be evaluated. Accordingly, the Hall conductivity assumes the form $\sigma_{xy} = (e^2/h)I/m$, where I is the sum of the contributions from all the subbands. In the limit of a weak periodic potential, this sum adds up to 1, i.e., $I = 1$. Hence, $\sigma_{xy}/(e^2/h) = 1/m$. In the opposite limit of tight-binding, $I = 0$, and localized magnetic states can be constructed. In fact, there is an intimate relationship between the Hall conductivity and magnetic localization, as Dana and Zak[205] have discussed. Thus, it is very interesting to study magnetic Bloch states. Note that the topological theory of Thouless et al. explains both the IQHE and FQHE.

Since the quantized Hall conductivity is associated with the fine-structure constant, its relation with quantum electrodynamics was sought by Widom et al.[206] based on Schwinger's equation for chiral currents. The chiral anomaly in quantum electrodynamics is related to the topological invariance of Thouless et al. Thus, Widom et al. derived the relation $\sigma_{xy}/(e^2/h) = \nu$ in a general way. However, their general approach did not clearly provide the conditions under which the Hall plateaus appear.

Returning to Laughlin's wave function, we note that the function was introduced intuitively. Therefore, Lee et al.[207] offered its interpretation based on a semiclassical treatment of closed-loop diagrams in the frustrated XY model with nearest-neighbor interactions. We have discussed that Laughlin's wave function contains essential features of the FQHE, but the function is approximate despite the exactness of the Hall quantization. For this reason, it is desirable to try to find the true ground

[205]D. J. Thouless, M. Kohmoto, M. P. Nightingale, and M. den Nijs, *Phys. Rev. Lett.* **49**, 405 (1982); D. J. Thouless, *J. Phys.* C**17**, L325 (1984); I. Dana and J. Zak, *Phys. Rev.* B**32**, 3612 (1885).

[206]A. Widom, N. H. Friedman, and Y. N. Srivastava, *Phys. Rev.* B**31**, 6588 (1985).

[207]D. H. Lee, G. Baskaran, and S. Kivelson, *Phys. Rev. Lett.* **59**, 2467 (1987).

state and its properties. it is also desirable to evaluate not only the energy but also the Hall plateaus as functions of impurity concentration, temperature, and filling factor. Moreover, the elementary excitations require further investigations. Furthermore, the interpretation that the state at $v = 1 - p/q$ is the electron-hole conjugate of the p/q state should be reexamined in view of the asymmetry mentioned in Section 16. Broken symmetry in such conjugates has been revealed more recently in a titled magnetic field and upon light illumination.[208] Even more fundamentally, the appearance of even q states requires further exploration. Willett et al.[28] have revealed a fractional state at $v = 5/2$, and not at $v = 1/2$ or $3/2$, in a relatively weak field (5 T). Hence, by using a "hollow-core" model, Haldane and Rezayi[209] attribute the 5/2 plateau to a spin-singlet state of mixed polarization. Such a state may not be expected for the lowest Landau level because of the form of the wave function. In contrast, the form factor of the next Landau level has a node at the center of the cyclotron orbit, resulting in less mutual repulsion for overlapping two electrons. The energy cost for pairing these electrons at zero angular momentum can then be substantially small.

However, such a pairing must be shown in the presence of Coulomb repulsion, inviting further studies. Nevertheless, it is safe to say that the basic role played by Laughlin's wave function is unaltered. In this respect, we might add that the ground state of the frustrated Heisenberg antiferromagnet is well described by the $m = 2$ Laughlin's wave function.[210]

To summarize, the FQHE has provided a great deal of new information on electron correlations, but its thorough understanding will require more time and effort. Meanwhile, the effect will continue to be an important source for new dicoveries and challenge.

Appendix
Glossary of Symbols

Except for special cases, the following notations are generally used.

N: Total number of electrons g: Lande's g-factor
n: Density of electrons $g(\varepsilon)$:   Density of states

[208]R. J. Haug, K. von Klitzing, R. J. Nicholas, J. C. Mann, and G. Weimann, *Phys. Rev.* **B36**, 4528 (1987); R. J. Nicholas, R. G. Clark, A. Usher, J. R. Mallet, A. M. Suckling, J. J. Harris, and C. T. Foxon, *in Proc. Euro. Conf. Cond. Matter* (Pisa, 1987), to be published.

[209]F. D. M. Haldane and E. H. Rezayi, *Phys. Rev. Lett.* **60**, 956 (1988).

[210]V. Kalmeyer and R. B. Laughlin, *Phys. Rev. Lett.* **59**, 2095 (1987). See also D. P. Arovas, A. Auerbach and F. D. M. Haldane, *Phys. Rev. Lett.* **60**, 531 (1988).

k_F: Fermi wavenumber = $(2\pi n)^{1/2}$

ε_g: Ground-state energy

$u(q)$: F.T. of the Coulomb potential

$v(q)$: F.T. of the impurity potential

c_A: Specific heat per area

a_0: Bohr radius

ρ: Resistivity

γ_0: $= \mu_B H/\varepsilon_F$

γ: $= \mu_B H/\mu$

r_s: $= 1/[a_0(\pi n)^{1/2}]$

m: Band effective mass. $1/\nu$ in Laughlin's state

μ_B: Bohr magneton

A: Total area

n_i: Impurity concentration

g_V: Valley degeneracy

κ: Average dielectric constant

ε_F: Fermi energy

μ: Chemical potential

B: Bulk modulus

ϕ_0: Flux quantum = ch/e

c_v: Specific heat per volume

ν: Filling factor = $n/(eH/ch)$

σ: Conductivity

l: Magnetic length = $(c\hbar/eH)^{1/2}$

α: $= kT/\mu_B H$

H: Magnetic field

m_0: Bare electron mass

ω_c: Cyclotron frequency

Author Index

Numbers in parentheses are reference numbers and indicate that an author's work is referred to although his name is not cited in the text.

A

Abe, R., 298(44)
Abe, Y., 262(190), 264(190)
Abraham, D. W., 94(8), 104(29), 105(30)
Abrahams, E., 241(113), 267(113), 350(134)
Abstreiter, G., 339(109)
Aburto, S., 187(254)
Adachi, S., 192(277, 279)
Adams, G., 286(15), 287, 304(62), 305, 307, 309, 311
Adler, D., 214(7)
Adler, J. G., 128(73)
Adrian, F. J., 248(138)
Adroja, D. T., 191(272), 192(272)
Aharony, A., 218(23), 239(110), 248(110), 249(110)
Ahn, B. T., 149(72), 151(72, 83), 152(83), 154(72), 155(96, 97)
Akachi, T., 174(195), 187(254)
Akinga, H., 175(206, 209, 211, 212), 176(206, 209, 211, 212), 179(229, 231, 232), 180(235), 182(231), 184(231, 232), 191(231)
Aksay, I. A., 163(127), 169(127)
Alario-Franco, M. A., 151(84, 86), 152(86), 153(86, 89), 154(89), 159(989), 174(194)
Alben, R., 61(94)
Albers, J., 251(146), 258(146), 261(146)
Albers, R. C., 268(214)
Alder, B. J., 292(35)
Alder, D., 227(7), 234(7), 235(7)
Alford, N. McN., 145(52), 163(52)
Allen, J. W., 20(49), 88(49), 234(82), 239(82, 107), 240(107), 243(124), 244(107, 124), 247(107), 251(124), 252(82), 253(124), 254(124), 255(124), 256(124), 257(124), 258(124), 259(124), 260(124)
Allen, P. B., 228(61), 287(16)

Allen, R. E., 228(63), 229(66), 232(71)
Allen, S. J., 342(122), 347(128), 348
Allen, S. J., Jr., 283(11), 339(109), 341(114), 343(123), 344, 345
Allred, D. D., 192(283)
Alp, E. E., 262(185), 263(185), 264(185)
Altshuler, B. L., 356(142), 357(144), 358(144)
Alvarez, M. S., 161(114, 115), 196(327), 197(327), 219(25), 234(25), 236(25), 237(25)
Ambegaokar, V., 95(9), 109(40)
Amelinckx, S., 147(69, 70, 71), 149(70, 71), 163(69), 164(69, 70), 165(136), 166(70), 169(155, 157), 197(345, 346, 347), 199(345, 346, 347, 353, 355), 200(347, 353), 201(346, 347, 353, 355), 202(380), 203(380)
Andersen, H. C., 8(16), 47(73)
Andersen, J. C., 51(80), 55(80)
Anderson, J., 260(181)
Anderson, J. S., 199(350), 200(350), 202(378), 203(378)
Anderson, O. K., 221(28a), 232(28a), 233(76), 234(76), 236(76), 242(76), 256(28a)
Anderson, P. W., 103(27), 122(58), 133(84), 214(6), 225(6), 235(6), 236(6, 87), 237(6, 94, 96, 97), 238(6, 97, 104), 239(94, 105), 244(87), 249(87), 350(133, 134)
Ando, K., 199(357), 201(357)
Ando, T., 274(3), 320(83), 331(96), 334(99), 335(100), 341(116), 349(131), 365(169), 366(169), 368(169)
André, J.-P., 337(103)
Andrei, E. Y., 312(76)
Angel, R. J., 137(13), 138(13), 143(13), 197(337), 198(332), 199(332, 337), 200(332, 337), 201(332), 202(363), 204(363), 206(363)

403

H

O

W

Wachs, A. L., 268(213)
Wada, H., 150(76)
Wada, N., 132(83)
Wada, T., 192(277, 279)
Wadati, M., 298(45)
Wagener, T. J., 257(171, 172)
Wagner, N. J., 16(30), 20(30)
Wagner, R. J., 345(125)
Wainer, J. J., 337(106), 339(106)
Wakabayashi, J., 331(92), 353(137),
   356(137), 360(148), 362(155, 157),
   363(157), 365(166), 366(157), 368(157),
   377(181)
Waki, S., 137(19), 138(19)
Wakiyama, T., 174(193)
Walker, E., 139(38), 268(213)
Walker, J. C., 187(251, 253)
Wallenberg, L. R., 199(350), 200(350),
   202(378), 203(378)
Walstedt, R. E., 122(59)
Wang, C. S., 234(74), 268(214)
Wang, G., 155(95)
Wang, H. S., 191(268)
Wang, K., 168(152), 169(152)
Wang, L., 136(7), 137(7)
Wang, T. Q., 197(332), 198(332), 199(332),
   200(332), 201(332)
Wang, W. I., 374(176, 177)
Wang, X. W., 233(74), 234(74), 242(118)
Wang, Y. Q., 92(4), 136(3), 173(184),
   176(184), 196(318), 197(331), 201(331),
   213(2), 221(31), 229(31)
Wang, Z. Z., 216(18), 218(18)
Wanner, M., 308(68)
Warren, W. W., Jr., 122(59)
Waseda, Y., 20(49), 88(49)
Washburn, S., 357(143)
Wassdahl, N., 262(189)
Waszczak, J., 122(59)
Waszczak, J. V., 124(64), 125(64), 131(64),
   144(48), 145(48), 151(88), 175(204),
   176(204), 177(204), 191(275), 192(275),
   197(340), 199(340), 236(91), 238(101),
   265(200), 266(101, 200), 267(101)
Watanabe, I., 132(83)
Watanabe, N., 139(37), 175(206, 211),
   176(206, 211), 179(225), 180(225)
Watanabe, Y., 161(113)
Watson-Yang, T. J., 222(33), 225(33),

226(33), 227(33), 228(33)
Weaver, J. H., 257(171, 172), 258(177),
   260(177)
Webb, D. J., 127(70)
Webb, R. A., 337(106), 339(106), 357(143)
Weber, T. A., 52(81)
Weber, W., 121(55), 196(310), 225(50),
   227(58), 245(58), 247(58, 136), 248(58,
   136, 137), 266(137), 267(58, 137)
Weber, W. J., 163(127), 169(127)
Weger, M., 228(65)
Wegner, F., 365(169), 366(169), 368(169)
Wei, H. P., 362(157), 363(157), 366(157),
   367, 368(157)
Wei, J. Y. T., 121(53), 194(301)
Wei, Y. N., 139(40), 143(40), 194(304)
Weigmann, P. B., 238(103)
Weijs, P. J. W., 251(149, 151), 252(149,
   151), 253(149), 257(149, 151), 258(149,
   151), 259(149), 260(182), 261(151),
   262(149, 195), 264(149), 265(149, 195),
   266(149)
Weimann, G., 337(103, 104), 345(126),
   364(126), 365(168), 374(176), 401(208)
Weimann, R., 365(167)
Weinberger, P., 252(167), 253, 255(167)
Weinert, M., 234(78)
Weis, J. J., 379(183), 380(183, 195),
   383(195), 397(183, 195)
Welch, D. O., 155(98)
Weller, M. T., 177(215)
Wen, X.-G., 226(54), 236(54), 237(54)
Wendin, G., 250(142), 255(142), 256(142),
   258(142), 260(142), 264(142), 265(142)
Werder, D., 220(26), 221(28), 232(28),
   256(28)
Werder, D. J., 145(51), 151(82, 88),
   153(90), 163(51)
Westra, C., 243(125), 255(125), 260(125)
Wetzler, K. H., 268(213)
Whangbo, M.-H., 225(51)
Wheatley, J. M., 238(104)
Wheeler, J. A., 41(68), 42(68)
Wheeler, R. G., 295(38), 337(106),
   339(106)
White, A., 120(50), 193(293)
White, A. E., 121(55)
White, D., 192(287)
Whitfield, H. J., 145(54), 163(54), 164(54),
   165(54)
Whithers, R. L., 202(378), 203(378)

X

Y

Subject Index

A

Activation energy
  Hall plateau, 391
  magnetic field dependence, 377–378
Aluminium, $YBa_2Cu_3O_{7-\delta}$, substitution, 192
Ambegaokar–Baratoff relation, 95
Anderson lattice Hamiltonian, 242–245, 268
Anderson localization, 350
Angular deficit, 42
Antiferromagnetic insulator, 217–218
Auger electron spectra, copper-oxide superconductors, 261–262

B

Bardeen–Cooper–Schrieffer coherence length, 92
Bardeen–Cooper–Schrieffer theory
  interactions, 118–119
  jump in specific heat, 132
Barrier height, current-dependent, 112
bcc crystal, 38
  free energy, 85
  tetragonal structure, 215–216
Bernal holes, 18
Bi2122, 199–201
Binary alloys, frustration and glass formation, 55–56
Binary glasses, icosahedral order, 51–52
Bismuth-containing superconductors, 197–201
  Bi2222, 199–201
  incommensurate modulation, 200
  unit cells, 198
Bohr radius, effective, 298
Bond-orientational order, 45–47
  long-range, 49
  rotationally invariant, 47–48

Bragg peaks, metastable icosahedral crystal, 85–86
Bremsstrahlung isochromat spectroscopy, 257–258

C

$CaLaBaCu_3O_x$, diffraction pattern, 161–162
Calcite oolite, 18–20
Calcium, $YBa_2Cu_3O_{7-\delta}$ substitution, 191
Charge density wave distortion, 225
Cluster CI model, 255–256
Cluster variation method, 157–158
Cobalt
  vapor-deposited, structure factors, 82–84
  $YBa_2Cu_3O_{7-\delta}$ substitution, 185–186
Coherence length
  Bardeen–Cooper–Schrieffer, 92
  Ginzburg–Landau theory, 99
  superconductors, 99
Coherent potential approximation, 331, 333–335
Condensed matter, 1–5
Conductivity
  Drude-type, correction, 352
  dynamic, 340
  transverse, 372
Conductivity tensors, 331
Configurational entropy, 61
Continuum model, superconductors, 103–104
Cooper pairs, 119
Copper, enrichment and stacking faults, 169
Copper-oxide superconductors, 213–215, 267–269
  anisotropy, 264
  antiferromagnetic structure, 234
  auger electron spectra, 261–262
  before hybridization, 245
  body-centred tetragonal structure, 215–216